THE LONG WAY HOME

THE
LONG WAY HOME

An American Journey from Ellis Island to the Great War

David Laskin

HARPER PERENNIAL

NEW YORK • LONDON • TORONTO • SYDNEY • NEW DELHI • AUCKLAND

HARPER ● PERENNIAL

Grateful acknowledgment for permission to reproduce illustrations is made to the following: *Library of Congress*: insert page 1, top; 3, bottom, photo by Lewis W. Hine; 4, bottom; *National Archives and Records Administration*: 5, bottom; 12, original poster by Charles Edward Chambers; *Photograph collection, Miriam and Ira D. Wallach Division of Art, Prints and Photographs, The New York Public Library, Astor, Lenox and Tilden Foundations*: 1, bottom; 2, bottom; 3, top; *Milstein Division of United States History, Local History and Genealogy, The New York Public Library, Astor, Lenox and Tilden Foundations*: 2, top; *Institute for Regional Studies, North Dakota State University, Fargo*: 4, top; *Nellie C. Neumann*: 5, top; *John Ricci*: 6, left; *Leonard Epstein*: 6, top right and bottom right; *John Chimelewski*: 7, top; *Dorothy Vancheri*: 7, bottom right; *John Riggs*: 7, bottom left; *Affatato family*: 8, top; *Ralph Madalena*: 5, center right; *Christy Leskovar*: 8, bottom; *Rick Pierro*: 9, left; *Philip Dorian*: 9, right; *El Paso County Historical Society*: 10, top; *The Pictorial Record of the 27th Division*: 11, top and bottom; 15, bottom; *Mole and Thomas*: 13; *The Absolute Truth, U.S. Army Signal Corps*: 14, top and bottom; 15, top; *Marine Corps History Division*: 16, top; *Michelina Rizzo*: 16, bottom.

A hardcover edition of this book was published in 2010 by HarperCollins Publishers.

P.S.™ is a trademark of HarperCollins Publishing.

FIRST HARPER PERENNIAL EDITION PUBLISHED 2011.

Designed by Jennifer Daddio / Bookmark Design & Media Inc.

The Library of Congress has catalogued the hardcover edition as follows:

Laskin, David, 1953–
   The long way home : an American journey from Ellis Island to the Great War / David Laskin.
      p.   cm.
   Summary: "Narrative account of how twelve immigrant soldiers became Americans through fighting in World War I"—Provided by publisher.
   ISBN 978-0-06-123333-3 (hardback)
   1. World War, 1914–1918—Biography. 2. Soldiers—United States—Biography. 3. Immigrants—United States—Biography. 4. United States. Army—Biography. I. Title.
D507.L26   2010
940.4'8173—dc22

ISBN 978-0-06-123334-0 (pbk.)

11   12   13   14   15   OV/RRD   10   9   8   7   6   5   4   3   2

FOR MY MOTHER,

daughter of immigrants,
niece of a World War I veteran

At its core, perhaps, war is just another name

for death, and yet any soldier will tell you,

if he tells the truth, that proximity to death

brings with it a corresponding proximity to life.

—TIM O'BRIEN, *The Things They Carried*

Oh the army, the army, the democratic army,

All the Jews and Wops, the Dutch and Irish cops

They're all in the army now.

—CORPORAL JOHN MULLIN, LYRICS TO
MARCHING SONG SUNG BY THE
77TH "MELTING POT" DIVISION

CONTENTS

Twelve Who Served   xi

Introduction   xv

CHAPTER 1   OLD COUNTRIES   1

CHAPTER 2   JOURNEYS   23

CHAPTER 3   STREETS OF GOLD   44

CHAPTER 4   THE WEAK, THE BROKEN, AND THE
            MENTALLY CRIPPLED   65

CHAPTER 5   THE WORLD AT WAR   79

CHAPTER 6   THE ARMY OF FORTY-THREE LANGUAGES   122

CHAPTER 7   I GO WHERE YOU SEND ME   157

CHAPTER 8   JULY 4, 1918   173

CHAPTER 9   THESE FOUGHT IN ANY CASE   194

CHAPTER 10 THE JEWS AND THE WOPS AND
            THE DIRTY IRISH COPS   221

CHAPTER 11 THE ARC OF FIRE   242

CHAPTER 12 BREAKING THE LINE   255

CHAPTER 13 BLANC MONT   271

CHAPTER 14 WHY SHOULD I SHOOT THEM?   296

CHAPTER 15 POSTWAR   312

Acknowledgments   351

Sources   359

Index   373

TWELVE WHO SERVED

This is the story of twelve immigrants from Europe who served in the American armed forces during World War I. Of the humblest origins, these men and their families came to this country at the turn of the last century in search of freedom and opportunity. When the nation went to war in 1917, the twelve returned to Europe in uniform and fought in the front lines. The survivors came back different men, transformed in the ways people are transformed by sorrow and sacrifice beyond words. Transformed in the way all of us are transformed when history catches us in its rough current.

EPIFANIO AFFATATO

Born Scala Coeli, Italy, January 3, 1895; emigrated with his brother 1911; joined his father and worked as a laborer in Brooklyn, New York, and briefly on railroads in Des Moines, Iowa; entered the army April 1, 1918; served as private first class with Company C, 107th Infantry, 27th Division.

JOSEPH CHMIELEWSKI

Born Russian Partition of Poland, 1896; emigrated 1912; joined his brother and worked as a coal miner in South Fork, Pennsylvania; entered the army June 17, 1917; served as private with Company A, 16th Machine Gun Battalion, 6th Division.

ANDREW CHRISTOFFERSON

Born Haugesund, Norway, April 14, 1890; emigrated with his sister-in-law and her children 1911; worked as a farm laborer in Larimore, North Dakota, and homesteaded in Chinook, Montana; entered the army June 25, 1918; served as private first class with Company M, 321st Infantry, 81st Wildcat Division.

MAXIMILIAN CIEMINSKI

Born Polonia, Wisconsin, October 11, 1891 to immigrants from Kaszubia, Prussian Partition of Poland; worked as a miner and night watchman in his brother-in-law's brewery in Bessemer, Michigan; entered the army November 19, 1917; served as private with Company C, 102nd Infantry, 26th "Yankee" Division.

SAMUEL DREBEN

Born Poltava, Ukraine, June 1, 1878; emigrated 1899; enlisted U.S. Infantry 1899 and fought in the Philippines, where he was dubbed "the Fighting Jew"; fought as soldier of fortune in Central America; enlisted February 12, 1918; served as sergeant with Company A, 141st Infantry, 36th Division.

MEYER EPSTEIN

Born Uzda, Russian Pale of Settlement, 1892; emigrated on the *Lusitania* 1913; worked as a hauler and plumber, New York City; entered the army April 27, 1918; served as private with Company H, 58th Infantry, 4th "Ivy" Division.

SAMUEL GOLDBERG

Born Lodz, Russian Pale of Settlement, March 19, 1900; emigrated with his mother and siblings 1907; lived in Newark, New Jersey, and later worked in an automobile dealership in Atlanta, Georgia; entered the U.S. Cavalry May 6, 1918; served as private with Company M Troop, 12th Cavalry Regiment.

MATEJ KOCAK

Born Gbely, Slovak section of Austria-Hungary, December 30, 1882; emigrated 1907; enlisted U.S. Marine Corps, October 15, 1907, and reenlisted twice; served in World War I as sergeant with 66th (C) Company, 5th Marine Regiment, 2nd Division.

TOMMASO OTTAVIANO

Born Ciorlano, Italy, May 1896; emigrated with his mother and siblings 1913; worked as a machine operator in Lymansville, Rhode Island; entered army April 27, 1918; served as private with Company I, 310th Infantry, 78th Division.

ANTONIO PIERRO

Born Forenza, Italy, February 15, 1896; emigrated with a cousin 1913; worked as a laborer in Swampscott, Massachusetts; entered the army October 4, 1917; served as private with Battery E, 320th Field Artillery, 82nd "All-American" Division.

PETER THOMPSON

Born County Antrim, Ireland, September 4, 1895; emigrated 1914; worked in the copper mines in Butte, Montana; entered the army in summer 1917; served as private first class (later promoted to sergeant) with Company E, 362nd Infantry, 91st "Wild West" Division.

MICHAEL VALENTE

Born Sant'Apollinare, Italy, February 5, 1895; emigrated 1913; worked as an orderly in a mental hospital, Ogdensburg, New York; enlisted in New York National Guard, 1916; served as private with Company D, 107th Infantry, 27th Division.

INTRODUCTION

Antonio Pierro—a dapper, dark-eyed young private in the field artillery—spent the morning of October 17, 1918, feeding shells packed with phosgene gas to the big guns in the Argonne forest in France. Tony's unit—the 82nd "All-American" Division's 320th Field Artillery—opened fire on the tiny village of Champigneulle at 6:10 A.M., and they kept it up until they had laid down twenty-six hundred rounds of phosgene. When that cloud of poison proved ineffective against the German occupiers, the All-Americans fired off an additional twelve hundred phosgene rounds just before noon. The second barrage did the trick—or seemed to. The Germans left Champigneulle and streamed into the nearby scrap of woods, the Bois des Loges, where they proceeded to slaughter the faltering, inexperienced American infantry.

Before the battle, Tony had transported artillery shells to the front with a horse and cart. Now the cart was piled with the bodies of men who died that day trying—and failing—to seize that bit of woods. Hundreds of American soldiers would perish in the Bois des Loges in the final weeks of October 1918, but Tony was one of the lucky ones. I know to the last decimal just how lucky because eighty-eight years later I sat down with him in the sunny back garden of his house in the seaside town of Swampscott, Massachusetts, and prodded him to ruminate on his life and times. It was July 8, 2006, and Tony Pierro was halfway through his 111th year. *One hundred and ten years old.* To me it seemed inconceivable to be face-to-face with someone who had gone to war when Woodrow Wilson was the commander in chief.

But service and survival were not the only extraordinary things about Tony Pierro. The very fact that he was living out his days in this lovely, prosperous, quintessentially American setting was in itself a remarkable feat, the final chapter in a humble epic that had begun in an impoverished hill town in the south of Italy. For Tony was not only a soldier but an immigrant. Though he fought in France with the All-Americans, at the time of his service he was not technically an American at all. Born in the far south region of Basilicata in 1896, Tony had emigrated to Massachusetts in 1913 at the age of seventeen. Like millions of other immigrants in the first decades of the twentieth century, he passed through Ellis Island, moved in with relatives who had come before him, and went to work at the first job he could find. Four years later, when the army mailed him a letter ordering him to report for duty, Tony went to war. Even though he was still a citizen of Italy, Tony fought for the United States. Some half a million other immigrants from forty-six different nations did the same. At the height of the nation's involvement in the worldwide conflict that became known as the Great War, fully 18 percent—nearly one in five—of the 4.7 million Americans in uniform had been born overseas.

Tony Pierro didn't say much about how fighting with the All-American Division had changed his life or his relationship with his adopted country. He didn't have to: the facts spoke for themselves. Nearly nine decades later he still had his discharge papers; he was still proud that he had chosen to serve with the American Expeditionary Forces instead of returning to Italy (our ally in that war) to serve in the Italian army; he still remembered his joyous disbelief when the Armistice was declared on November 11, 1918; he still loved his country, by which he meant the country in whose military he had served.

Tony did not choose to fight in the Great War. It was not a conflict he had a stake in or really understood. He had no great love for the discipline or privations of army life. Nonetheless, he fought bravely and loyally. In fighting for the United States of America, he and thousands of immigrants like him became Americans.

G od knows, military service was the last thing most of these men had in mind when they and their families came to this country. Many, in fact, had emigrated expressly to avoid mandatory military service. Before the war the United States had no draft; its army was tiny compared to the behemoths massing in Europe, and its military culture quiescent. Had Tony Pierro remained in Italy, Meyer Epstein in the Russian Pale, Andrew Christofferson in Norway, Joe Chmielewski in the Russian section of Poland, all of them would have faced compulsory military service. They came to America for freedom, and freedom from the army was a big part of it. They came to America not to fight but to work—and America obliged, however grudgingly, with dirty, backbreaking, unskilled jobs. Tony dug a rich man's garden; Meyer hauled radiators through the streets of New York; Andrew reaped wheat on the prairie; Joe mined coal in the hills of western Pennsylvania. Americans gave them work—but as more and more of them poured in, Americans began to doubt the wisdom of keeping the golden door open. They worried about what all these foreigners would do to the strength and purity and complexion of the population. By the early twentieth century, some 14 percent of the country was foreign-born—and every year hundreds of thousands of fresh immigrants were arriving from the ghettoes of eastern Europe and the blasted villages of southern Italy. The likes of Tony Pierro and Meyer Epstein and Epifanio Affatato and Peter Thompson were fine to build and dig and haul—but what if they were called on to fight? Would they? Could they?

The questions took on a new edge when Europe went to war in August of 1914. Most of the immigrants came from the belligerent nations. How would they react? Would Slavs, Italians, Poles, and Germans return "home" to fight for their native lands? Or would they import the conflict into the streets of New York, Chicago, Buffalo, Boston? Anti-immigrant sentiment had been intensifying as the numbers of aliens rose, and now it exploded. Politicians insisted that hyphenated Americans must choose—100 percent American or not American at all. After the Russian Revolution broke out in 1917, fear of foreigners fused with fear of Bolsheviks. Wild rumors circulated of an alien fifth column poised to poison reservoirs, blow up munitions supplies, undermine the government. It was Europe's war, but Europe's wretched refuse was smuggling the horror into the cities and towns of America.

Everything changed, as it always does, when the nation declared war on April 6, 1917. The United States needed an army—a sizable army—in a hurry, and immigrants overnight went from being a dangerous threat to a valuable resource. Valuable, but unstable. The fundamental issue was, would they fight? But the more pressing question was, would they understand orders? When Tony arrived at Camp Gordon for training in the fall of 1917, three-quarters of his fellow recruits did not speak English. The enlistees from New York pouring into Camp Upton on Long Island spoke forty-three different languages. To talk of cannon fodder was distasteful in time of war, but these swarthy, brutal, jabbering aliens did not even know what a cannon was. If they weren't cowards, they would be traitors. Or spies. How would they fight when they couldn't even drill?

Everything changed, as it always does, when the men went into battle together. Tony Pierro had never wanted to be a soldier. Neither did Meyer Epstein, Tommaso Ottaviano, or Max Cieminski. But all of them shipped out to France in the spring and summer of 1918. All of them got crammed in boxcars, transported east to the front line, marched down roads deep in mud and strewn with corpses, handed rifles. And when they were told to go over the top, they did it—and so did the overwhelming majority of other foreign-born soldiers. Most of them didn't give a damn about making the world safe for democracy. God and country were the last things on their minds. They fought not for an idea but because the sergeant ordered them to fight, because their buddy was fighting, because they were part of a platoon. But in the end, they also fought because they were Americans. Maybe in the grand scheme of things they *were* cannon fodder, another 150 pounds in the avalanche of flesh that the generals were piling on the enemy— but to the amazement of their officers, and sometimes themselves, they fought like American soldiers.

"Our minds were becoming warped," said Italian immigrant Giuseppe (Joe) Nicola Rizzi—Woppy to his buddies—of what happened to him and his comrades in the 35th Division after weeks of bloodshed in the Argonne. "I had become as vicious as the rest. Our nerves were mighty strained. We were crabbing about everything in general—hunger, cold and fatigue. Still, the last puff of a cigarette would be split up; the last bit of chewing tobacco was passed around; the last can of corned willie shared. You see, we were all buddies." War has its own strange alchemy. Soldiers fear and hate and grouse about every minute of it—and yet nothing else in their lives compares to the intensity, the selflessness, the significance of combat. "Combatants live only for their herd," writes war correspondent Chris Hedges in his searing book *War Is a Force That Gives Us Meaning*. "Those hapless soldiers who are bound into their unit to ward off death. There is no world outside the unit. It alone endows worth and meaning. Soldiers will die rather than betray this bond. And there is—as many combat veterans will tell you—a kind of love in this." In World War I, this bond became especially powerful for the foreign-born. To their fellow soldiers they were kikes, wops, micks, hunkies—no matter how the War Department tried, they couldn't stamp out these ethnic slurs. But after the battles fought at Soissons, Blanc Mont, Montfaucon, the

Hindenburg Line, and the Bois des Loges, the slurs became terms of camaraderie. As Joe Rizzi said, "You see, we were all buddies."

"To go 'home' to the USA means more to us as immigrants fighting for our adopted country," wrote Morris Gutentag, a Jew who had emigrated from Warsaw in 1913 and enlisted in the 77th Division (nicknamed the Liberty or Melting Pot Division because it drew so heavily on immigrants living in New York). "I was proud that I fought for and we won the war and all my family shared in that. I never regretted it." Many felt the same way. No man could explain why he was proud of being forced to endure madness, atrocities, filth, hunger, cold, mud, disease, and the unspeakable horror of killing or being killed. But the hell was an essential part of the alchemy of war—an alchemy with the power to turn strangers, even despised aliens, into comrades.

Not that they had it easy when they were shipped back across the Atlantic after the war. *Woppy* may have become a term of endearment in the Argonne, but in the Red Scare era that followed hard on the heels of the war, immigrants became the target of vicious attacks and discrimination. It didn't matter that you'd won a medal for bravery; if your name was Cieminski, Rizzi, Dreben, Kocak, or Valente, you were dangerous, subversive, potentially Bolshevik and anti-American. Immigrant soldiers came home from the war to discover that someone else had been hired to do their jobs, that the resurgent Ku Klux Klan openly advocated their deportation, that in the popular press and back-street mutterings they were being lumped together with the "Huns" they had fought in France and Belgium. Pogroms erupted once again in eastern Europe—but now the Jewish victims had nowhere to flee. In 1921 and 1924, Congress voted overwhelmingly to cut the flow of immigration from eastern and southern Europe to a trickle. The parades had barely ended when the doors started slamming shut.

But the pride that Morris Gutentag and thousands like him brought back from the war proved to be a most valuable commodity. Immigrants had learned to stand up for themselves in the army. They had picked up American slang and American swagger. They had mingled for the first time with people from outside their groups—and there was no going back to the way life had been. No one was going to convince these men and their families that they weren't real Americans, that their pride and patriotism didn't count.

The same holds true today. Currently about 5 percent of the troops on active duty in the U.S. Armed Forces were born overseas. "Their service is steeped in pride, but also in the paradoxes of allegiance inherent in serving under a foreign flag," reports Patrik Jonsson in the *Christian Science Monitor*. Jonsson quotes a senior spokesman for the Department of Homeland Security saying that foreign-born soldiers "identify with the ideals of the United States and they are willing to fight and protect those ideals, even before they've secured all the liberties of citizenship." Two key differences between our current wars and World War I are worth noting: first, today all soldiers, foreign-born and native-born alike, serve voluntarily, while the majority in both groups who fought in 1917 and 1918 were drafted; second, the largest groups of immigrant soldiers are no longer Italians, Poles, Irish, Germans, and Jews but Filipinos and Mexicans.

Different names, different faces, but the issues and feelings have altered little. The journey from alien to citizen—something American immigrants and their descendants carry in their DNA—is both hastened and skewed by war. We all bear the scars and the rewards of this journey, though for those who have fought, who are fighting now, the scars are deeper, the rewards more precious. The journeys that unfold in this book are unique, peculiar to the circumstances of the individuals and the pressures of their time and place. But the outcome is familiar. All of us have come this way; many more are coming still.

Change is one of the great imponderables in the life of an individual, a culture, or a nation. We crave change, or think we do, but rarely do we control or comprehend the forces that bring it about. Great fortune or misfortune, love and loss, inspiration, revelation, a truly new idea, natural or human-induced disaster, friendship, war, relocation: these are among the prime movers of personal change and, in the aggregate, social change. The immigrant soldiers who fought in the Great War experienced two of these fundamental changes almost simultaneously. In many cases just a few years or even months separated their arrival at Ellis Island from their induction in the American Expeditionary Forces. The coincidence profoundly altered the course of their lives. Some were decorated for heroism, passing cherished medals on to their sons and daughters. Some came home broken men, maimed by artillery shells or machine-gun bullets, permanently disabled by poison gas, shell-shocked, alcoholic, unreachably depressed. Some marched with chests thrust out in Memorial Day parades but awoke night after night screaming from combat nightmares. Some—too many—never came home or returned in coffins that had been carefully exhumed from battlefields or foreign cemeteries and shipped back under military escort. Every story is different. Even the ones with happy endings bring tears to the eyes. Every story concerns a life—and often the life of a family—that took a sharply different course because of the changes wrought together by immigration and war.

Tony Pierro was born in Italy in 1896, emigrated to the United States the year before Europe went to war, and entered the U.S. Army six months after Congress voted to declare war on Germany. It was a classic trajectory for immigrant men of Tony's generation—still numerically the greatest generation of American immigration. Half a million other foreign-born soldiers shared the same fate. In the pages that follow I recount the stories of twelve of them—twelve men who epitomize what this generation of immigrants endured and how they changed in the course of their journeys from immigrant to soldier to citizen. For each of the twelve, I begin in Europe, going as far back as memory and family lore penetrate. I describe the journeys, almost always unforgettably traumatic, from village to port, from port to Ellis Island, in the reeking steerage of the immigrant ship. As the young men and their families spread out—to Boston; Brooklyn; Butte, Montana; Polonia, Wisconsin; South Fork, Pennsylvania; El Paso, Texas—their first priority was inevitably to make a living. But when the world went to war, that priority was rocked by more pressing concerns—concerns for the fate of the countries they had left behind, for family now living and fighting in the war zone. The nearly three years that passed between the outbreak of war and the United States' entry was a period of intense strain and conflicting loyalties for these men—and that strain only increased after April 1917, when their adopted country became one of the belligerent powers. Of the dozen men I follow—three Jews, four Italians, two Poles, one Irishman, a Norwegian, and a Slovak—six were drafted, four enlisted, two were career soldiers who had spent almost their entire American lives in uniform. Three of these men died in France—two on the battlefield, one of wounds sustained in battle. Two won the Congressional Medal of Honor, the nation's highest military award for valor. Together the twelve fought in every major engagement that the American Expeditionary Forces, the AEF, pursued in Europe—Belleau Wood; the Aisne-Marne offensive; Belgian Flanders; St. Mihiel, the first battle planned and executed solely by the Americans; the breaking of the Hindenburg Line at the end of September 1918; and the Meuse-Argonne offensive that won the war. The combat experiences of these dozen men—the hours in which they tossed grenades, shuddered under the pounding of artillery shells, crouched in shell holes while the air sizzled with machine-gun fire, died of raging fevers from infected wounds—do not add up to a comprehensive military history of the war. But their actions under fire do frame some of the most critical and proud moments of the war. Moments that changed both the outcome of the fighting and the outcome of the lives I have been privileged to follow. Moments that have proven to be impossible to forget.

The narrative also includes the stories of two men who died not in uniform but nonetheless in combat directly related to the war. Their deaths—the result of persecution to the point of torture of German Americans and German-speaking conscientious objectors—are also a part of the immigrant experience of this war.

Each of the men who appears in this book is remembered, loved, and honored to this day—I know this for a fact because their families have come forward to share stories, letters, diaries, grief, and, inevitably, pride. The descendants of these immigrant soldiers—children, grand-children, nieces and nephews—cherish photos and medals, display dis-charge papers framed on their living room walls, preserve uniforms and helmets, pass down memories of how their ancestor came to this country, when and why he fought in the war, and what happened to him. They weren't all heroes, these immigrant soldiers, but they all fought bravely and they are all remembered with love and with honor. This is a story that continues to unfold in the fabric of American life.

THE LONG WAY HOME

CHAPTER 1

OLD COUNTRIES

At the end of the nineteenth century, the Jewish population of the Pale of Settlement at the western fringe of
the Russian Empire was the largest in the world—over 5
million Jews confined by the czar's decree to a province
that reached from the Baltic to the Black Sea and that comprised territory that the coming wars would carve into Poland, Ukraine, Belarus,
Lithuania, and Moldova. Crowded into shtetls of wood and dirt and
brick, barred alike from farmland and the major cities, the Jews of the
Pale inhabited a vast ghetto of sorrow. For the men there was never
enough work. Whether they were tailors, carpenters, shoemakers, ped-
dlers, shopkeepers, or butchers, always there were other men clamoring
for the same few jobs, the same scarce kopeks. "There were ten times
as many stores as there should have been," one Jewish girl remembered,
"ten times as many tailors, cobblers, barbers, tinsmiths." Sons as young
as twelve were swallowed up in the "military martyrdom" of service in
the Russian Army for terms of twenty-five years—and either died or
disappeared or returned with their Jewish identity beaten out of them.
To avoid the virtual death sentence of conscription, boys sliced off
their own trigger fingers. Mothers and daughters were heckled openly

in the street, cheated in the Gentile shops, excluded from the solace of Talmudic study that kept the men spiritually alive. When pogroms erupted—violently during Easter week 1881 after the assassination of the czar; even more violently in 1905 when Cossacks opened fire in Bialystok—no Jew was safe. For Jews in the Pale of Settlement, sorrow was a guest that arrived uninvited and never left.

Sorrow came to lodge swiftly in the household of a young Jewish couple named Yehuda and Sarah Epstein. Yehuda Epstein and Sarah Lotwin married sometime around 1890 in the town of Uzda in the province of Minsk in what is now Belarus and, obeying the most basic impulses of love and desire, started a family. But in a region teeming with poor Jews starved for hope, love and desire were expensive, even dangerous luxuries.

In 1892, Sarah gave birth to a son—a round-faced, gray-eyed child she and her husband called Meyer. Though small and pale, Meyer was stronger than he looked—a survivor. Four years later, Sarah became pregnant again—but this was when sorrow began to sink its teeth in. By the time she was ready to give birth, Sarah's husband had left for New York. It was the fork in the road that every Jew living in the Pale at the end of the nineteenth century arrived at sooner or later. To stay in the land of the pogrom or to make a new life in America? Over the next fifty years, as two world wars laid waste to Europe and to Europe's Jews, the answer to that single question would decide the fate of millions of families—not just the difference between prosperity and poverty, freedom and oppression, shtetl and tenement, democracy and autocracy, but between life and death. But of course no one knew that in 1897.

Between 1881 and 1914, some 2 million Jews would choose to leave the Pale, and Yehuda Epstein was among them. Leaving Uzda, leaving his son and his pregnant wife, Yehuda sailed to New York to find work as a butcher. He believed he would prosper. And when he did, he would send for his family, as hundreds of thousands of other men were doing. That was what Yehuda intended and what his family expected. None of them could know that when they separated, splitting the family between two continents, it was for good. Husband and wife, father and son would never set eyes on each other again.

Sarah gave birth to her second son in 1897. Someone named the baby Alexander—someone, but probably not Sarah, since she died giving birth. With their mother dead and their father a continent and an ocean away, the two boys were left to the mercy of relatives. Who had the means to take in two orphans? The Epstein family, divided once by emigration, was divided again, as the boys went to live with different aunts, in different parts of the Pale. The aunt who took Meyer was married to a butcher, and the couple had no children of their own. In theory, a four-year-old orphan nephew would bring joy to the household, but with money so tight, and the business of butchering so bad, joy was beyond their means. The uncle began to cast a cold eye on the growing boy who had been palmed off on him. One day, before his tenth birthday, Meyer came home from school to find his aunt in tears. Quick and intuitive, Meyer sensed the tears concerned him. "We can no longer feed you," the aunt told him between sobs. "There isn't enough money." Meyer noticed that the uncle's eyes were dry.

No matter how poor, no matter how mean, Jews in the Pale did not toss one of their own out on the street to starve. The tradition of *tzedakah*—a Hebrew word connoting a life steeped in justice and righteousness that came to be used for a communal obligation to be charitable—was too strong in Jewish communities. In Lithuania and the Ukraine, *tzedakah* supported nearly one-quarter of the Jewish population—money raised primarily by poor Jews to feed and clothe even poorer Jews. As one Jewish scholar wrote, "Feeding the hungry, clothing the naked, visiting the sick, burying the dead and comforting the mourners, redeeming the captives, educating the orphans, sheltering the homeless and providing poor brides with dowries" were among the duties prescribed by Talmudic law, and in the Pale, Jews clung to Talmud because they had little else. So something would be done for little Meyer Epstein. The aunt and the butcher uncle found another more distant relative willing to take the boy in. It wasn't asking so much—after all, he was ten, in a few years he'd be working. The arrangements were made, Meyer's meager possessions were packed, the small scrappy boy—he was destined never to be tall or burly—was placed on a train, admonished to be good, waved to, and watched through damp or dry eyes until the train disappeared. Someone else's burden now.

The train conductor came through the car demanding in Russian to see the passengers' tickets. Meyer understood what was expected and handed his ticket over like everyone else. He was stunned when this provoked a torrent of angry Russian. The aunt and uncle had put Meyer on the wrong train. The conductor informed the Jewish boy that he must leave the train at the next stop. With no money and the wrong ticket, how was he to return and find the correct train? If the other passengers felt sorry, they kept their feelings to themselves. The conductor escorted the boy off the train and watched to make sure he did not sneak back on. The iron doors clanged shut, the whistle blew, and in an explosion of noise and smoke, the train pulled out. The ten-year-old orphan was on his own now.

Meyer was not a boy prone to panic or despair. He had never had that luxury. Being an orphan had taught him to be resourceful, a trait that would stand him in good stead all his life, and even at the age of ten he had a good head on his shoulders. Meyer started to walk. He figured that if he followed the tracks, eventually he'd come to another village, another shtetl, and that's exactly what he did. When he reached the next town on the train line, he asked someone who looked Jewish where the synagogue was located. At the synagogue, Meyer humbly approached the men and told them his story. He asked for shelter. The men took him in—what else could they do?

For two weeks Meyer lived in the synagogue, sleeping on a bench, subsisting on bread and herring. Then he had his first lucky break. A man named Brevda, the richest Jew in that shtetl, learned of the boy who was living in the temple and offered him a job. Mr. Brevda had grown rich, at least by the standards of the Pale, on junk. Metal, fabric, bits of furniture, pieces of clothing that others had tossed—Mr. Brevda scavenged junk from all over the region, hauled it in his cart, sorted it, sold it, and salted away the profits. Mr. Brevda was rich but he was also kindly. He offered to take Meyer in and keep him—but not for nothing. Meyer must work. Brevda would pay the boy a ruble a year and a suit of clothes, and in exchange Meyer got to sleep in Brevda's barn and go to work as an apprentice junkman.

Meyer was not only smart and resourceful, he was also cute and personable. Brevda took a shine to him, and so did the women who worked for him, Jews and Gentiles both. The women looked after the boy, even quarreled over who got to care for him. Brevda noticed how well the kid was getting along and soon offered him a sweeter deal: he would stake the boy to horse and wagon and take him in as a kind of junior partner. Meyer would now be the one traveling around the countryside collecting junk, while Brevda acted as a kind of clearinghouse for whatever the kid scavenged. When he was on the road, Meyer slept under the wagon. When he was back at Brevda's, he had the women to fuss over him. For an orphan who had in essence been abandoned, it was a pretty good setup. Meyer began to prosper, a little businessman, a successful junkman.

In his years with Brevda, Meyer had not forgotten about his brother Alexander. There must have been some contact between the boys during the years when Meyer still lived with the aunt. Now that he had a horse and wagon of his own, he visited Alexander—Zender, he called him—every once in a while. Someday he hoped that both of them might be reunited with their father.

Meyer was working too hard to go to school—but still he remembered he was Jewish. Mr. Brevda was religious, even something of a scholar, and one of his sons was studying to become a rabbi. This son took Meyer under his wing and taught him Jewish law and scripture. In Brevda's household, Meyer learned the obligation of *tzedakah*—a lesson that would guide him through life. For him, religion was always expressed by doing good deeds, by giving to those in need, by helping others, by standing up for what he believed in.

Meyer was quick-witted and good with language, speaking Yiddish, Russian, and Polish fluently and later adding English and bits of German, French, and Italian. Under the tutelage of Brevda's son, he had learned to be an observant Jew. But the life of scholarship and prayer that all Jews aspired to was not in his future. Meyer was never destined to bend his head over the Talmud with other learned men or to argue poetry and philosophy in smoky cafés or to declaim on street corners to the poor and oppressed. He would make his way in the world by the sweat of his brow and the strength of his back, which, lucky for him, was amazingly strong despite his small stature.

A decade passed under Brevda's roof. Meyer had grown to his full height of five feet, four inches; he had filled out to 135 pounds of compact muscle. He was a nice-looking young man with a clear, steady blue-eyed gaze, brown hair, a high-bridged nose. At an age when most boys are just entering the world of work, Meyer was already a successful small businessman with a tidy sum saved up. It was enough to start him on a new life. He decided to go to America and find his father.

When, in 1861, Giuseppe Garibaldi finally realized the dream of uniting Italy's disparate provincial kingdoms and mini-republics into a single nation, the map of the peninsula was radically redrawn. But the boundary that was, and remains, Italy's most important—the line scoring the peninsula from Rome to Pescara that divides north from south—never appeared on any map. North of the invisible line—in essence north of Rome—were jobs, schools, factories, modern cities, prosperous farms. To the south, there was poverty, superstition, illiteracy, unemployment, epidemic diseases, absentee landowners, primitive secretive villages, overflowing slums, and exhausted degraded land. *Benvenuti in Africa* reads a famous sign on the side of a highway heading down into Italy's forlorn lower third. If anything, the rebirth of a unified Italy made the plight of the south even more desperate. Newly decreed taxes, now the highest in Europe, and compulsory universal military service drained the peasant south of cash and labor. "We plant and we reap wheat but never do we eat white bread," peasants of the south lamented. "We cultivate the grape but we drink no wine. We raise animals for food but we eat no meat."

Il Mezzogiorno, the midday, Italians called the sun-scorched territory of the south—and a hundred years ago it was a name for human misery. As if poverty, overpopulation, and spiraling taxes weren't bad enough, in the decades immediately following the Risorgimento—the resurgence, as Italians termed their national unification movement—the Mezzogiorno suffered repeated plagues of disease and natural disaster. Fifty-five thousand died of cholera between 1884 and 1887. An aphid-like sucking insect known as grape phylloxera swarmed south from the vineyards of France and Lombardy and laid waste to the region's wine grapes, one of the few reliable crops. In December 1908, a 7.5 magnitude earthquake followed by a tsunami destroyed the medieval city of Messina, Sicily's third largest city, and leveled coastal towns and cities in the southern province of Calabria, killing as many as two hundred thousand. "The earth opened and threw stones at us," one peasant said in despair after the earthquake.

But even despair has its shadings and gradations. In the Mezzogiorno, the shades deepened the farther south you went and the higher you climbed into the mountains away from the coast. Most stricken of all was the remote mountainous region pinched between Italy's heel and toe that is known today as Basilicata but in ancient times was called Lucania. "Christ never came this far," wrote Carlo Levi of malarial, godforsaken Lucania in his 1945 memoir *Christ Stopped at Eboli*, "nor did time, nor the individual soul, nor hope, nor the relation of cause to effect, nor reason nor history. . . . No one has come to this land except as an enemy, a conqueror, or a visitor devoid of understanding." It was in the deep harsh interior of Basilicata in a stone-and-stucco hill town called Forenza that the Pierro family lived, or tried to.

Rocco Pierro and Nunzia dell'Aquila made a handsome couple—vigorous, reserved, hardworking people, both of them broad-faced, dark-eyed, with fine chiseled mouths, broad straight noses, and thick dark hair. Even in youth they harbored neither lofty expectations nor undue fear of the years before them. Nunzia, born in 1866, had a dark stern dignity to her gaze that almost looked Native American. Rocco, four years older, carried his shoulders square, his spine straight, his hair cropped close like a Roman emperor. In photos that have come down through their family, they face the world straight on, without illusions or demands—ready to take what life dealt them, ready to survive. Ready to prosper, if only there was work to afford them prosperity.

Sweethearts from childhood, they waited to marry until Rocco was nearing thirty, Nunzia in her midtwenties—whether from prudence, poverty, or some other obstacle no one any longer remembers. Husband and wife were healthy and in love, and they set out together to make the most of what their situation offered. Forenza, though certainly no gem in a country blessed with breathtaking beauty, had a fine situation atop a lofty but gentle rise. Sunstruck vineyards and orchards swept up to the outskirts of town like a kind of patchwork tent; where agriculture left off the town abruptly began—a maze of narrow streets, arched portals, alleys that climbed between tight-packed one- and two-story stone houses. Some of the alleys were so narrow you could stand in the middle and practically touch the buff or pale gray stones of the houses on either side; some of the inclines were so steep that the streets ended in flights of steps. In Forenza the people lived close and volubly, as Italians had lived for millennia. By day the stone streets echoed with the sound of voices, footsteps, basins of water or waste flung from the doors and windows. On Sundays and feast days the churches and piazzas were mobbed. Privacy in Forenza was as scarce as summer rain. There was plenty of opportunity to be by yourself under an open sky in the fields and vineyards. In town they expected to be rubbing up against their own kind.

At the top of Forenza's hill, rising straight up over the tile rooftops like a finger pointed at heaven, was the square stone bell tower of the town's main church, the Chiesa Madre di San Nicola—the mother church of Saint Nicholas. When San Nicola tolled its bells on Sunday, Christmas Eve, Easter, and All Saints' Day, all Forenza made the stony ascent. For the Pierros, however, San Nicola was only for special occasions. Usually the family went to a tiny church just a hundred feet from their house on the north side of town—the neighborhood church of San Vito. The patron saint of dancers, actors, comedians, dogs, young people, and those who suffer from epilepsy, San Vito—Vitus in Latin—was the kind of stubborn, rebellious saint who appealed to an oppressed rural people. His feast day, June 15, is still a major public celebration in Forenza.

Saints and stories Forenza had in abundance—if only there had been work that paid wages. There was work, all right—the grueling relentless work of scratching out a living from the earth, cooking and preserving what they grew, keeping a growing family clean and fed— but jobs with salaries were all but nonexistent in the south. Like most of the Forenza families, the Pierros owned some small fields a couple of miles outside of town. In the growing season, the townspeople went out to the land on foot or on donkeys—trekking two miles, four miles. The land became so hot and dry in summer that the ground cracked. The Pierros had a hut on their land where they slept in the summer to avoid the four-mile ride to and from town. On their plot west of town, they grew wine grapes and melons; they tended a small orchard of fruit trees—cherries, pears, figs—and kept some pigs and chickens.

Everybody worked, but to make money, you went to America. It was the only way to survive. Entire villages from Calabria and Puglia up and left—family by family, first the husband, then the sons, then the wife and daughters. The money that trickled back in worn, stained envelopes to those left behind altered the economy of the Mezzogiorno—dollar by dollar, it watered the parched soil of the south and raised a meager crop of bread and wine and eventually fares for passage to America. By the time war broke out in Europe in

1914, three-quarters of a billion dollars had been transferred from the United States to Italy.

Rocco left Nunzia for the first time in 1890 when she was pregnant with her first child. He didn't intend to stay—and the family made no plans for Nunzia to join him. He would work for a salary, send home the money he could spare, and return to Italy when he'd earned enough. Tens of thousands of men from the Mezzogiorno were doing the same thing. Rocco made his way from Forenza to the pretty seaside village of Swampscott, north of Boston, a summer resort for well-heeled Yankees in those days. He was in the United States when his first child was born in 1891, a boy named Michele.

For two decades Rocco swung back and forth between Swampscott and Forenza. The fruit of every trip to Italy was another baby. Guarino, their second son, arrived in 1894. Two years later, there was another son, Antonio, born February 15, 1896. Rosa came in 1904, Daniele the following year, then Nicola in 1909, and finally Vito in 1911. Six sons and a daughter.

Growing up in Forenza, the children rarely saw their father—but when he was home, they hung on his stories of the big hotels fronting the ocean in Swampscott, the harsh snowy winters and humid rainy summers. The kids found it strange that their father didn't farm—no grapes, no orchard, no pigs. How could a man live without growing his own food? But that was not the way in *l'America*. Rocco worked not for himself but for money, picking up whatever work he could get. One of his jobs was with a company that put up telegraph poles. The children took it all in, but to them it seemed more alien and strange than the miracles of the saints. The boys understood that someday they'd go to America with their father to earn money too, but meanwhile Forenza was all they knew, and what happened in its streets and houses and fields was all that mattered. And of course they worked, all of them worked on the land when they were old enough to hold a shovel or control a donkey.

Antonio, the third son, hated snakes. There were snakes in the family vineyard and it made Antonio shudder just to think about them. A story made the rounds that a boy was sleeping outside on a hot summer afternoon when a snake slithered into his mouth and got stuck there. Antonio made sure to keep his mouth closed when he snoozed outdoors with his brothers. He was a good-looking kid, not as dark as the two older boys, and always careful about his appearance. But it was hard to keep clean and neat the way people lived in Forenza. Nunzia stabled her mule inside the house and put the boys out to sleep in the shed: the children were safe from thieves—who would want another kid in Italy?—but if someone stole the mule they'd all starve. The more fortunate families had houses with two stories: a downstairs room for the mules, the cow, and the pigs; a loft upstairs where the family slept—all in the same room, as many to a bed as would fit, often all in the same bed. They put the babies in cradles suspended from the ceiling just above the family beds. Carlo Levi was struck by the way the peasants in Lucania divided their houses into "three layers . . . animals on the floor, people in the bed, and infants in the air." There was no such thing as a kitchen. At the side of the room was a stove and a pile of sticks that the women and children collected every day from the scraggly hillsides. At that end of the room the walls and ceilings were black with soot. No windows—windows were not for the poor. For light they opened the door; when it was cold or wet they sat in the dark. If the town had no water source, someone went down to a spring or fountain or stream every day and hauled water up in jugs carried by the donkey. The Pierro children remember their mother dragging the family washing to a stream outside of town, scrubbing the clothes on flat rocks, spreading them out on dusty shrubs to dry. Food was mostly whatever vegetables were in season, bread, sometimes rubbed with garlic or tomato, pasta when they were lucky, meat rarely. A beef heart delighted the children; a chicken would be slaughtered for a special occasion or when someone fell sick. "Once in a while we had an egg," remembers an Italian immigrant. "Sugar was only for Christmas and Easter."

Still, for the Pierro kids, there was the sweetness of childhood. In the long hot sunny summers they fished in the streams and raided the neighbors' orchards. Every June came the Feast of San Vito. A capon for Christmas, a kid roasted for Easter. Ripe fruit in summer and nuts in the fall. And the weather, month after month the glorious cloudless skies of the Mediterranean, free for everyone. "My father lay in the fields and looked up at the sky," said one son of a Calabrian immigrant. "That was all they had—beautiful weather."

In the Mezzogiorno, the sweetness of childhood didn't last long. Antonio came to hate the vineyard with its snakes and its inexhaustible crop of rocks. Every year come spring the boy was given the job of grubbing the rocks out of the vineyard's soil—and every year there were more rocks. He began to wonder if rocks grew in the ground. Once there was an incident with a neighbor's dog. The dog kept messing around in the vineyard and one day Antonio got his father's gun and shot it. The owner found the dog's body and was enraged; word got out that he suspected the Pierro boy. There could be trouble—blood feuds flared up with far less provocation in the south of Italy. "They told me to get a job," said Antonio. But what job was there to get? His older brothers had already left for America—Michele joined his father in Swampscott in 1910; Guarino tried his luck in Argentina, as many Italians were doing then, leaving home in 1912, when he was eighteen. Antonio was next. In February 1913, he turned seventeen years old. That summer, after the fleeting season of green hills and perfumed orchards, Antonio and a cousin left Forenza and set out for Naples.

For more than a generation now, as they left the villages of the Mezzogiorno, the men and boys sang "Il Canto degli Emigranti"— "The Song of the Emigrants":

> *In time of peace we sickened in hospitals or jails*
> *In time of war we were cannon fodder*

We harvested bales of grass, one blade for us, the rest for the wolves . . .
No disease can be more horrible than hunger from father to son . . .

Magnus Andreas Brattestø was born on April 14, 1890, in a house near the sea on the west coast of Norway. Haugesund was the closest town, a small port with a harbor full of fishing boats between Bergen to the north and Stavanger to the south. This is fjord country, island country, mountain and sound country, where the craggy land takes its time before finally surrendering to the open salt water. Rain country, too, especially in winter when the westerlies rake the North Sea and deposit load after load of liquid on the first steep flanks of terra firma. Waterfalls, dark dripping conifers, pastures of thick velvety green, lingering summer twilight—Haugesund and the farms and islands around it was a country of an almost heavenly beauty for those who could care about beauty.

Magnus Andreas Brattestø was the firstborn son in a family that would quickly swell to a dozen children—ten boys and two girls in all, though one of the boys did not survive childhood. With blond hair, round blue eyes, a finely cut mouth, and straight high nose, the boy was thin and small-boned and seemed a little delicate—but he wasn't. Never one to roar or raise his fist unprovoked, he was nonetheless stubborn. Take the matter of his name—Magnus Andreas Brattestø. Magnus for obvious reasons didn't suit him, so the boy dropped it and called himself by his middle name Andreas. As for Brattestø, it means, roughly, "The Farm of the Steep Boat Landing." It wasn't really a family name at all but the name of the large farm on which a number of families lived and worked. It was the custom in rural Norway for families to take the name of the land, so everyone from the same valley or from the fields on one side of a lake shared the same surname even if they weren't related. For whatever reason, Andreas objected to dragging around Brattestø as his last name—and when the time was right he shed it too. It took a bit longer to accomplish than shedding Magnus, but he seized his chance when it came. Stubborn.

Actually, the name Brattestø said a lot about what life was like on the west coast of Norway. The land here was rocky and sloping, the weather uncertain. Over the years, families had divided the farms into smaller and smaller parcels, so by the time Andreas was growing up it was all but impossible to make a living from farming alone. A farmer needed to have a boat in the water to supplement his income by fishing. And even then, many of them barely made enough to feed their families. With twelve kids in Andreas's family, they couldn't afford more than a few years of childhood for each. Andreas left school before he was ten and went to work on a fishing boat. The man he worked for lived on the rugged North Sea island of Utsira eleven miles west of Haugesund and had a fleet of boats. Winter was not a pleasant time to be fishing in the North Sea, but that's when the herring began to spawn, so Andreas, small and slight as he was, went out with the men. The crew slept in the hold and were expected to bring their own food—potatoes, beans, a bit of meat or fish, and hardtack when those ran low. The men and boys netted the schools of herring that whales chased through the deep channels. The catch was dumped into rowboats and then transferred to the holds of waiting steamers. Little boys like Andreas stood up to their waists in squirming silvery herring as they emptied the rowboats. Andreas earned a bit of extra money by taking buckets of fish off the boat and selling them door-to-door in Haugesund. Often he brought his little brothers along as assistants.

The town was full of child workers—little girls cleaning the houses of the well-to-do, boys doing odd jobs for bakers, blacksmiths, cobblers, all manner of shopkeepers. Always a good worker, Andreas put in long days on the fishing boat, but it was never enough. The earth was niggardly, the farms small, the sea fickle and dangerous. "It required such a bitter battle against the elements to get a living from the sea," a fisherman from the west coast of Norway said. One of

Andreas's brothers died at sea—went out fishing and never came back. By the time he was a young man, Andreas was getting weary of the bitter battle. He could see for himself how his life in Norway would go. The stony farm, the short summers, the cold wet winters on the North Sea—and then, if he survived into his late twenties, five years of mandatory military service. There were too many other young men like him jostling for the same scanty living. Despite the poverty and the subdivision of the farms, Norway's population grew steadily in the second half of the nineteenth century—from 1.7 million in 1865 to 2.4 million in 1910. A stunning rate of growth considering that 600,000 Norwegians emigrated during those same years, most of them bound for North America. Tens of thousands more moved within Norway— mostly to the west coast or to the north for the fish.

It was a restless, anxious time, especially for the young. With so few opportunities and so little to look forward to, young Norwegians were susceptible alike to the promise of new beliefs and the threat of old superstitions. Stories of mountain trolls and *huldrer* (seductive, elusive woodland fairies) still haunted the countryside. It was whispered in the farms around Haugesund that *huldrer* disguised themselves as beautiful girls who lured unsuspecting children into the crags of the mountains and bewitched them. The *huldrer* wore skirts to hide their long tails— so if you saw a bit of tail peeping out, you had to run for your life.

Lutheranism had long dominated religious life in Norway—but now an austere, exacting strain took hold in the impoverished coun- tryside. Children barely old enough to prattle were taught that their only hope of salvation was a sudden, overwhelming encounter with God—like a Christian thunderbolt that clove their soul in two. "It was pretty well taken for granted that no one, children, teenagers, or adults, was a Christian until such an experience had taken place," one boy from Haugesund recollected. "My mother told me once that she had a religious experience or conversion when she was nine years of age. My father, no doubt, had a similar experience at a later age. . . . I can recall very vividly that at the age of five I felt very definitely that if I died, I would be eternally lost. . . . As an unconverted person, I had no right to approach God."

In the northland's long summer evenings and the endless dark of winter, some found God, some caught America fever—and some were taken by both. The first Norwegian immigrant ship—a craft called the *Restauration* and nicknamed the "Norwegian *Mayflower*"—had departed the western port of Stavanger on July 4, 1825, and in the decades that followed, an ever increasing flow of Norwegian farmers and fishermen and laborers crossed the ocean. One young Norwegian wrote of how the fever spread, especially among the young, whose blood "stirs easily": "Very attractive tales of the broad stretches of productive soil, that just lay waiting for people to guide the plows and erect homes were sent from mouth to mouth. It was something different than stones, stumps, steep mountainsides. It sounded good."

In 1911, Andreas Brattestø turned twenty-one. He was a fine young man, well-shaped, clear-eyed, wiry, thin, and strong. While working on the fishing boat, he had fallen in love with his boss's daughter, Juline Ostrem, and hoped some day to marry her. But first he would have to do military service. And after that? How could he and Juline raise a family on what Haugesund had to offer? The year 1911 was when Andreas caught America fever. A letter came from his brother John—from Nebraska. John was homesteading on the prairie and doing all right. Doing well enough to send back to Norway for his wife and children. John made an offer: if Andreas would escort his wife and children to Nebraska, John would help him find a homestead of his own. Free land on the American prairie. Something different than stones, stumps, steep mountainsides.

Andreas bade farewell to Juline on the island of Utsira. Then he and his sister-in-law and her children made their way to Stavanger, where the Norwegian *Mayflower* had set out to the west eighty-six years before.

The Chmielewski brothers were born in a country that had not officially existed for a century. Frank and Joseph Chmielewski, sons of a Polish father and a Lithuanian mother, had a Polish surname, they spoke Polish, called themselves Poles, worshipped at a church with a Polish priest—but in the late nineteenth century Poland itself was like a body that had been operated on, dissected and sutured so many times and in so many different ways that it had long since given up the ghost. The first amputations came in 1772 when Prussia, Russia, and Austria agreed to partition the defeated and demoralized Polish-Lithuanian Commonwealth and take control, each in its own way, of the three unequal pieces. Subsequent partitions in 1793 and 1795 redrew the boundaries, altered the size and shape of the partitions, dealt further humiliating blows to Polish resistance movements, and tightened the grip of the partitioning powers. By the time Frank Chmielewski was born in 1885 and Joseph in 1896, the Polish people had grown inured to poverty, oppression, and political paralysis at home. As repeated uprisings were crushed, Polish revolutionaries, visionaries, and artists like Tadeusz Kościuszko, Frédéric Chopin, Adam Mickiewicz, and Joseph Conrad fled abroad to foment revolution in art, music, and politics throughout the West. When the Polish people rose against their oppressors once again, as they had in 1830–1831 and 1863–1864, there would be a ready supply of soldiers and martyrs. The Poles made natural freedom fighters. Meanwhile small landholders like the Chmielewskis worked with clenched teeth and prayed.

To the extent that the idea of a free and independent Poland survived in Poland itself, it was the Roman Catholic Church that kept it alive. Though conservative and repressive in some ways, the church became the major safeguard of Polish nationalism and the most powerful symbol of Polish resistance. During the century of the partitions, to be Polish meant to be Catholic: the church was the one institution that the three severed pieces had in common, the one source of hope that they might some day be reunited.

The Chmielewskis came from the Russian Partition, or Congress Poland as the sector became known after yet another reorganization was made in 1815 in the wake of the Napoleonic wars. They lived in the region's northeast corner, not far from the current Polish-Lithuanian border. Frank always maintained that he had it good in Poland. A quiet, gentle man, fair-haired, blue-eyed, and husky with a round face and a shy reluctant smile, Frank went to work for Lithuanian missionary priests, first as their houseboy and later as their carriage driver. He already knew how to speak Lithuanian from his mother, and from the priests he learned French as well. The priests took him along when they traveled abroad, and Frank picked up something of an education in the cities of western Europe. Bright and responsible, Frank made a good impression and the priests were happy to employ him and help him in any way they could. His biggest worry was his brother Joseph, his only sibling. Eleven years Frank's junior, Joseph was a passive, unassuming kid who looked a lot like Frank only paler, less forceful, somehow less substantial. Frank felt responsible for him, more like a father than a brother.

Frank never could explain why he decided to leave. There was something in the air, some impulse that young men could not resist. He was doing all right working for the priests, but somehow he came to believe he could do better in America. All around him, he saw small farmers suffering—plots had become too small to support families, grain prices fluctuated unpredictably with world markets, too many mouths to feed and never enough bread. Young Polish men were going to Germany and Denmark in search of seasonal farmwork, but they grumbled at being treated little better than slaves. Word had spread that in America wages were eight times higher than in Poland. So what if they had to dig coal out of the hills or stand for twelve hours a day in factories next to blast furnaces? Wages were wages. "I left for America because I got tired of eating turnips," one Polish farmer told his daughter. In America it would be different—better. Between 1899 and 1913, some 750,000 people staked their future on this hope and departed Congress Poland for the United States. They left for the wages, for the chance to eat something better than turnips, for the idea of returning someday with money in their pocket, for the dream that when they returned, Poland would be Poland once again. And even if they never returned, so many Poles were now living in Chicago, Buffalo, Milwaukee, Detroit, Hamtramck, Pittsburgh, and the towns adjoining the coalfields of Pennsylvania, that it was almost like home. When the day came again to fight for Polish independence, they'd raise an army of Polish soldiers in America.

In 1907, when he was twenty-two years old, Frank Chmielewski joined the exodus. Before he left, he vowed that he'd send for his younger brother Joe. He'd have a house big enough for a family—his own house, not rented—and Joe would live there with him. Frank would look after Joe and teach him how to be an American. That's what brothers were for.

Peter Thompson, born in County Antrim in 1895 and raised on the back streets of north Belfast, went to work at the age of twelve sorting flax fibers at the Whitehouse linen mill. The place was full of children. Two of his sisters, Nellie and Mary, put in nine-hour days moving spools of thread and getting showered with water that flew off the spinning flax. The girls worked barefoot, their feet immersed in dirty puddles. When the Thompson children walked home in the gloaming, kids in the neighborhood who did not have jobs teased and taunted them. Peter was a bright, eager boy, a good student at the Christian Brothers school, an avid reader. The Christian Brothers headmaster told Peter's mother Rose to let the boy quit working and continue with his studies, but Rose refused. As long as the mill offered paying jobs, her kids would work there. The family could not survive without the wages the kids brought in—it was as simple as that. At forty-one, Rose had twelve living children, including a newborn baby daughter. But not all of them were at home anymore. Mary had been liberated from the mill and sent off to the United States, joining an uncle in the mining town of Butte, Montana. Peter remembered how their father had pinned a sign to her overcoat as the girl stood weeping on the pier: TO BUTTE. After that, a book and a letter arrived for Peter every now and then from Montana, bringing a gust of fresh air. In the summer of 1911, when he was sixteen, Peter wrote to thank Mary for the latest book: "I am still working at Whitehouse and I guess I am just tired of it. . . . At present I don't see any loophole of escape out of the spinning room for me."

Two and a half years later, in the winter of 1914, Peter Thompson found the loophole and left the spinning room, left Belfast, and sailed to America.

Every family departed with a store of memories, fears, terrible regrets, wild expectations, and pangs of anguish that would forever haunt their dreams. Samuel Levin was ten when his father died in a pogrom in the Ukraine south of Kiev. The Gentile neighbors tried to help by hiding the Jewish family—Samuel, his ten brothers, his parents. But when the father left the hiding place to go look for food, he was attacked in the street. He died at the age of forty-two, leaving a widow and eleven sons. It was 1905, the year when nearly a thousand Jews died in pogroms in the Pale.

Samuel Goldberg was four years old when the anti-Semitic violence swept through the explosively growing textile city of Lodz in Congress Poland. More than a century later he could still remember the blood and the sound of that terrible word *pogrom*. "I was standing in front of our house and all of a sudden I saw a group going by me with bandages on their heads and bleeding wounds. I ran into the house and asked my grandmother what was happening. 'Pogrom,

pogrom,' she said, 'they were wounded in the pogrom.' My father went out with a gun."

"There was no life in our town," said one child of the Mezzo-giorno. "No doctors. If you got sick, you died or got well by yourself. Nobody ever left the town except to pass away or move to America. No one ever came from the outside."

"Among the various arguments for going to America, the strongest was the poverty among the common people where we lived," wrote a Norwegian immigrant. "Also, the hopelessness of ever amounting to anything."

They weren't refugees from war—that would come later—but they were refugees all the same. From hunger, passed like a disease from father to son. From villages with no water, no doctors, no schools, no hope. From state-sanctioned riots and systematic oppression and military conscription from which their sons never returned. Between 1880 and the 1920s, more than 23 million immigrants came to the United States—one of the largest population shifts in human history. Two million Italians emigrated in the first decade of the twentieth century, depleting the Mezzogiorno of more than a third of its popu-lation (though in many villages the loss was temporary since half or more of the immigrants returned). Jews from the Pale departed in comparable numbers, though unlike the Italians they rarely returned: 200,000 emigrated in the 1880s, 300,000 in the 1890s, 1.5 million from the start of the new century to the outbreak of war. In a little more than three decades, more than one-third of eastern Europe's Jews had fled to the United States, most of them bound for New York. Germans, Scandinavians, Irish, and Dutch—the primary immigrant groups in the decades immediately following the Civil War—con-tinued to arrive in large numbers; but by the turn of the century, the flow from eastern, central, and southern Europe had far surpassed anything the nation had seen before. Over a million and a quarter souls arrived in 1907 alone.

They came for work, and for freedom, opportunity, and the hope of better lives. They came to keep their sons out of the armies of kings and emperors, czar and kaiser. None of them dreamed that one day these sons would be transported back across the ocean, some in the same ships that had carried them to freedom, to fight in Europe's war.

CHAPTER 2

JOURNEYS

The journeys began the way journeys had begun for thousands of years—with a horse or mule, a cart, a loaf of bread, a bundle made of all that was necessary and the little that was precious. And of course with the wild dream that everything would be changed, changed utterly, at the journey's end. This much was familiar. But when the warmth of the last embrace faded, something cold and new closed over the travelers. People had migrated to strange lands since time began, but until the turn of the last century, they had never been *processed*. The endless shuffling lines; the sheaf of indispensable documents; the crowding and filth; the peering and prying at every body part; the bored official with the power over the fate, even the life, of families: it would all become numbingly routine in the tragic decades of war and slaughter that lay ahead. But for the millions of Europeans who set out for America in the great migration at the turn of the last century, this reduction of life to a number, a code, a stamp, a sum of money, was a shock to the system. No one had warned them that to get into the United States they first had to surrender their humanity.

The Affatato boys had descended the hill from their Calabrian village thousands of times in the daily water run. Three miles to the well and three miles back to the dry hilltop village with the enchanting name of Scala Coeli—the Stairs of Heaven. But this was different. This time when Epifanio and his brother Carmine reached the well, they did not turn back. It was just after Christmas of 1910 when the boys left for America. Epifanio, born on the day of the Feast of the Epiphany in 1895, was fifteen, Carmine twenty-three—unmistakable sons of Italy's Mezzogiorno with their black hair and deep brown eyes. Epifanio, not yet five feet tall, was still growing, but he wouldn't grow much more since he topped out at five foot two inches. Smooth olive skin, strong muscles, big heart, big dreams. Dreams most of all of New York City. Epifanio had never been outside Calabria, never been on a train or a boat, never been out of sight of Scala Coeli. But like all southern Italian kids of 1910, he'd heard so much about New York he could practically sell real estate. The money, the cars, the jobs, the beautiful women in their fur coats; Enrico Caruso singing at the Metropolitan Opera and, by the miracle of radio, booming out his arias right in your own kitchen. Everyone said that in New York you could do whatever you wanted; even a poor boy could become rich and live in a palace on Fifth Avenue—instead of leaving school at the age of ten and going to work on a road crew, as Epifanio had done, because his mother needed the extra money. Dreams so big they made his mouth water—but still, Epifanio had no idea what really awaited him in *l'America*. His father had gone before—every now and then a letter and some money arrived from a place called Brooklyn. At the end of 1910, Epifanio and his older brother left their mother behind in Scala Coeli and followed their father all the way to Brooklyn.

The long bumpy cart ride from Scala Coeli to the port of Reggio di Calabria would have been the pleasant part of the journey. From Reggio di Calabria at the toe of the peninsula's boot they boarded

a train to Naples, the first train ride of their lives. Grain fields, vineyards, olive groves, blinding views of the sea filled the gritty windows
as they rattled north through the Mezzogiorno. Naples was teeming
with immigrants in those years—"a continuous startling whirl," in the
words of the immigrant poet Pascal D'Angelo, who passed through on
his way to New York as a teenager around the same time as the Affatato boys. The ancient city on the bay was beautiful, wrote Gay Talese,
whose father emigrated from Italy around this time, but "boisterous,
dirty, overcrowded. Nothing new was being built; nothing old was
being renovated. Beggars were everywhere."

The moment Epifanio and Carmine hit Naples, chaos and confusion overwhelmed them—and they were in constant fear of being
robbed, a fear that would not subside until they were safe with their
father in America. They disembarked from the train directly into a
gamut of thieves and con artists and beggars. Back to the days of ancient Rome, literature is full of stories of wide-eyed country bumpkins
getting royally bamboozled by shrewd city rats—and the schemes and
cheats really hadn't changed much over the centuries. Shills sent out
by lousy flea-ridden hotels promised cheap lodging and plentiful food
for immigrants waiting for boats or trains. Crooks offered to exchange
their Italian lire for dollars, insisting falsely that it was better to change
the money in Italy. All manner of street vendors shouted at them with
something to sell—rosaries to beseech the Virgin's blessing for a safe
voyage; souvenirs and cheap postcard "views" of Italian scenes to bring
to their relatives in *l'America*; citrus and bread at exorbitant prices.

In the narrow streets between train station and port, the boys passed
all manner of makeshift enterprises specializing in emptying the pockets of unwitting immigrants—schools advertising instant proficiency
in English, street dentists hawking last-minute extractions that they
claimed were essential to passing the inspection at Ellis Island, scam-
mers offering to stamp their steamer tickets "inspected by the American doctor" for a favorable price. No stamp, no entry into America,
the con artists insisted. Many believed—and a few more precious coins

were left behind. "What makes the emigrant so meek in the face of outrageous brutalities," wrote Broughton Brandenburg, an American reporter who journeyed undercover from Naples to New York with his long-suffering wife in the steerage of an Italian immigrant ship, "so open to the wiles of sharpers, so thoroughly disconcerted and bewildered in the face of an examination, is his terrible dread of not being allowed to enter America. He would as soon think of cutting off a hand as doing anything that 'would get him into trouble.'"

As the hour of embarkation approached, a kind of human vortex drained into the Neapolitan waterfront, sucking passengers by the hundreds toward the office of North German Lloyd, the company that owned and operated the *König Albert* on which the boys were booked. The ship was not one of the line's more prepossessing vessels. With two funnels set rather close together in its middle and two ungainly masts at either end, the 10,643-ton *König Albert* was long (nearly 500 feet), squat, pinched and tapered like a submarine and rode low in the water. Epifanio and Carmine, of course, had never seen anything like it—though they saw precious little of it through the crowds of people and luggage that packed every available inch of harborside pavement. The ship had capacity for nearly eighteen hundred steerage passengers, and most of them converged outside the steamship broker's offices, shouting at the top of their voices and waving documents in the air. The 346 first- and second-class passengers were nowhere to be seen: they were either ensconced in their cabins already or, if they were running late, they would board calmly once the pandemonium at the North German Lloyd office subsided.

Iron railings enclosed the worn paving stones of the processing station of the Naples port authority—the Capitaneria. As soon as the Affatato brothers made it through the gate, they were told to separate their hand luggage from the larger bags or trunks they intended to stow in the ship's hold. The hold luggage had to be inspected by an agent of the American consul, inspected again by a port health department official, then registered by a North German Lloyd employee, who gave the boys a receipt for each bag they surrendered. Before any hand luggage could be carried on board, it had to be fumigated for noxious insects and then affixed with an official yellow label indicating it was safe. To accomplish this the boys had to board a creaky little boat and motor half a mile out to the fumigation station built on the harbor's breakwater—all the while glancing anxiously at the bulk of the *König Albert* and praying fervently that the ship would not embark without them. There were the usual scam artists hawking counterfeit labels for a couple of lire a bag.

Next came the human processing. Back inside the precincts of the Capitaneria the boys waited until it was their turn to stand before an official of the steamship company and answer a slate of twenty-nine questions prescribed by U.S. immigration law: Were they married? Could they read and write? Where did they reside in Italy? What was the name and address of the person they were joining in the United States? Had they ever been in prison or an almshouse? Who paid for their passage? Were they polygamists or anarchists? Did they have at least $50 with them? Carmine told the official he was holding the money for the two of them—a total of $42. They would be joining their father in Brooklyn. Epifanio was measured at four feet, eleven inches, his complexion was rosy, his hair dark. He was neither a polygamist nor an anarchist.

Then they were vaccinated against smallpox and had their health tickets stamped. Still trembling from their first encounter with a hypodermic needle, with their arms beginning to swell and go sore, the brothers passed into the jammed holding pen to wait their turn for the medical inspection. Policemen peeled off those who had been churned to the head of the crowd and directed them to a panel of doctors from the U.S. Marine Corps Hospital, the port authority, and the ship's staff. Epifanio stood rigid, his heart pounding, while a doctor folded back his eyelids to search for the redness and scarring caused by trachoma (an infectious bacterial eye disease that leads to blindness if not treated) and rubbed his head for signs of the crusted scabs of the chronic fungal disease favus. Trachoma and favus sufferers were sent back to their villages. Anyone with other obvious symptoms of illness was taken aside for further inspection. The brothers, fortunately, were robust. At last they were free to go on board.

A steep gangplank led from the paving stones of the Capitaneria to the deck of the *König Albert*. As the surging crowd carried the boys from solid land to the film of wood and iron suspended on salt water, they had no time to stop and reflect on the momentousness of the occasion. The last step on their native soil, the last breath of Italy's air, the last time they would turn their faces into the Italian sun. From now on they were aliens. The passengers pressing around the boys were nearly all peasants from the Mezzogiorno, most of them dressed in their finest clothes, the women in black shawls, the men wearing their church jackets or carrying them carefully folded over the crook of their arms, all of them burdened with as many parcels and bags as they could possibly drag on board. Even children had their hands full with baskets of food.

As soon as they were on board, the Affatatos added their bodies to the mad crush of passengers piled up at the narrow steps behind the mast. Every immigrant was bent on securing one of the more desirable berths in the cavelike steerage section sandwiched between the lowest deck and the cargo hold. On the *König Albert*, as on most immigrant ships, steerage was subdivided into large gender-segregated compartments. Children stayed with the women. The compartment where the Affatato boys ended up was dark, rank, low-ceilinged, and crammed floor to ceiling with blocks of iron-frame bunk beds. They flung their parcels down on the first empty mattresses they came to—lumpy burlap-sheathed sacks of straw set on a lattice of iron slats. No pillows or sheets. Bedding was a single blanket made of a wool-cotton-jute blend and freshly scrubbed of whatever its previous user had spilled, shed, vomited, or secreted into it. The brothers were too young and too green to reflect on how much the place looked like a prison—or a barracks.

"For such quarters and accommodations . . . the emigrant pays half the sum that would buy a first-class passage," remarked Brandenburg. "A comparison of the two classes shows where the steamship company makes the most money." The North German Lloyd company laid out about 60 cents a day on food for each steerage passenger. Profits for a single voyage ran as high as $60,000.

As the *König Albert* finally left the pier and crawled out past the breakwater, Epifanio and Carmine went up on deck to stand with the other passengers. Everyone waved frantically to the people on shore frantically waving handkerchiefs. The blasted cone of Vesuvius, still an active volcano, seemed to loom larger as the city receded and its buildings shrank into pink and white toys capped with tile roofs and gold crosses. Immigrants were sometimes treated to the sight of smoke rising from the volcano's crater. The vineyards and orange groves of Ischia and Capri glided by. After all the hours of preparation, delay, and chaos, it took only minutes for the ship to reach the open water of the fabled Bay of Naples, the filthy azure bay that had inspired so many fervid descriptions and so many pitched battles. Back to the time of the ancient Greeks, boys had died attacking and defending the cities around this bay—and they would die again. Musicians in the crowd—there were always a few—brought out guitars and tambourines. Italy receded to the wail of folk songs.

Had the Affatato boys remained in Scala Coeli, Carmine would have been drafted into the Italian army in five years, Epifanio in ten. But not in *l'America*. Everyone in the Mezzogiorno knew this for a fact. The steamship company circulars that got passed from hand to hand and village to village spread the word. There was always someone on hand in every village piazza to read to the illiterate the most important sentence in the text: in America, you went into the army only if you chose to.

And who would choose to be a soldier in a country where the streets were paved with gold?

U nlike most immigrants of the time, Meyer Epstein left the Pale of Settlement with money in his pockets. At the age of twenty-one, he had already put in ten years in Mr. Brevda's junk business. Meyer was a good worker, honest, careful, trusted by those who knew him. Brevda, his boss and patron and in some ways his surrogate father, had treated Meyer well and never begrudged him the money he made. By the standards of Jewish Belarus, Meyer had prospered. But he was convinced he could do better in America. And besides, he wanted to see his father—his real father, Yehuda Epstein, who had emigrated to the United States so long ago that Meyer could no longer remember anything about him. In the autumn of 1913, Meyer Epstein left Brevda's household and set out to seek his fortune and his father in America.

Before he started the journey, Meyer went to see his younger brother, Zender, the only immediate family he had left in Belarus. Meyer asked the boy to come to America with him. Why not? He had enough money to pay passage for both of them and then some. Zender may well have jumped at the offer—what sixteen-year-old wouldn't?—but the aunt he had been living with ever since their mother died wouldn't hear of it. "I raised him," she told Meyer, "and no one is taking him." So Meyer set out for America by himself, just as he had set out by himself on the Russian train that led him to Brevda's house a decade earlier.

The route of his passage from Belarus to the British Isles is no longer known—but somehow Meyer ended up in Liverpool. It was here, in the shadow of the soaring twin clock towers of the new Royal Liver Building, that he walked across the granite-rimmed dock, mounted the gangplank over the Mersey River, and boarded the *Lusitania*. Meyer never got over the quirk of chance that put him on such a fabulous, fateful ship. In fact, even before the disaster of its demise, the *Lusitania* was one of the most famous ships in the world—celebrated for its size, its opulence, and most of all its speed. The Cunard Line built the *Lusitania* and its sister ship the *Mauretania* in the first decade of the twentieth century as part of a fierce naval competition with the Germans. Since the days of Lord Nelson, Britannia had ruled the waves with the fastest ships afloat—but British naval pride took a blow in 1897 when North German Lloyd's *Kaiser Wilhelm der Grosse* crossed the Atlantic in two hours shy of six days and set a new record for speed. It was the first time that the Blue Riband—the prize pennant awarded by commercial shipping companies to the fastest transatlantic steamer—flew from the topmast of a non-British vessel. The British could not let this stand. Not only was losing the Riband a blow to national prestige, it was also a matter of potential military significance, since in times of war commercial ships were commonly pressed into naval service. In 1903, Cunard chairman Lord Inverclyde secured a hefty loan from the Balfour administration to construct ships capable of retaking the Blue Riband, and four years later the sister ships *Lusitania* and *Mauretania* were christened and ready for service. The *Lusitania* won the Blue Riband back for Britain on its second commercial voyage in October 1907, crossing from Queenstown to Sandy Hook in four days, nineteen hours, at an average speed of 23.99 knots. Ruling the waves again gratified Britannia, and the government paid Cunard £150,000 a year to maintain the swift sister ships in a state of readiness for war. Just in case.

For those who were able to afford it, the *Lusitania*'s luxury was legendary. The 550 first-class passengers lived like kings in public rooms fashioned after the great courts and country houses of Europe. First-class cabins had up-to-the-minute amenities like hot and cold water and telephones. Sumptuous dinners of turtle soup and jellied ham, lobster and roast beef, brandy and cigars, lasted hours. Tycoons, opera singers, diplomats, and royalty preferred the "Greyhound of the Seas" because they knew they would encounter only their own kind in the tightly segregated first-class quarters.

Barely a whiff of this elegance crossed the barrier to third class, where Meyer Epstein and some eleven hundred other immigrants, most of them from the Russian Pale, rode out the voyage in the ship's bow. The best that could be said for the *Lusitania's* third-class accommodations was that passengers slept in cabins with up to six berths rather than in huge crowded dormitories, ate food prepared in the same galley that served the ship's 850 crew members, and relieved themselves in flush toilets (a bit of a mixed blessing since the Cunard designers, working on the assumption that most steerage passengers would be unfamiliar with modern plumbing, installed toilets that flushed continuously as long as someone was sitting on them).

Meyer was lucky to have ended up on the world's fastest ship. For most immigrants, the voyage across the Atlantic was a week or more of hell. Even those inured to the primitive, waterless huts of southern Italy or the muddy shtetls of the Pale were appalled by what they encountered belowdecks. In time, they would forget the nuances of their mother tongue, the geography of their native village, the stories of wars and wolves their grandparents told them on winter nights, the names and faces of beloved friends. But no immigrant ever forgot the crossing. "I remember everything," remarked Samuel Goldberg, ninety-nine years after he sailed from Liverpool to New York on board the *Campania*, a Cunarder whose first-class luxury rivaled the *Lusitania's*. It was December 1907 and Goldberg, a little blond-haired blue-eyed Jewish boy just three months shy of his eighth birthday, was traveling with his mother and three siblings. "We traveled steerage, naturally, and all of the kids slept in the same bed with our mother. The five of us in one bed. The smell of that steerage was something I could not get rid of for a long goddamn time. It took us seven days to cross the Atlantic—no bath, we kept the same clothes on the entire time. Before we left, my mother's parents gave her a loaf of black bread and a few buttons of garlic. She would slice the bread and rub it with the garlic. You have no idea what it was like."

The stench of the unventilated quarters was what immigrants found hardest to endure. Decades after he sailed from Italy with his family when he was nine, Angelo M. Pellegrini could still describe the smell of steerage in vivid, nauseating detail: "We were packed in filthy bunks like herring in a barrel.... [The stench] was something very palpable and substantial. We felt as if we could touch it, lean against it, move it from place to place; but we could not escape it. It originated in the galley: a heavy, warm, humid, sour odor of desecrated food. It fused with the smell of acid vomit. It gathered overtones from the exhalations of sour stomachs and of dirty sweat-drenched, peasant flesh. Then, whichever way we turned, it blew into our nostrils in thick, pulsating blasts." Seasickness became epidemic as soon as the ships hit the open sea. "Hundreds of people had vomiting fits," wrote a Russian Jew named Israel Kasovich of his voyage out of Liverpool, "throwing up even their mother's milk.... The confusion of cries became unbearable, and a hundred persons vomited at one and the same time. I wanted to escape from that inferno, but no sooner had I thrust my head forward from the lower bunk I lay on than someone above me vomited straight upon my head. I wiped the vomit away, dragged myself onto the deck, leaned against the railing and vomited my share into the sea, then lay down half-dead upon the deck." Sleep was often impossible for the cries and moans of the suffering.

Rare was the steerage that had a dining room. On most ships, third-class passengers ate their meals on their beds or standing at shelves that ran alongside the sleeping compartments or, when weather permitted, squatting or sitting on bundles on the cramped third-class decks (usually situated behind the smokestacks). One passenger noted that the Italian immigrants on board a German ship failed—or refused—to understand the instructions to toss the remains of their meals over the sides of the ship or into the scuppers. Instead they cleared their plates onto the deck, which, after the first dinner, became covered with "unctuous filth that made footing very uncertain." Fights broke out when children climbed on strangers' berths with their foul sticky shoes. An entire family received a single cup of fresh water a day to share for washing themselves. Broughton Brandenburg, the American journalist traveling incognito with a group of Italian peasants on board the *Prinzessin Irene*, quailed when he overheard first-class passengers disparaging the slovenliness of immigrant families: "What dirty little imps they are." "Terrible to think of admitting such people wholesale to the United States." "There ought to be a stop put it: they are a menace to our civilization." "How," demanded Brandenburg, "can a steerage passenger remember that he is a human being when he must first pick the worms from his food?"

And yet there were moments of calm, dreamy reverie, even euphoria. Some remembered sitting on deck in the sun when the weather was tranquil and talking endlessly of the marvels that awaited them in America. At night the haunting music of Russian or Ukrainian folk songs rose from huddled groups. As a Russian song faded, another would begin in Polish or Yiddish or Norwegian. The Scandinavians loved to dance. Jewish girls who had never touched a man before found themselves dancing wildly with grinning deckhands who didn't speak a word of their language and didn't need to.

As the end of the voyage neared, anxieties mounted. "Who can depict the feeling of desolation, homesickness, uncertainty, and anxiety with which an emigrant makes his first voyage across the ocean?" wrote one Jewish immigrant of the final hours of his crossing. "The thumping of the engines was drumming a ghastly accompaniment to the awesome whisper of the waves. I felt in the embrace of a vast, uncanny force. And echoing through it all were the heart-lashing words: 'Are you crazy?'" Rumors circulated in multiple languages. Questions tormented the travelers. Was their English good enough to get past the inspectors at Ellis Island? What if relatives who had promised to meet them failed to show up? Little Samuel Goldberg had had trouble with his eyes. It wasn't trachoma—but the child's blue eyes were red and inflamed. Back in Poland, his mother Sarah had been putting in eyedrops every day—and she kept it up during the journey. What would happen when Samuel had to face the medical inspector on Ellis Island? What if he was detained—or deported? What then? What would Sarah do—and what about the other children? The Goldbergs talked about little else as the *Campania* neared New York.

Are you crazy? Meyer Epstein was coming to America with the hope of finding his father, but he had no idea where the man lived—or indeed if he was still alive. It had been so many years Meyer couldn't even remember what he looked like. Folded into his bag was a slip of paper with the name of an aunt he'd never met and a house number on New York's East Broadway. His plan was to go there and beg a room of her while he looked for his father.

Andreas Brattestø had an easy crossing compared to most. He was supposed to look after his sister-in-law and her kids on the voyage over, but since the sexes were segregated in steerage and children bunked with their mothers, there really wasn't much for him to do. Seasickness wasn't a problem, since he was accustomed to winter fishing on the rough waters of the North Sea. So Andreas spent the long empty hours at sea thinking about his sweetheart Juline back in Norway and wondering what it would be like to farm 640 acres of free land. Free American land.

At some point in the course of his journey, Andreas decided to change his name. Magnus Andreas Brattestø had boarded the ship in Stavanger; but Andrew Christofferson would disembark in the New World. When he finally reached the free open land everybody talked about, Andrew Christofferson was the name he intended to sign on his homestead papers.

Epifanio could not believe how cold New York was. A cold gray city rising from the cold gray water of an enormous bay, even dirtier than the Bay of Naples. No mountains, no leaves on the trees, no red tile roofs—just gray buildings and black chimneys spewing clouds of soot into the cold gray January sky. The great hulking ship he'd grown to hate seemed to be stuck—close to land but not going anywhere. Where was the great fabled city he'd been dreaming of? Epifanio didn't know that he was gazing not at Manhattan but at Brooklyn and Staten Island: the *König Albert* was stopped temporarily in Lower Bay at the entrance to New York Harbor while doctors came on board to inspect the cabin passengers for contagious diseases. Small wooden and brick houses huddled close together—houses that looked flimsy and makeshift after the stone buildings of Calabria. This was not how Epifanio had pictured New York. But maybe it would look different once they finally got off the ship and set foot on dry land. The waiting, the endless freezing waiting, was driving both the brothers crazy.

Finally the ship's engines rumbled to life again and the steamer began to creep through the Narrows between Brooklyn and Staten Island past scores of other anchored ships awaiting quarantine clearance and on into Upper Bay. As the New Colossus came into view—La Statua della Libertà—passengers in their hundreds crowded the port side railing to gaze out in silence. She stood on her island robed like a saint, but a saint too cold and high and severe to hear their prayers. "No one spoke a word," recalled one immigrant of the vision of Liberty, "for she was like a goddess." Just a glimpse and then the goddess and her gilded torch and her stone-girt island were behind them and Ellis Island came into view and just as quickly disappeared. Smokestacks, grime-covered warehouses, enormous tawdry billboards with strange English words. The *König Albert* steamed past Jersey City and the tip of lower Manhattan and kept moving into the wide slate gray mouth of the Hudson River. An endless gaping rank of docks and piers hung from the western shore like a jaw full of rotten teeth. A tugboat eased the *König Albert* beside one of these piers and the engines subsided. They had arrived—not at New York—but at Hoboken.

The hours that followed were a blur of unclean bodies pressing too close, cavernous warehouses, chaotic piles of bags and trunks, crying babies, yelling parents, orders barked in strange languages, long snaking lines. Trunks and rope-bound suitcases and wicker baskets were unlocked or untied so customs officials could rifle through their contents. Vendors of apples and cakes worked the crowd. Men with sticks kept the lines moving and the exits guarded. Epifanio and Carmine shoved their way through the throng, dragged their bags through customs, shuffled step by step closer to the end of the pier. They were grateful when their turn came to be loaded onto the barge and shipped back down the river to Ellis Island.

By the time they disembarked at the Ellis Island pier, the brothers were dazed by exhaustion and hunger. It had been hours since they had had anything to eat or drink. The din of foreign languages was relentless. Somehow as they entered the portal of the gorgeous palace of red brick and white limestone and carved taloned eagles and towers capped with spiked hemispheres, they were made to understand that they must surrender all of their bags and stow them in the cavernous baggage room. The only thing they kept on hand were their landing tickets marked with their numbers on the ship's manifest—Carmine was 8, Epifanio 9. Despite the sea light dazzling at the banks of high windows, it was dark inside the ground floor of the palace of immigrants—dark and unbelievably noisy. Shouts, cries, murmurs, wails, baby screams, barked commands, whined demands—everything but relaxed laughter—bounced off the tile floors and ceilings, collided, echoed, merged, amplified, and resonated in deafening waves.

After the baggage room they were herded to the back of another line, a long one that continued up steep flights of stairs and disappeared at the top. The doctors standing at the head of the stairs were surveying the flock for culls. Those too lame to mount the steps without help, those whose panting or sweating might be a sign of heart disease, anyone who seemed unduly bewildered or disoriented—all of the sick and halt had letters chalked on their shoulders. Through an interpreter, a doctor told Epifanio to hold still while he examined his face, neck, hands, and hair. Nothing wrong with him, nothing with Carmine. They continued.

The stairs were just the prologue—the real medical inspection was still to come. Seven-year-old Samuel Goldberg, his three siblings, and his mother Sarah were ushered into a kind of corridor delineated by metal pipes and bars. Doctors worked the crowd in teams. Sarah was asked to remove her hat so the doctor could see if scabs or sores lurked at the roots of her long blond hair. The children had to stop while the doctor examined each of their dirty hands and faces in turn. At the end of the corridor, another doctor performed the dreaded eye exam. Sarah held her breath while the doctor took a long hooked stick—a buttonhook of the kind that was once used for lacing up boots—and peeled back Sam's eyelids. She was frantic at the idea that the boy might be detained, sent to the hospital, and then sent back to Poland. But no. It was all over in two minutes. The family had gotten through. On to the next stage.

Trachoma, conjunctivitis, hernia, goiter, venereal disease, leprosy, ringworm, favus, dysentery, tuberculosis, mental retardation, insanity, drunkenness, impudence, surliness, obvious stupidity: any of these was grounds for detaining an immigrant for further examination and possible deportation. Several thousand were treated every year at the Ellis Island Marine Hospital, the largest percentage by far suffering from trachoma. Despite the rumors and horror stories that circulated through steerage, despite the fear and trembling with which new arrivals faced the examiners, deportation on medical grounds was rare. Some 98 percent of those who passed through Ellis Island were ultimately admitted.

The final verdict was delivered upstairs in the celestial Registry Room. Rust-colored tiles on the floor, tan tiles covering the vaulted ceiling, immense arched windows set high above an encircling balcony—it was like the nave of a church, even more crowded than Easter Sunday, vibrating with a chorus of five thousand voices, practically bursting with the convective energy of prayer. Wooden benches filled the center of the space; dusty semicircles of light slanted down from the clerestory windows. Inspectors and interpreters stationed in rows at the far end of the room called out names and the chosen would rise and rush to the altar. Immigrants knew that they would be asked to corroborate the information that had already been recorded on the ships' manifests—their final destination in the United States, how much money they had, name and address of the relative they were joining, whether they were anarchists or polygamists—but still the fear of failing to clear this last hurdle knotted their stomachs. In fact, the great rustling ceremony in the Registry Room cathedral was more or less a formality. Unless you were a criminal, a contract laborer, clearly immoral or unable to support yourself and your family, you were in. Ellis Island was a form of purgatory—but for most it was a swift, transient purgatory. For all the dread, the average time of processing was just five hours. It had to be swift—in these years an average of five thousand immigrants arrived at Ellis Island each day, and on busy days it could be twice that. The high-water mark was reached on April 17, 1907, when 11,747 newcomers passed through Ellis Island in one day.

The end went fast. Dazed or delirious, stamped papers in hand, the arrivals passed through an arch at the far end of the Registry Room and descended back to ground level on three sets of steep staircases—the Stairs of Separation. At the bottom, the river of bodies divided into three streams. "New York Outsides" turned left and exited through a door marked PUSH, TO NEW YORK that opened to the landing dock for the ferry to lower Manhattan; the "New York Detaineds" continued straight ahead to a crowded room where they sat killing time until they were met by a husband, father, cousin, friend, or some other kind soul from the Old Country willing to help them get established in New York; those stamped "Railroads" went to the right to a rail ticket office and baggage check area at the rear of the building, where agents booked their passage on one of twelve railroads or three steamship companies.

There was no time to say good-bye. Nobody told them anything. They turned right or left and disappeared forever into the vastness of America.

W here was I to go? What was I to do?" a twenty-year-old Italian immigrant named Bartolomeo Vanzetti wondered as he finally set foot on the pavement of New York City in 1908. "Here was the promised land. The elevated rattled by and did not answer. The automobiles and trolley sped by, heedless of me."

S am Goldberg, still glowing with joy at having passed the eye exam, was eating soup and bread in the "New York Detained" area with his mother and siblings when the announcement came that Mrs. Goldberg's husband was waiting for them. "They didn't even let us finish the soup," recalled Sam decades later. "For that son of a bitch." It was December 21, 1907, the shortest day of the year, when the seven-year-old Polish-Jewish boy was reunited with the father he had always hated. Sam started his life in America with a half-eaten bowl of soup and a gnawing ache in the pit of his stomach.

D omenico Affatato was there at Ellis Island to meet his sons Epifanio and Carmine just as he said he'd be. January 27, 1911. They embraced and kissed and the boys shared the family news from Italy. Then they went out to the ferry landing to catch a boat to lower Manhattan, where they would get on a train for Brooklyn. Domenico could not understand how his wife could have sent Epifanio from

Scala Coeli to America in the middle of winter without an overcoat. The boy shivered all the way to Brooklyn.

The newly renamed Andrew Christofferson shepherded his sister-in-law and her children to the railroad ticket office and counted out the money for their fares to Nebraska. Andrew's brother, waiting for them in Nebraska, had told him exactly what to say and how much the tickets should cost. The clerk gave Andrew a tag with big square letters on it—abbreviations of the names of stations where they would be changing trains en route to Nebraska. Andrew had no idea that in 1911 he was already a generation too late to claim a Nebraska homestead. The good farmland on the prairie was long gone. The wide open spaces of the heartland were not as wide or as empty as they once were.

Tony Pierro and his cousin trudged through Ellis Island in the company of hundreds of other southern Italians who had come over on the Italian liner *Stampalia* in August 1913. They found out fast that August in New York was almost as bad as the malaria-ridden Mezzogiorno. But Tony and the cousin were not staying in New York long enough to care. In Swampscott, where they were bound, there would be sea breezes and big white hotels facing the blue-gray water. And there would be jobs. Lots of jobs for young immigrants from Italy. When they emerged at the bottom of the Stairs of Separation, Tony and the cousin headed for the railroad ticket office and bought two one-way tickets to Boston.

The arrival of the *Lusitania* in New York was always an occasion, even when it failed to break another speed record, and so it was on October 10, 1913, when the noble ship delivered the sixty-four-year-old painter William Merritt Chase, celebrated for the rich flattering color of his portraits and the flamboyant style of his dress, to the Cunard dock. Chase went on shore in a small stately procession of first-class families—Blackstone, Dempsey, Williams, and Barrows. The steerage passengers, most of them from the Russian Pale, with a few families from Belgium and Germany mixed in, followed. Meyer Epstein disembarked from the *Lusitania* with a thousand other immigrants. While William Merritt Chase made his way back to his Manhattan town house, Meyer Epstein, his pockets lined with his modest savings and the address of his aunt, joined the procession of Jews that had been flowing for decades now from Ellis Island to the Lower East Side.

From Belfast to Liverpool, from Liverpool to New York, from New York to Butte, Montana. In the first weeks of 1914, eighteen-year-old Peter Thompson traveled halfway around the world so he could quit working in a linen mill and start working in a copper mine. Peter was the kind of boy people used to call "likely"—quick-witted, beguiling, magnetic, confident, eager. Though small and slight for a laborer, he had the wavy glossy dark hair, fair complexion, and arresting blue eyes of a matinee idol. Being handsome doesn't count for much in a copper mine, but it would help Peter in every other part of his life.

The date was February 22, 1914, not quite the dead of winter but close to it, when Peter and a buddy disembarked at Ellis Island and boarded the train for Butte. The great tough back of North America was frozen solid under the wheels of the train. Butte was awash in labor disputes, but that didn't stop immigrants from converging there. They came for the jobs, and as long as the bosses were hiring, they'd keep coming.

Generals in Germany already knew that war was inevitable. Had he stayed behind in Belfast, Peter would have been in the trenches in

France or Belgium by the following winter, another Northern Irish lad fighting for the English King. But not in America. In America he'd be burrowed deep in the richest hill in the world doing his bit to supply armies and industry with copper wire. And who knows—maybe one day he too would strike it rich.

CHAPTER 3

STREETS OF GOLD

On his first morning in America, Epifanio Affatato got up early and went out to look for gold. Even in the eyes of a sixteen-year-old boy from a poor town in Calabria, Park Avenue in Brooklyn—a street of modest row houses in a working-class neighborhood between Ridgewood and Bedford-Stuyvesant—did not look like a promising place to go prospecting. The buildings were small, brick or wood, low and close together; skinny trees stood naked in the winter wind; the people on the street looked almost as poor as the villagers back in Italy. But Epifanio had heard that in America the streets were paved in gold—and why would so many people say it if it wasn't so? So much was strange in this big crowded country—anything seemed possible. Though he had to admit that nothing so much as glittered when he stepped out of his father's place at 472 Park Avenue at Skillman Street and gazed at the pavement.

"Try again tomorrow," Carmine told his kid brother when Epifanio came back empty-handed.

Maybe Park Avenue in Brooklyn wasn't paved with gold, but Madison Avenue a few miles away in Manhattan where J. P. Morgan had his palace might as well have been. At the time Epifanio and his brother Carmine landed at Ellis Island, Morgan's companies controlled some $22 billion in capital—capital that in turn controlled steel, railroads, electrical production, farm machinery, insurance—making his financial empire the single biggest entity in the U.S. economy, including the government. All of it powered by cheap immigrant labor.

In J. P. Morgan's America, nearly 2 million children under sixteen— Epifanio's age—were working full-time and earning pennies a day. On March 25, 1911, two months after Epifanio and Carmine disembarked from the *König Albert*, some 146 garment workers, most of them young Jewish and Italian women and girls, perished when the Triangle Shirtwaist Factory caught fire in lower Manhattan. Immigrant kids put in ten-hour days in the coal mines of Pennsylvania and Ohio for 50 or 60 cents a day, while their fathers, who didn't earn much more, died by the score in mine explosions. At a time when a dozen eggs cost 75 cents and steak sold for $1 a pound, female workers counted themselves lucky if they earned $4 a week.

"Alexander's Ragtime Band." The debut of Jell-O and the dime novel. Baseball unchallenged as the national pastime. Two-reelers cranked out by D. W. Griffith, Mack Sennett, and, starting in 1914, Charlie Chaplin. The IWW—the International Workers of the World, better known as the Wobblies—recruiting workers at the bottom of the economic ladder—Slavic miners in Pennsylvania and Colorado, migrant farm workers in California, lumbermen in the Pacific Northwest, unskilled factory workers in grimy midwestern cities—and urging them to rise up and seize control of the means of production. NAACP-founder W. E. B. DuBois speaking out against "the crying evil of race prejudice." An average life expectancy of fifty years, four months, for American men, fifty-five years for women—though of course if you were black or poor, lived in a tenement or worked in a coal mine, your days were likely to end much more quickly. Such was the America where Epifanio Affatato went prospecting for gold in Brooklyn on a cold January morning.

In the great cities, trolleys had begun to replace horse cars as the prime mode of transport, and electrical wires sprouted and branched over the packed streets. Urban homes and businesses rapidly connected to the electrical grid. The first subway trains began rumbling under the streets of Manhattan in 1904. Trains sped from New York to Chicago in eighteen hours. The number of automobiles on American roads surged from 8,000 registered in 1900, to 77,988 in 1905, to 458,000 in 1910 (two years after Henry Ford introduced the Model T at a sticker price of $850), to 2,490,932 in 1915, to some 8 million in 1920. When William Howard Taft became the twenty-seventh president in March 1909, his considerable bulk was transported to the White House by horse-drawn carriage. Four years later, Woodrow Wilson rode through the streets of the capital in the back of an automobile. The inaugural horse-and-carriage had vanished forever.

The technology of killing was also changing fast. The year Epifanio arrived, a U.S. Army colonel named Isaac Newton Lewis perfected a lightweight machine gun that would be widely used in the coming war and proved to be especially well suited to being mounted on aircraft. It was a banner year for firearm advances. In March, the U.S. Army designated a semiautomatic pistol designed by John Browning as the standard issue military sidearm—and Browning's M1911 Colt pistol would remain the army sidearm of choice through the next four major wars until it was retired in 1985.

All of this newfound national wealth, speed, illumination, connection, power, and glory arose as record numbers of immigrants poured into the country. "Through this metal wicket drips the immigration stream," wrote H. G. Wells after visiting Ellis Island in 1906, "all day long, every two or three seconds an immigrant, with a valise or a bundle, passes the little desk and goes on ... into a new world." By 1914, after decades of this steady drip, one out of every three Americans was either an immigrant or the child of an immigrant. In the course of thirty years, the new arrivals changed the face of America—just as radically and profoundly as America changed them. It went beyond the proliferation of foreign languages, the darkening of complexions, the pungent smell of city streets at suppertime. It was more than a matter of available muscle to dig tunnels, extract ore, wash clothes, man production lines. Through the golden door streamed a hungry army—hungry for food and freedom, but even hungrier for work. The immigrant's willingness—indeed, need-driven compulsion—to get down to work made modern America possible. New sources of labor and new technology came together perfectly to foment America's industrial revolution.

When war came, the same reciprocal forces went to work. Only in war the formula became infinitely more complicated because it was no longer just sweat and money but now also blood that was the medium of exchange.

Epifanio perceived almost nothing of this upheaval in his first days in *l'America*. He had sailed to the New World—but landed in a makeshift replica of the old. The streets of his Brooklyn neighborhood were crowded with Italian families and Italian shops; Italian newspapers were for sale in the newsstands; Italian was spoken at home, in fact in just about every home on every street around his. After the shock of lower Manhattan, with the tall buildings "seemingly shooting out of the water like so many sky rockets," as one Italian immigrant wrote, and the electric trolley cars rattling down the middle of the avenues, and the people covered in wool and fur jostling each other—Brooklyn had a familiar feel and smell. An Italian smell.

The America that the newcomers saw in their first hours after Ellis Island was mean and poor and wretched—so much uglier than anyone had led them to expect. "Noise is everywhere," one Italian immigrant wrote of his first impression of New York. "The din is constant and it completely fills my head. . . . The streets were full of horse manure. My town in Italy, Avellino, was much more beautiful. I said to myself: 'How come, America?' On hot days when the manure dried, the wind lifted it into the air like confetti and breathing became difficult." Maria Valiani, who emigrated from Tuscany as a little girl in the same year as Epifanio, remembered her mother's horror at all the shabby little wooden houses they saw on the train from New York to Chicago. "Ugly wooden houses. We always had stone houses in Italy. You don't see no wooden house over there. My ma says, 'What kind of a house is that? People live in those houses? Ugly. All ugly cottages.' Oh, I'm sorry we came here. Our little town was better. We could see that beautiful Viareggio, the ocean. Mama mia, how we gonna live over here? What kind of a place is this?" Maria's mother was also stunned as she gazed out the train window at the vast quantities of salt that Americans hoarded. "Angelino, will you tell me why you need so much salt in America?" she demanded as soon as she was reunited with her husband. "How many houses of salt do they have over here?" She was mortified when her husband explained that "sale"—the Italian word for salt—had a different meaning in English.

Other Italians were amazed to see the word "Ave" written on signs hung all over this godforsaken Protestant nation—Ave A, Fifth Ave, Atlantic Ave, Commonwealth Ave—though for some reason the Americans never followed Ave with Maria.

The typical immigrant left Ellis Island with $8 to his or her name. Shelter, food, and work were urgent crises that quickly pushed everything else from their minds—and for most, the only recourse was to get the help of family. Never were the bonds of kinship, no matter how distant, more important than in the immigrant's first days in America.

By the time Meyer Epstein found his way to his aunt Dora Gornitz's flat at 225 East Broadway on Manhattan's Lower East Side, the grandeur of the *Lusitania* and the clamor of Ellis Island had receded before an onslaught of new impressions. The crowds Meyer passed in the street were unlike anything he had ever seen before—unlike anything anyone had seen before in this country. Even by New York standards, the density of the Lower East Side in the first decades of the last century was unparalleled. In 1910, some 2.3 million people packed Manhattan—nearly 800,000 more than in the year 2000—and the most densely packed part of the island was the Lower East Side. Some blocks housed nearly a thousand people, almost all of them foreign-born. When Meyer Epstein arrived in October 1913, three-quarters of New York's Jews lived south of Fourteenth Street and east of the Bowery; half the Italians lived in three wards bordering Canal Street. It was "a gray, stone world of tall tenements," according to Yiddish writer Leon Kobrin, "where even on the loveliest spring day there was not a blade of grass. . . . The air itself seems to have absorbed the unique Jewish sorrow and pain, an emanation of its thousands of years of exile. The sun, gray and depressed; the men and women clustered around the pushcarts; the gray walls of the tenements—all looks sad." The cacophony of the streets, the unyielding stone and cement, the press of bodies, the smells rising from the pushcarts would have been overwhelming to a young man who had spent his days traveling the Pale from village to village with a horse and cart. It was as if every shtetl in Russia had been swept up and dumped on a few blocks of lower Manhattan. "America was . . . noise," recalled Jewish immigrant Aaron Domnitz of his most dominant memories of the Lower East Side. "Centre Street . . . was at that time paved with rough, uneven stones, so the constant clopping of horse hooves and the hard wheels of freight wagons drifted overhead. It hummed, whistled, and clamored. . . . [I] wondered, is this America?" Everyone was on the make, selling, buying, haggling, yelling, working or looking for work.

It was probably from his aunt Dora that Meyer learned that his father Yehuda had died two years earlier. The father he had not seen for fourteen years and could not remember—the father whom he dreamed would receive him with open arms in America, help him find

work, get him established, help him to bring his younger brother over
from Russia. Yehuda Epstein had left his wife, Sarah, pregnant back in
Russia in order to make a better life for himself, and presumably one
day for his family. But neither plan worked out. He ended up working
as a kosher butcher—though evidently not a very successful one for he
never sent his sons a nickel. In 1911, two years before Meyer boarded
the *Lusitania*, Yehuda Epstein died at the age of forty-three. Influenza?
Tuberculosis? Meyer never discovered what had killed his father; nor
did anyone ever tell him why Yehuda had died so poor that he had
to be buried in a pauper's grave funded by the charity of a German-
Jewish nobleman and philanthropist named Baron Maurice de Hirsch.
In time, Meyer would locate his father's grave in the Baron Hirsch
Cemetery on Staten Island and buy a stone to mark it. It always pained
him that his father's American years remained a blank.

Aunt Dora's apartment was in a tenement building with the bath-
room down the hall. The smell of chicken fat, cabbage, and human
waste permeated the place. Behind each door lived who knew how
many people, children sleeping on chairs or under kitchen tables,
greenhorn lodgers crowded into back bedrooms never touched by a
ray of light or a breath of air. Still, no matter how squalid the place
was, Meyer was grateful to Dora for taking him in, giving him a
toehold in America, sharing what little she had. And anyway, the real
life of the Lower East Side was on the street. Whenever he stepped
outside his aunt's building onto East Broadway, Meyer was caught up
in the swift current of the main artery of Jewish commerce, culture,
religion, debate, struggle, aspiration. A block away was the famous
Garden Cafeteria, where Yiddish journalists, writers, intellectuals,
labor leaders, and union organizers gathered. A few doors down was
the office of *Forverts*—the *Daily Forward*—the socialist Yiddish newspaper founded and edited by the brilliant, dynamic Abraham Cahan.
Artists Chaim Gross and Sir Jacob Epstein took art classes at the
Education Alliance at 197 East Broadway, one of the first settlement
houses. Synagogues were everywhere. East Broadway before the war

was the epicenter of Jewish life in New York—which really meant Jewish life in the world.

More than half of the eastern European Jewish immigrants in New York earned their living doing manual labor—and Meyer joined their ranks. He went to work in the scrap metal business, because that's what he knew. Hauling junk in Russia or scrap metal in New York—what was the difference? Meyer's employer provided him with a horse and cart and he soon got to know the city up and down. One day his boss told him to deliver a load of 200-pound radiators to a construction site way up town. Meyer took the horse onto Broadway and headed north. By the time he found the worksite, it was lunch hour and the foreman told him to wait—that he'd get a couple of guys to help him unload after lunch. Either Meyer didn't understand the man's English or he chose to do things his own way. When the crew got back, they found all of the radiators stacked up neatly by the side of the building. "Who unloaded this wagon?" asked the plumbing contractor in charge, a fellow by the name of Sam Minskoff from the Lipsky and Rosenthal plumbing outfit. "I did," Meyer replied. "What do they pay you?" Minskoff wanted to know. "Why?" asked Meyer. "Whatever it is, I'll double it." And that's how Meyer Epstein got into the plumbing business.

In this country immigrants of the same town stick together like a swarm of bees from the same hive," wrote one Italian immigrant with a flair for metaphor. Family by family, entire Sicilian, Calabrian, and Sardinian villages more or less reconstituted themselves in American cities and towns. In their American churches, *paesani* staked out an altar for their own patron saint and celebrated the saint's feast on the same day and in the same way as it was done back in Italy—the same statue, the same procession, the same music. Villagers lived together, prayed together, braved the strangeness of the New World together by making it less strange. In Boston, Chicago, New York, Philadelphia,

Pittsburgh, and New Haven, Little Italies sprang up with all the regions of the Mezzogiorno compressed into a few blocks. The biggest difference, aside from the American setting, was the preponderance of males. In the Italian colonies, it was always the men who came first and made up by far the largest portion of their population. Fathers, uncles, sons, and brothers were the pioneers. Once they established a beachhead, the men went back to Italy to father more children and bring more young men back with them. It was often years or decades before the women joined them.

And so it happened with the Pierro family. In 1890, four months after marrying, Rocco left his wife Nunzia pregnant with their first child in Basilicata, sailed to America, found his step-uncle Vito living in a boardinghouse in Swampscott, moved in with him, and went to work. Swampscott was not, on the face of it, a likely place for an Italian immigrant to wash up. Settled in the seventeenth century by Yankee fishermen, the coastal village north of Boston had become a posh summer resort soon after the Civil War—a velvety enclave of clipped estates, vast resort hotels, manicured golf courses, vine-trimmed porches, and mansions set back from rocky promontories or fronting long sandy beaches spread before the Atlantic. All those rich leisured Boston Brahmins needed workers to build their beautiful houses, tend their gardens, drive their carriages (and eventually their cars), pave their streets. Who better to make things beautiful than the Italians? By the time Rocco showed up, there were a couple of distinct Italian islands in the elite WASP sea of Swampscott. Rocco moved in as a boarder with the Buffalino family at 10 Shelton Road and took what jobs he could get. For a while he worked setting up telegraph poles alongside the train tracks—hard, dirty work, not that different from squeezing grapes and melons out of the dry stony soil of Forenza. At some point around the turn of the century, he went to work in neighboring Lynn for a florist named Archie Miller. Miller was a good man. Eventually he helped Rocco get a job with the town of Swampscott working for the parks department, a job he kept for years.

Rocco was lucky. He had steady work. He always made enough to support himself and to send back money to his wife and kids in Italy. Swampscott was a clean, safe, beautiful place to live compared to Boston's North End or New York's Lower East Side or Chicago's Little Hell. Most Italian immigrants had it a lot tougher. The men typically lived in packed boardinghouses, slept three to a bed, six or seven beds to a room. Six days a week they worked—long, backbreaking, monotonous days usually spent outdoors with a pick and shovel in their hands. Italian workers laid track for the Pennsylvania Railroad and excavated tunnels for the New York City subways; they dug coal and talc in the mines of the interior East and the Midwest; they shined shoes on street corners and cut the hair and beards of other men more fortunate than they. The Ashokan Dam in New York, the Brooklyn Navy Yard, the iron mines of Wisconsin and Minnesota, the stone quarries of New England, the cigar factories of Tampa, the stockyards of Chicago and Kansas City: all were built and run on the muscle of men from the south of Italy. The *paesani* arriving on steamers from Naples found out soon enough that not only were the streets of New York not paved with gold, many were not paved at all—and it was their lot to pave them. "Everywhere was toil," one immigrant wrote, "endless, continuous toil, in the flooding blaze of the sun, or in the slashing rain—toil." Same for the women who raised the families, ran the boardinghouses, took in piecework, or stood all day in factories. Toil.

Rocco had been living in America for nearly a quarter of a century, off and on, with trips back to Italy every couple of years, when his third son Antonio joined him in 1913. Tony, as he became known in America, boarded with his father at the Buffalinos' house on Shelton Road—$3 a week for a place to sleep, decent food, and clean laundry courtesy of Mrs. Buffalino. The house was nothing special—gray shingles, white railings, hip roof, grapevines growing in the yard—but compared to Forenza it struck Tony as quiet and orderly and it was better than living with the livestock. A store on the corner sold candy; a bookie lived upstairs, so there was always coming and going when the horses were running; every once in a while the knife-sharpening man would come through on his bicycle yelling his trade up at the houses.

The Pierro men were an odd combination of genial and fastidious. They liked their possessions, especially their houses and yards, to be immaculate; they were proud of where they'd come from and proud of where they ended up. But at the same time they were relaxed, warm-hearted guys who enjoyed shooting the breeze over a plate of pasta and a glass of wine. Tony was the most fastidious one in his family—and also the sweetest-tempered. At nearly five feet, five inches, he was slimmer than his older brothers and his dark brown hair was a shade lighter. His posture was perfect, his gaze steady, and he liked to look good—clean and snappy. Everyone always said he was a perfectionist. Soft-spoken, stubborn when he had his mind made up about something, Tony was also quite the charmer. Girls were drawn to him, and he was always gallant and flirtatious in their company—a real gentleman, as they used to say.

So it was something of a blow to Tony that his first job in the United States was working as a laborer on a rich man's estate. Eugene H. Clapp, one of the WASP *prominenti* of Swampscott, hired Tony and his brother to work on the grounds of his glorious seaside villa at Galloupes Point about half a mile from where the Pierro men boarded. The brothers built beautiful stone walls around sunken rose gardens, on terraces, along the shore. In the summer they dug, pruned, mowed, weeded, clipped, staked, watered, and harvested. Tony was not one to complain, but he made it known that he wasn't happy working as a gardener. It wasn't a job fit for a man with dignity, even if the man in question was only eighteen years old.

Andrew Christofferson arrived at the town of Larimore in eastern North Dakota with just one word of English at his command: hello. But he didn't need to know English to figure out fast that the American Midwest wasn't all it was cracked up to be back in Norway.

The first disappointment had come in Nebraska right after he delivered his sister-in-law and the kids to his brother John, as promised. "Come to America—free land!" they all said in Norway. But Andrew was twenty years too late for anything free and arable in Nebraska. If he wanted a homestead with decent soil, he'd have to keep moving—farther west. He had another brother named Tollef working in Larimore, just east of Grand Forks and the Red River Valley, and Andrew decided to try his luck out there. Word was that the bonanza farms in the Red River Valley were as large as counties, with plenty of work for young able-bodied men from Norway or Sweden or the Ukraine. Andrew got a job on one of the sheep farms, saved what he could and kept his ears open. He soon picked up more English than "hello"—though whatever he said always came out with a strong Norwegian accent. Whenever there was talk of homesteading, Andrew paid close attention. He knew there had to be free open land somewhere in all this expanse of grass.

Growing up in Norway, Andrew had not been especially religious. His family was Lutheran, more out of habit and expectation than from deliberate choice, but it wasn't something they dwelled on much. The pietistic religious revival sweeping through the west coast of Norway swept past his family's farm in Brattestø and kept moving. So it was something of a surprise to Tollef when Andrew went off to a Nazarene camp meeting in Larimore one summer day—and even more of a surprise when he came back practically incandescent with grace. "That man preached what I believed," Andrew told his brother.

Religious conversion, when it comes, takes everyone in a different way. The way it took Andrew was to instill in him a simple but unshakable belief that he was sanctified in Christ and could no longer sin. "Andrew, that will never last," Tollef told him. But it did last. Andrew Christofferson was a changed man. He had always been humble, gentle, unassuming, honest—a person who loved animals and whom animals loved back—but henceforth Andrew was a devout Christian. A simple strict Christian who devoted himself to living a holy life and who shunned strong drink, cards, gambling, dancing, swearing, or taking the Lord's name in vain. A man of God. "That man preached what I believed"—and he believed it for the rest of his life. Andrew became a Christian under the tent at Larimore, and a Christian he remained.

Not long after his conversion, Andrew learned that there was open land to be had in Montana. Dry short-grass prairie that greened up in a good year in May and then quickly burned to gold and brown and tan under the relentless summer sun. Land about as different as land can be in character and climate and topography from the rain-washed rocky hills tumbling down to the sea that Andrew had known back in Norway. America's short-grass prairie looks like ideal farm country— flat, clear of trees, "just tickle it with a plow and it would laugh with a beautiful harvest," one Norwegian farmer claimed. But as Andrew and thousands like him would discover, this western prairie was only marginally arable. You could tickle all you wanted, but if there wasn't enough rain, harvests were miserable. Still, in the first decades of the twentieth century, lots of it was free to any homesteader who cared to stake a claim—and to Andrew, that was all that mattered.

In June 1913, Andrew rode James J. Hill's Great Northern Railway across North Dakota and kept going west until he reached the town of Havre, about thirty miles south of the Canadian border in pretty much the dead center of Montana. At Havre he found the land office, and on June 18 he filed on a quarter-section of land, a 160-acre square cut out of the sea of grass north of the tiny town of Chinook and south of the even tinier town of Elloam. When his papers were duly submitted and stamped, Andrew took the train back to Chinook, got off, and walked the twenty-two miles to his new home. No one was around to witness his arrival, and he didn't speak of it later—but it seems likely, considering what kind of man he was, that Andrew fell to his knees when he reached his land and thanked God for this blessing.

Tommaso Ottaviano was seventeen years old when his mother Antonia sold everything she owned in Italy and moved with her five children from the southern Italian hill town of Ciorlano to Providence, Rhode Island. Young as he was, Tommaso was already the man of the family. His father, Ascanio, had died of diabetes some years earlier, and Tommaso as the firstborn and the oldest son was expected to take over where his father had left off. It was a role that suited him well. With his straight black hair slicked back from his high forehead, intense dark eyes, shoulders squared like a soldier, Tommaso had the air of someone who could be counted on. Though he was slender and handsome, with thin arching eyebrows like Rudolph Valentino, he had substance.

Tommaso stood by his mother through all the chaos and upheaval of the voyage out of Naples, the frantic attempts to connect with relatives once they landed in America, the search for a place to live in the Italian community of Lymansville, north of Providence. And of course once they got settled into the house on Emmanuel Street, he went to work to support the family. Most of the Italian guys in the neighborhood worked in one of the textile mills that ringed North Providence—it was pretty much the only game in town for an Italian immigrant unless you wanted to sell canned tomatoes and salami in a corner grocery—so that's where Tommaso looked for a job. He got hired as a machine operator at Esmond Mills, a blanket manufacturer famous for its supersoft Esmond Bunny Baby Blanket.

Those first years in Lymansville were hard ones for the Ottaviano family. Five kids, no husband, a strange country—Antonia thanked God she had a son like Tommaso who went to work every morning and turned his modest pay over to her every week. Antonia was a tough, resourceful lady—and she would get tougher in the years to come in America—but with Tommaso beside her, she would make it. As for Tommaso, he was proud to have his mother's love and respect, proud to be treated like a man, proud that he had landed a job so quickly. In a photograph taken soon after they came to the United States, the Ottaviano family faces the camera stiff and unsmiling in their starched ironed Sunday clothes—Antonia and Tommaso alone are seated and they take the exact same pose, right elbow resting on the arm of the chair, left hand on left knee, while the four younger children (three sons and a daughter) stand deferentially beside them. Tommaso in a dark suit and high lace-up boots looks at ease, slim, dignified, intelligent, reliable—a young prince, summoned to rule before his time, who has put youth behind him.

South Fork, Pennsylvania, had two claims to fame when Frank Chmielewski came to settle there. The first was the town's association with the devastating Johnstown flood of May 31, 1889, in which some twenty-two hundred people died when the dam maintained (badly) by the exclusive South Fork Hunting and Fishing Club burst. The second was the extraordinarily deep seams of coal running inside the hills that hemmed in the town. South Fork the club was the haunt of millionaire tycoons like Andrew Carnegie, Andrew Mellon, and Henry Clay Frick. South Fork the town, a dozen wooded miles away, was the home of eastern European immigrants willing to devote their lives and their health to mining coal. Tucked deep in a valley carved by the Little Conemaugh River, South Fork was all about coal—extracting it from the hills, loading it onto trains, sweeping and wiping and washing its dust out of one's clothes and house, and, for the men who mined it, coughing it out of their lungs.

In 1907, when Frank Chmielewski moved here from Congress Poland, the town had a couple of blocks of stone-and-brick shops, a bank, an opera house, half a dozen hotels, and a few fine homes belonging to the Welsh and Dutch Protestant mine owners. A few years later, the stately dome of the United Methodist Church put a cap on South Fork's splendor. Across the Little Conemaugh River, on the other side of the tracks, stood the satellite village of Fifficktown—and it was here that the Polish and Slovak and Greek and Lithuanian immigrants who dug the coal had their cramped wooden houses and squat wooden churches.

A small shallow river was all that separated Catholic immigrant Fifficktown from Protestant second- and third-generation South Fork, but it might as well have been an ocean. It would take a major social upheaval—or a war—to bridge the divide.

Frank lived in Fifficktown near the Oak Street bridge on a narrow street that ran perpendicular to the river and steepened as it climbed. He rented a spare room from another Polish family and worked six days a week at the Argyle Coal Company. Six days a week he'd cross the river, walk in the damp morning beside the train tracks, grab a small bucket of beer at the South Fork Brewery, and keep going west out of town and up past the "boney piles"—heaps of tailings discarded outside the mines—until he disappeared into the side of a hill where the coal was hidden. Two and a half miles, twice a day, on foot, rain or shine. What with the smoke from trains and the soot of coal dust that blew through the air and the yeasty smell of the brewery and the sour reek of garbage and bottles piled behind the hotels that lined Railroad Avenue, it was not an uplifting walk, even in the best of weather. But Monday to Saturday, if he wanted to work, Frank had no choice. Sunday was for church, though the women, when there started to be Polish women in Fifficktown, tended to go every morning. Two years after Frank arrived, St. Anthony's—the Polish church—up on Mountain Avenue burned down. The next year, 1910, the parishioners built a new and better one, right beside the river on Portage Street, just a few doors down from where Frank boarded. It cost a pretty penny for a poor immigrant mining community—$10,000—but the steeple was a lovely sight and the windows were colored glass and Father Jastzemsky was one of their own and gave the sermon in Polish. Young as he was, Frank quickly emerged as a leader in the St. Anthony's congregation. When there was work to do on the church building, when parishioners were in need, when there were letters to be written or official documents to be deciphered, Frank could always be counted on. The education and polish he had acquired from those priests back in Poland stood him in good stead. Even small, poor communities have pillars, and Frank was a pillar of Fifficktown.

Oak Street, where Frank lived, pretty much mapped the soul of Fifficktown. It started at the river, ran by the train tracks and the church, climbed past the miners' houses and the tiny yards bursting in summer with cabbage, beets, and tomatoes, and finally petered out at the Catholic cemetery that overlooked the bony pile of yet another mine. Train, church, houses, mine, cemetery. And taverns—five in Fifficktown alone, a couple of dozen more in South Fork—little islands of fleeting comfort in a deep valley where the sun rose late and set early and the coal dust darkened the windowpanes and sprinkled the laundry hung out to dry. "More taverns than churches," the women said and shook their heads. Frank got used to starting his evenings at the Blue Mirror Tavern with the coal miner's round of choice: a shot of whiskey chased down by a beer. "To clear the throat and wash down the dust," they said.

In 1912, five years after he settled in Fifficktown, Frank managed to get his kid brother Joseph out of Poland. In Poland, it had just been the two of them—no other siblings—and after Frank left he worried about how his quiet, withdrawn younger brother would get by without anyone to look out for him. So Frank put aside what money he could spare from his wages until he had enough to pay the kid's passage from Poland. Joe Chmielewski was fifteen years old when he arrived in America, already broad and stocky like his brother, plenty strong enough to go to work. Frank got the kid a job at the Argyle Coal Company—what else was there in Fifficktown?—and without really thinking about it or making a choice, Joe slipped into the stream of Polish life in the small central Pennsylvania mining town. He boarded with the family of Walter Yablonsky, taking over a back room of a little house a couple of doors away from where Frank lived. Weekdays Joe joined Frank on the long walk to the mine. Sunday mornings the brothers stood outside St. Anthony's with the other guys to talk and smoke until the second bell rang, whereupon they all stubbed out their cigarettes and cigars and filed in to join the women. On Saturday night

there were dances in the Polish hall—loud thumping polkas and ober-
eks that made you work up a sweat no matter how cold it was outside.
Young as he was, Joe fell into the habit of hitting the tavern on the way
home from work, though he was too shy to join the rowdy drinkers and
card players who kept the place going strong to all hours. He never really
developed a taste for alcohol—just enough to get the coal dust out of his
throat. Joe and Frank became close despite the eleven-year difference in
age and the years they had spent apart.

Joe saw right away how much respect Frank commanded in Fifficktown, how much the other Polish guys looked up to him. At twenty-
five years old, Frank was a mature, solid man—a good worker, a devout
Catholic, admired by all who knew him. In short, Frank Chmielewski
was an excellent prospect for a husband. It was finding the wife that
was the problem, since there were so many more men than women in
the Polish communities. That problem ceased when Frank met Mary
Yablonsky, a nineteen-year-old beauty with dark hair, eyes sparkling
like wet coal, a high forehead, a full generous mouth. The daughter
of a Polish career soldier, Mary emigrated in 1912, the same year as
Frank's younger brother Joe. "In Poland you'll be the wife of a farmer,"
her father had told her; "you'll milk cows and dig in the ground. But
in America you can learn, you can be more than a farmer's wife." So
Mary came to Fifficktown to join her sister Josephine and brothers
John and Walter, and in no time she had every eligible Polish bachelor
in town lining up for her. Frank always said that the day Mary chose
him over the other fellows was the proudest day of his life. The two
were married in January 1913. The fact that they barely knew each
other and that there was a seven-year difference in age struck no one
as an impediment. Love would come after marriage. They knelt side by
side in St. Anthony's and Father Francis Piltz, who had just taken over
as priest, blessed them and solemnized their union. A year later their
first child was born, a son they named Chester.

"They have their own churches, their own stores and business places, their own newspapers," a report written in 1910 noted of Polish immigrants in Buffalo. "They are content to live alone, and the rest of the population generally knows little about them and cares less." That was pretty much how it was for Frank and Mary and Joe Chmielewski. Joe at fifteen was young enough to pick up English easily, but he never bothered. There was really no need or opportunity. Everyone he knew or had to deal with at the mine, at home, in the tavern, in church spoke Polish. This was not uncommon in Polish American communities. "I find very little use for it," one Polish immigrant wrote of the language of his adopted country. "My fellow workers are Polish, my landlord is Polish, I hear Mass on Sundays in Polish, I read the Polish newspaper and I even buy my food from a store owned by a fellow countryman." The Chmielewskis and the Yablonskys found that Fifficktown, for all of its strange dark hills and sooty air, was a little Polish world unto itself in the anthracite wilds of central Pennsylvania. Mary cooked stuffed cabbage, cabbage fried with noodles, and mushroom soup for Frank—the same food her mother cooked in Poland. Each year on the Saturday before Easter she made a basket of food—ham, hard-boiled eggs, rye bread, babka, horseradish, butter carved into the shape of a lamb, cheese, bread, salt, and pepper—and brought it to church to be blessed by Father Piltz. On Christmas Eve they fasted until the first star was spotted; only then did they break off pieces of the Christmas wafers—the *oplatki*—and begin the feast. Between Christmas and Easter, christenings and birthdays, there were long stretches with nothing but work. Every night, when Frank came home from the mine, he stripped off his clothes and stood in a galvanized tub while Mary washed the coal dust off his back. Every morning, he and Joe went back there to crawl in the dark. In winter, when the men walked back from the mine on cold nights, their clothes, soaked from the damp underground, would be frozen to their bodies by the time they got home.

What was America anyway? Where was the opportunity? To Joe

Chmielewski, America sometimes seemed like a sliver of Poland set down under a darker, dirtier sky.

W e went by ferry from Ellis Island to the Jersey City dock," recalled Samuel Goldberg of his first free hours in America. "And then by horse-drawn cab to Newark. My father had found a place in a crummy tenement *upstez* in a Jewish neighborhood. Nobody else there but Jews. If a dog came around he'd have to prove he was Jewish before they let him in—that's how Jewish it was."

At home, at 149 Howard Street, Sam spoke Yiddish, and Yiddish was all he ever heard on the street. As the third of four kids, he had it rough. Sam hated Newark. He hated the packed Jewish ghetto. He hated the school where all the kids were Jewish and all the teachers were Gentile. Most of all he hated his father, Asriel. "He was a bad, bad man," Sam said with rage in his voice nearly a century after he joined his father in America. A bitter domestic tyrant who, as a young man, had managed to survive five years in the Russian army, Asriel had come to the United States alone in 1902 with $3 in his pocket. When he had earned enough, he sent ·for his wife Sarah and their four children—but once the family arrived, Asriel did nothing but make their lives miserable. Sam remembered looking on in horror as his father knocked his younger brother Leib unconscious at the dinner table. "I was scared of him. He never hit my mother. If he had, I might have gotten a knife and taken his gizzard out."

In 1912, when Sam was twelve, the family moved to an Irish neighborhood. It was his first contact, outside of schoolteachers, with people who hadn't been born in the Russian Pale. He made friends with kids named Mickey Sayers, Tim Maloney, Martin Moore. The Yiddish accent faded; he picked up a bit of an Irish brogue and a tough, scrappy, cocky manner. Four and a half years he had lived in America—but it was only after he left the Jewish ghetto behind that Sam felt he had truly arrived in the New World.

That was how it was for most of the newcomers. They found out soon enough that it wasn't enough to be in America. If you lived in a neighborhood where even a dog had to prove it was Jewish—or Italian, or Polish—you were never going to become American. If you went to different schools, ate different foods, spoke a different language, worked at different jobs from *them*, you were never going to assimilate, no matter how much you wanted to. Of course there was also the question of what *they* wanted—the native-born who held the keys to the golden door. How could you become American if *they* looked at you as something stupid, dirty, scarcely human? *They* might have let you in—a decision many of them were increasingly coming to regret—but they weren't about to let you take over. It was one thing to have you mine their coal, press their pants, cut their hair, dig their gardens, mold their steel, lug their radiators, loom their blankets, build their bridges and tunnels. But it was something else again to grant you a voice, a say in how the country was run, or a uniform and a gun to defend it with.

But what if war came? How could they expect you to fight for God and country when God only knew what you believed in and which country you were loyal to?

CHAPTER 4

THE WEAK, THE BROKEN, AND THE MENTALLY CRIPPLED

One day, soon after he got to the United States, when Rocco Pierro was still on the job of putting up telegraph poles next to train tracks, one of the non-Italian workers asked him if he would come home to meet his father. "My dad, he's never seen an Italian before," the guy told Rocco. When the two of them got to the house, they found the father in bed—whether sick or feeble or just old Rocco didn't know. The son introduced Rocco as a genuine Italian, and the father looked him up and down with a wide, surprised gaze. "What do you know," the old man said. "And I thought all Italians were black." The Pierro family still talks about it.

That old guy was not the only one to harbor strange ideas about immigrants. The massive influx spawned suspicion, anxiety, curiosity, fear, loathing, and outright hatred at all levels of society. Henry James, returning to the United States in 1904 after an absence of twenty-one years, was stunned at how the nation, and especially New York, was being transformed by "this visible act of ingurgitation on the part of our body politic and social." In *The American Scene*, the book in which he recorded the impressions of his journey, James wrote at length of the "dense Yiddish quarter" of the Lower East Side: "There is no swarming like that of Israel when once Israel has got a start, and the scene here bristled, at every step, with the signs and sounds, immitigable, unmistakable, of a Jewry that had burst all bounds." Aghast and amused, James felt himself wading helplessly through "some vast sallow aquarium in which innumerable fish, of overdeveloped proboscis, were to bump together, for ever, amid heaped spoils of the sea." On the "electric cars" that plied Broadway and the Bowery he found nothing but foreign faces—"a row of faces, up and down, testifying, without exception, to alienism unmistakable, alienism undisguised and unashamed." "What meaning," James demanded, "in the presence of such impressions, can continue to attach to such a term as the 'American' character?—what type, as the result of such a prodigious amalgam, such a hotch-potch of racial ingredients, is to be conceived as shaping itself?"

In James's imagination, "the great inscrutable answer" to such questions hung obscurely "in the vast American sky." Others of his class and background, however, had no trouble making it out. "These Southern and Eastern Europeans are of a very different type from the Northern Europeans who preceded them," wrote New York educator Ellwood P. Cubberley in 1909. "Illiterate, docile, lacking in self-reliance and initiative, and not possessing Anglo-Teutonic conceptions of law, order and government, their coming has corrupted our civic life." Woodrow Wilson, before taking office as president, wrote in disgust of the southern European countries that were "disburdening themselves of the more sordid and helpless elements of the population." Harvard zoologist Charles Davenport, founder of a research center into human heredity called the Eugenics Record Office, wrote in 1911 that the mixing of southeastern Europeans with America's old British and northern European stock would result in a strain of smaller, darker, inferior Americans prone to murder, theft, and sexual immorality. It's worth noting that eugenics grew out of the Progressive Movement as a biological theory aimed at social reform, and that many eugenicists, including early birth-control activist Margaret Sanger, were motivated by a desire to ameliorate the conditions of the poor. The horrors advocated under the banner of eugenics, however, have overshadowed these distinctions. In a hugely popular bestseller of 1916 entitled *The Passing of the Great Race,* New York lawyer Madison Grant, chairman of the New York Zoological Society and trustee of the American Museum of Natural History, argued that a "dark Mediterranean subspecies" was diluting the "splendid fighting and moral qualities" of the old stock and pushing the nation to the edge of a "racial abyss." With unchecked immigration, the soldiers, sailors, rulers, explorers, and aristocratic adventurers of the "great race" would be replaced by "the weak, the broken and mentally crippled of all races drawn from the lowest stratum of the Mediterranean basin and the Balkans, together with hordes of the wretched, submerged populations of the Polish Ghettos." Uncannily anticipating Hitler (who called *The Passing of the Great Race* "my Bible"), Grant warned: "If the valuable elements in the Nordic race mix with inferior strains or die out through race suicide, then the citadel of civilization will fall." "The most influential tract of American scientific racism," Stephen Jay Gould later wrote of Grant's book.

Madison Grant's venom did not reach Meyer Epstein on East Broadway or the Pierros on Shelton Road. Grant's close friend Teddy Roosevelt, however, took up some of these ideas, gave them a manly patriotic twist, and broadcast them on the national stage. Unlike Grant, Roosevelt was not concerned with the taint of inferior racial strains on America's Nordic purity: the president firmly believed that the new immigrants could become good Americans, even good warriors—but *only* if they cleansed themselves of any allegiance to their countries of origin. "There can be no fifty-fifty Americanism in this country," said Roosevelt. "There is room here for only 100 percent Americanism, only for those who are Americans and nothing else.... The foreign-born population of this country must be an Americanized population—no other kind can fight the battles of America either in war or peace. . . . We have room for but one flag, the American flag. We have room for but one language here, and that is the English language . . . and we have room for but one sole loyalty and that is a loyalty to the American people." All this thundering about battles and allegiance and loyalty might have seemed odd, even a touch paranoid, when the nation was at peace. But war was coming. Roosevelt, practically alone among his countrymen, could smell it. And relish it. England, Germany, and France, having carved up the world, were armed to the teeth and waiting for an excuse to march on each other. How could the United States remain on the sidelines if it wanted to fulfill its destiny as a great—as *the* great—world power? But when war came at last, woe betide the nation if its armies had to rely on the unassimilated scum from the Mediterranean basin and the ghettoes of Poland. Grant warned that these inferior "races" were genetically too weak and defective to fight for America; Roosevelt countered that even if they developed the physical capacity to be soldiers, they would lack the backbone and the stomach to fight like real Americans until they had been purified of all trace of their alien origins. But the two agreed that the vast pool of unassimilated immigrants from the south, east, and center of Europe posed a serious risk to what we would now call national security. In time of peace, these newcomers were a drag on the nation's vitality; in time of war, they would hobble and corrupt the army, unless they could be magically transformed into "100 percent" Americans.

The bottom line was, America wouldn't stand a chance in the next big war if it was reduced to fielding an army of "undisguised and unashamed" aliens.

When the test of actual battle comes," wrote Madison Grant, "it will, of course, be the native American [i.e., the old stock from Northern Europe] who will do the fighting and suffer the losses. With him will stand the immigrants of Nordic blood, but there will be numbers of these foreigners in the large cities who will prove to be physically unfit for military service." Yet in fact, even before the test of battle came, there were "foreigners" who were serving with perfect physical fitness in the American armed forces. They may not have been "100 percent," but there they were in uniform, loyal, brave, even reckless soldiers in the nation's small ill-equipped prewar military.

Two of the bravest and most reckless were Matej Kocak and Sam Dreben—a Slovak marine and a Jewish soldier of fortune. Career soldiers both, they each came into their own the day they signed on to be soldiers in the U.S. Armed Forces. When the test of battle came in the Great War, Kocak and Dreben found themselves by chance on the same battlefield, one of the bloodiest and most fiercely contested in the war; in the heat of that battle, the two of them became heroes and inspired heroism in those they fought beside.

Kocak and Dreben were living proof, if proof were needed, of what poisonous rubbish Madison Grant was spouting. On the eve of war in 1914, they already had more than twenty years of military service between them. Three years later, when the United States entered the fight and a generation of immigrants was called on to serve, Kocak and Dreben led the vanguard into action.

Fair-haired, blue-eyed, strapping, ruddy-complexioned, Matej Kocak was a pretty close match to Madison Grant's Nordic ideal. The only problem, aside from his middling height of five feet, six and a half inches, was that Kocak was a Slovak and thus, in Grant's schema, not a Nordic at all but a stubby rude Alpine peasant. Far better than being a Jew, of course, and a decided rung up the racial ladder from the squat coarse Mediterraneans, but still a far cry from the elegant godlike Nordics who carried the true torch of civilization.

Alpine peasant he may have been by birth, but by temperament and preference Kocak was a warrior—and he proved to be a natural fit in the American military. Not that he was especially wrathful or bloodthirsty. Rather, it was the discipline, the conditioning, the camaraderie, the outdoor life, and the opportunity to advance that suited him to the U.S. military. When war came, he displayed the born warrior's instinct for decisive action—the instinct that wins battles, rouses comrades, and sometimes, when you're unlucky, gets you killed.

Matej was born on the last day of 1882 to a family of farmers and small landowners outside the town of Gbely (also called Egbell) in the northwest tip of what is now the Slovak Republic but was then an interior region of the vast, fading Austro-Hungarian Empire. In the ethnic stew of Mitteleuropa—Czechs, Croats, Slovaks, Ukrainians, Carpathian Germans, Gypsies, Vlachs, Jews, and Magyars all simmering together under the same benignly reactionary monarchy—the Slovaks stood out for their vibrant folk culture, pride in their Slavic origins, and resentment at the duration of their subjugation by neighbors. As far back as the twelfth century, Slovaks had been squeezed by Hungarian wars and social upheavals. After 1867, when Hungary became Austria's coequal in the dual monarchy of the Austro-Hungarian Empire, Slovaks were subjected to an official policy of Magyarization—a campaign to suppress their culture, folk traditions, language, and identity and assimilate them into Hungary's dominant Magyar culture. Matej's native town of Gbely was especially hard-hit since it had been a center of Slovak culture, and hundreds of local families fled to North America.

Matej's own departure was impelled as much by personal passion as by politics. He had grown up with a sister and a brother in a traditional Slovakian Catholic household where the word of the father was akin to the word of God. Schooling ended with the primary grades, and then Matej went to work alongside his father and brother on the family farm. The sister married, evidently with the father's blessing. Matej, seething under the lash of paternal tyranny, kept his head down, worked on the farm, and drank. It was inevitable that father and son would fight and—given Matej's personality, slow to anger, but once angry quick to lash out—that the fights would be violent.

When he was twenty-three, Matej fell in love with a girl who lived on a nearby farm. Matej didn't care that the girl stood a rung lower on the social ladder and had no dowry, but his father did. Father and son exploded—again. This time the rift between them was irreparable. Matej packed a bag and left his family and his girl. He arrived in the United States in 1906, fluent and literate in Slovak, Czech, and Russian but with nothing else to show for himself. Through the immigrant grapevine he heard there was work to be had in the mining towns of western Pennsylvania, so after he ran through the little bit of money he'd brought over from Gbely, Matej drifted out to Lodi. Instead of digging coal underground like so many hardworking, broad-backed Catholics from the Austro-Hungarian Empire, he got a job as a cook in a mining camp. That didn't last long. In 1907 he turned up in the small town of Sturgeon southwest of Pittsburgh and went to work as a laborer—what kind of labor is not specified on his papers, though it was probably in a steel mill or coal mine or digging ditches or laying train track, because those were pretty much the only kinds of labor open to immigrants in western Pennsylvania in 1907. But Matej was not cut out to be a laborer.

On October 15, he showed up at the U.S. Marine Corps recruiting office on 412 Third Avenue in downtown Pittsburgh and in his broken but adequate English informed the guy at the desk that he wanted to enlist. He was escorted to the examining room and presented to E. J. Trader, the surgeon in charge. Trader told Matej to strip and went over him inch by inch. On a form labeled "Marks, Scars, etc." with a line drawing of a bald stocky naked man, front and back, Trader noted that Kocak had perfect 20/20 vision, a fair complexion, light brown hair, and blue eyes. He weighed in at 157 pounds; he was 24 years old; his chest measured 36 inches around and expanded 2 more inches when he took a deep breath; he had a scar on his knee and another on his backside. Aside from that Matej was a perfect specimen. The Marine Corps signed him up the following day—and that was that. On October 16, 1907, Matej Kocak ceased to be a laborer and became a private in the U.S. Marine Corps.

He was posted to the Marine barracks at the Philadelphia Naval Shipyard on League Island and assigned to an aging naval sloop named the *Lancaster* that dated back to before the Civil War. Kocak's year and a half on the *Lancaster* passed largely without incident, and in the spring of 1909 he was transferred to the USS *Georgia*. At the beginning of November 1910, the *Georgia* was slated to go to France, but for some reason, Matej was dead set against shipping out to Europe, and he hatched a scheme to avoid it that nearly scuttled his military career. On the morning of October 31, Matej went ashore without leave in Norfolk and he and a buddy remained AWOL for a little over forty-eight hours, during which time the *Georgia* embarked for France without them. The men received a hasty deck court-martial and were convicted of desertion, but the charge was subsequently overturned as "erroneously made," evidently because of some procedural misstep. In punishment, Matej was docked $10 in pay and given thirty days of extra police duties.

After this one blot on his record, Matej shaped up and settled down. Discharged on October 16, 1911, he spent a couple of months with his brother and sister-in-law in Binghamton, New York, and then, a few days before Christmas, went down to New York City, made his way to the Marine recruiting office at 112 East 23rd Street, and signed up for four more years.

Since his first enlistment, Matej had bulked up to 179 pounds and his complexion had turned from fair to ruddy bronze. He had learned to deal with military discipline and carry himself with a jaunty swagger, and he got on well with his comrades and superiors. In short, he had become a marine—and his second enlistment sealed this fate.

In 1912, Matej requested a three-month leave so that he could return to Europe to visit his mother, Roza, in Gbely. No mention of the father, though he was still alive. Back in the United States in June, Matej was posted to the naval prison at the Boston Navy Yard.

In April 1914, after nearly seven years with the Marines, Matej finally saw some overseas action, though it was not one of the more glorious military episodes in U.S. history. Early that month, a misunderstanding in the Mexican port city of Tampico led to a standoff that ended with the Mexican army taking nine American sailors prisoner. Although Mexico promptly released the American sailors and issued a written apology, the U.S. government insisted that it would not be satisfied until the American flag was raised and a twenty-one-gun salute fired. The fact that this incident took place during the chaotic period of the Mexican Revolution when various factions were vying for power and frequently toppling the government made the situation more combustible. When the Mexicans refused to meet Washington's demands, President Wilson sent the U.S. Navy's Atlantic Fleet down to Mexico to bolster American forces there. On the morning of April 21, 1914, the Atlantic Fleet began bombarding Veracruz, a city to the south of Tampico on the Gulf of Mexico, and a contingent of nearly eight hundred marines and sailors stormed the city's customhouse. A short nasty bout of urban street fighting and shelling from U.S. gunships ensued, and then the Americans moved in to occupy the city.

By the time Matej Kocak arrived with a contingent of marines aboard the SS *Morro Castle* on April 29, the fighting had been over for five days. Nonetheless, the United States saw fit to maintain a sizable garrison at Veracruz, and Kocak remained there on active duty until November, 23, 1914—by which time Europe was mired deep in the mud of war and the U.S. occupation of a sullen Mexican port had receded to the shadowy wings of the world's stage.

S hoot de woiks" was Sam Dreben's favorite expression, whether in a crap game or a battle, and in the course of his incredible life he had plenty of opportunity to act on it in both venues. Confronted with injustice, with an enemy, or simply with an opportunity to jump in and blaze away, Dreben did not pause to think. Like Kocak, he was born to be a soldier. And yet, as a comrade in arms pointed out, "He was the last man you would pick out of a crowd to be a soldier.

Short, heavyset, with a huge comedy nose, a stomach always straining at the belt, he was a walking vaudeville routine. Always kidding, always cheerful, always carefree and happy, he could make a bunch of men laugh when they were ragged, starving, and facing violent death." To look at him you'd think he was a store clerk or a peddler, when in fact, even before the Great War made him a hero, Dreben was a crack shot with a machine gun, a fearless infantryman, a gun runner, a crafty spy and raider behind enemy lines—and always a world-class character.

In April 1914, as Matej Kocak was steaming to Veracruz with a company of marines on the SS *Morro Castle*, Sam Dreben was living in El Paso, waiting for his application for U.S. citizenship to be approved and engaged in the highly dangerous enterprise of supplying arms to Pancho Villa, the most colorful and powerful of the Mexican revolutionary leaders. Had Dreben been caught, he would have been deported and his chance of becoming a U.S. citizen would have been shot to hell. But Dreben's customary luck was with him.

How a Jew from the Russian Pale ended up, via a string of wars and revolutions, being the only American Pancho Villa really trusted is a tale too strange to be invented. Born in 1878 to devout Jewish parents in the beautiful city of Poltava (in today's Ukraine) and raised in Kiev, the teeming regional capital, Sam was a short, stocky kid with dark piercing eyes and a prominent nose. As a boy he was shy and "almost painfully polite," a friend wrote later; though he had a wild streak, he kept it under wraps.

Sam's parents hoped their son would become a rabbi, but he had other ideas. His first dream was to join the army and become an officer with big shining buttons down his chest. That dream ended when he learned what happened to Jewish boys in the Russian army: no shining buttons, twenty-five years of grueling service, no hope of ever being an officer. While still in his teens Sam ran away from home twice—and then at eighteen he left home for good. He got himself to England by stowing away on a ship; in London he kept himself alive with a job delivering vegetables to markets, but he got fired for eating the merchandise. From London to Liverpool and more odd jobs, as a dock worker and a tailor's assistant for a penny a day. Sam picked up some English and somehow managed to scrape together the fare for a ticket to the United States. He arrived in New York in steerage—what else?—in January 1899.

Sam had an aunt and uncle in Philadelphia, so he went to Philly in hopes of finding work that paid and suited his temperament. The uncle, a tailor, took on Sam as his apprentice. By day he hunched over trousers and jackets; at night he went to classes to improve his English. It was the life that tens of thousands of Jewish immigrants were living in America's cities, but Sam hated it. When an opportunity came along to trade tailoring for soldiering, he jumped at it.

Sam's chance to fight arose because the United States had just declared war on the Philippines—one of those trumped-up late-nineteenth-century wars that gave an imperial power an excuse to flex its muscles and tighten its grip over a client nation. On February 4, 1899, when three armed Filipino men tried to cross a bridge into an American military compound near Manila and ignored (or failed to understand) an order to halt, American soldiers shot them. Since the Philippines was at that time an American territory—purchased from Spain in the aftermath of the Spanish-American War—President McKinley deemed the incident an insurgent attack and demanded that the nation go to war to quell the "insurrection." And so the brief, inglorious Philippine-American War began and Sam Dreben got his first chance to fight.

With plans to ship eleven thousand American soldiers to Manila by August 1899, the U.S. Army sorely needed recruits. The story goes that Sam encountered a recruiting sergeant in Philadelphia and was incredulous when he learned that if he volunteered to fight in the Philippines he'd get $15 a month and three meals a day. "Do they give the uniform too?" Sam wanted to know. Not only uniforms, the sergeant assured him, but medical care and a free funeral if he died. Sam signed up on the spot.

Whether or not it happened this way—there's a mythic element to a lot of the stories that have come down about Dreben—Sam was duly enrolled in the army, assigned to Company G, 14th U.S. Infantry (the so-called Golden Dragons), and dispatched to Bacoor outside Manila by way of San Francisco. The first legend of Private Sam Dreben dates from his initial hours of combat. Ordered to go after Filipino rebel leader Emilio Aguinaldo (the nation's future president), Company G was approaching a stone bridge when suddenly the rebels let loose a terrific blast of black powder, nails, rivets, and scrap iron. Eleven American troopers, including the captain, were killed or injured in the explosion. The Company G survivors took cover as best they could and tended their wounded—all except one man.

One of Dreben's comrades later described what happened: "As I lay watching this slaughter only a few yards away, I suddenly saw one soldier emerge from the smoke, still trotting forward toward the bridge. He was the loneliest figure I have ever seen, jogging along like a boy running an errand. There were several thousand insurrectos in those trenches and the bullets were snapping around him, but he didn't seem to notice. Down the road he went, over the bridge, and into trenches as if he were taking part in a drill on the parade ground." Sam's explanation, when the rest of the company joined him after American reinforcements arrived and put the rebels to flight? "Vell, I heard the captain say 'Forwards!' and I don't hear nobody say 'Stop.'" From then on, Dreben was known in the U.S. Army as the Fighting Jew.

Sam spent the next decade and a half in and out of the army—or rather a series of armies—chasing coups, uprisings, revolutions, and civil wars around the globe. In the course of his first enlistment he was dispatched to China to help quell the Boxer Rebellion and then returned to the Philippines in 1901 to fight Islamic insurgents on the southern island of Mindanao. After his discharge, he went back to the States and bummed around California, picking up money as a tailor in Los Angeles, a streetcar conductor, a fruit picker, a lumberman, a teamster. In the summer of 1904, he reenlisted. Stationed at the cavalry outpost of Fort Bliss in El Paso, Texas, Dreben mastered the weapon that would transform modern warfare—the machine gun. "Handling a machine gun," wrote his biographer, "was to be his only real trade." Honorably discharged as a sergeant three years later, at loose ends again, Dreben ended up bankrupt in Panama as the half owner of a saloon in the canal zone. Eventually, he drifted back to warfare—only this time as a soldier of fortune, first in a revolt in Guatemala, then as a freedom fighter in Nicaragua, then in Honduras (where he freed a general from a prison on the eve of his execution by emptying his rifle into the door of the stucco jailhouse), and finally, inevitably, in the revolution that was raging in Mexico. Dreben teamed up with another colorful American soldier of fortune named Tracy Richardson whom he had gotten to know in Nicaragua, and the two of them taught Mexican revolutionaries how to use machine guns and throw advancing troops into confusion by crisscrossing machine-gun fire in their path. "Those poor Federal soldiers were marched up against us in close formation," Richardson wrote later. "Rank after rank Sam and I mowed them down till it sickened us." Dreben and Richardson had many bloody adventures in Mexico and many near disasters after the faction they backed was overthrown in February 1913 and the new president Victoriano Huerta put a price on their heads.

Which was how Dreben ended up in El Paso, supplying arms and information to Pancho Villa, while Kocak was doing his bit in the U.S. occupation of Veracruz in the spring of 1914. On the face of it, the Fighting Jew and the Slovak marine might seem to have been on different sides in a messy border war. But these two immigrant career soldiers had one fundamental thing in common: both were unshakably loyal to their adopted country. When the time came to choose between Pancho Villa and the American force raised to chase him down, Dreben chose without hesitation.

It was a testament to the transforming power of the idea of America at the turn of the last century—the power to remake identity and even instinct—that Sam Dreben and Matej Kocak became American sol-

diers before they became American citizens. The Jew from the Russian
Pale and the Slovak Catholic both submitted to this power at the same
time and in the same way. In the war to come, this idea of America
would lead an Italian laborer to save his wounded buddies from a hail
of German hand grenades and a Jewish barber to deliver the message
that rescued a battalion trapped behind enemy lines; it put a Polish
farm boy on a wheat field raked by exploding artillery shells, and it
won a medal for an Irish kid who had enlisted to escape the mines.

Dreben and Kocak had taken the step from immigrant to soldier
as soon as they arrived in the United States; in the coming war, Tony
Pierro, Meyer Epstein, Joe Chmielewski, Andrew Christofferson, Epifanio Affatato, and half a million other immigrant soldiers followed
them.

CHAPTER 5

THE WORLD AT WAR

The Great War began on the afternoon of June 28, 1914, with two bullets fired from the gun of a tubercular nineteen-year-old Serbian nationalist in the Bosnian city of Sarajevo. One bullet lodged in the neck (near the heart, according to some accounts) of Archduke Franz Ferdinand, the heir to the Hapsburg throne, killing him within minutes. The other bullet entered the abdomen of his pregnant wife, Sophie, killing her and her unborn child simultaneously. Four and a half years later, when the firing finally ended in Europe, nearly 10 million people had joined the royal family in premature death.

For nearly a month after the assassinations, apparent calm prevailed in Europe while generals and ministers secretly assessed the situation and weighed their options. Three weeks passed before the Austro-Hungarian Empire responded to the loss of its heir with an ultimatum to Serbia: the empire's intent was to back Serbia into an impossible corner and provoke a limited Balkan war. But instead the ultimatum and the preexisting alliances of the major European powers "let loose the irresistible onrush of events," in the words of one historian. Serbia, with long-standing ties to Russia grounded in Slavic solidarity, took the occasion of the Austrian ultimatum to secure Russian support in the event of conflict. Meanwhile, Austria-Hungary appealed to Germany to join it should Russia declare war on the empire—and Germany, spoiling for a fight and with a meticulously detailed battle plan at the ready, was only too eager to comply.

On July 26, a month after the killings at Sarajevo, Austria-Hungary rejected Serbia's reply to its demands, and on July 28 declared war. Russia duly rose to Serbia's defense as promised and began to mobilize its vast army. On August 1, Germany in turn mobilized its armed forces and declared war on Russia on the grounds that the Russian mobilization was an act of war. Russia's foreign minister and Germany's ambassador to Russia, conscious of the enormity of these actions, embraced in tears. Two days later, on August 3, Germany declared war on France.

On August 4, the kaiser's army entered Belgium with the intent of piercing through to France and making a rapid strike on the French capital. At midnight Britain, already mobilized, officially entered the war on the side of France and Russia. The Eastern Front opened two weeks after that, on August 16, when Russia sent its First Army through Poland into eastern Prussia.

As that preternaturally warm and sunny summer of 1914 rounded toward its close, young men all over Europe left their jobs, their schools, their wives and families and marched to a war that their leaders vowed would be swift and glorious. The flow of immigrants to the New World all but ceased. Europe suddenly wanted to hoard its human capital.

The two shots fired in Bosnia raised only the faintest reverberation in the United States. The assassinations made front-page headlines in the American dailies on June 29, and the *New York Times* ran an anxious story on page two titled "Tragedy May Alter Politics of Europe," but within a few days the situation in Europe had dropped

out of sight. All that summer, as the machinery of war creaked to life
in Europe, the United States was preoccupied with its own internal
crises: a bloody miners' strike in Butte, Montana; the premature explosion of a bomb in Tarrytown, New York, that killed several Wobblies
instead of its intended target, John D. Rockefeller; the endless, perplexing ebb and flow of revolution and counterrevolution in Mexico.
On July 26, Americans read of a raucous mob crying "To Berlin!" on
the boulevards of Paris and cheering crowds crying for war in the
streets of Berlin. But the *Times* devoted nearly as much space that day
to a story about a "crazed man" who opened fire on the New Haven
line and killed three passengers.

Initially, the events that culminated in war in Europe played in the
United States like so much political theater—at least for the native-
born population. It was different for immigrants with ties of love and
memory to the warring powers.

In South Fork, Pennsylvania, Frank Chmielewski and his younger
brother, Joe, read of the outbreak of war in the *Naród Polski*—the
Polish Nation—a weekly published by a Chicago-based Polish nationalist fraternal organization called the Polish Roman Catholic Union
of America. Frank, a devout and hardworking Catholic family man,
subscribed to the *Naród Polski* because the paper reflected his conservative views—*Bóg i Ojczyzna*, God and Fatherland, was the motto of
the organization that published it—though it seems unlikely that he
had much interest in the factional wrangling that filled its columns.
Men with soft hands and clean lungs might occupy themselves with
whether or not saying the Mass in Polish was heretical and how best to
silence the disgraceful antipapal rhetoric of the breakaway Polish National Catholic Church or the dangerous liberalism of KON (Komitet
Obrony Narodowej—the National Defense Committee), which
placed patriotism above religion in the struggle to free Poland and
opened its membership to socialists and Jews. With a job to do and a

family to feed, with neighbors and friends and his brother to look out for, Frank didn't have time to care about such things.

But with the start of war, the weekly issue of *Naród Polski* became electrifying. "WAR! WAR!" the headline blared on August 5. "For the past eight days, the world hasn't spoken about anything but the war that will be let loose upon all of Europe . . . the murder of the Austrian prince, and his wife, in Sarajevo, is the spark from which a gigantic fire may arise." In the third week of August, while the mainstream American press rose in shock at the German atrocities in Belgium, Frank read of a massive Russian push westward through the Lithuanian-Polish border country where his family lived. The pro-Russian editors of the *Naród Polski* reported breathlessly on the sweep of the Russian First and Second Armies into eastern Prussia that began on August 17, and cheered as the Prussian city of Konigsberg fell. "The Germans have tried to hold them," the paper reported on August 26, "but have not been able to. . . . The Russians have taken many German prisoners, who are being transported deep into Russia." But the euphoria was short-lived. By August 27, Field Marshal Paul von Hindenburg was called out of retirement to take charge of the routed German forces, and under Hindenburg the Germans staged a swift, skilled counterattack. In the disastrous Battle of Tannenburg, the Russian Second Army suffered some 30,000 casualties, and 95,000 more of its troops were taken prisoner. The Second Army's commander, General Alexander Samsonov, committed suicide rather than return to Czar Nicholas II with news of the catastrophe. The volatility of the first weeks of war continued on what would soon be called the Eastern Front. Unlike the entrenched paralysis of the fighting in France and Belgium, the Eastern Front was a fluid, fitful war of routs, pillage, seizure, and retreat.

"Battles on Polish Lands" read the headline in the September 2 issue of *Naród Polski*. This was what most deeply alarmed the Chmielewski brothers: while they were safe in South Fork, the country of their birth was a battlefield, and God only knew what was happening to their parents and other relatives still living there. The escalating violence in Europe put Poland in a painfully delicate situation. In the collapse of empires, Poland might at last slip free and regain its independence—but Poles would surely have to buy their freedom with mass suffering and death. Poles in America, reading their Polish language papers with increasing horror, wondered what outcome would benefit Poland most. A hasty victory? But for which side? Catholic Austria? Slavic Russia? And to achieve that victory and that long-prayed-for release into freedom, how many Poles must perish, how many would be left homeless?

Frank and Joe Chmielewski and Polish immigrants like them found no answers in the pages of the *Naród Polski* or *Zgoda* or *Kuryer Polski* or *Gwiazda Polarna* or *Dziennik Chicagoski* or any of the scores of other Polish-language journals they relied on for news. But they didn't need newspapers to understand that their futures—as Poles, as Americans— depended on the outcome on the Eastern Front. Of all the newly arrived immigrants, the Poles were the first to grasp that the war in Europe was their war too.

On August 19, 1914, when the extent and seriousness of the war in Europe was beginning to make Americans sit up and take notice, President Wilson made his Declaration of Neutrality in a message to Congress. "The United States must be neutral in fact, as well as in name, during these days that are to try men's souls. We must be impartial in thought, as well as action." These were the lines most frequently quoted; but the bulk of the address was actually a stern lecture aimed at America's immigrant communities. After pointing out that "the people of the United States are drawn from many nations, and chiefly from the nations now at war," Wilson warned these immigrants not to try to drum up support for their country of origin. Divisions fomented by ethnic partisanship would, said Wilson, "be fatal to our peace of mind and might seriously stand in the way of the proper performance of our duty as the one great nation of peace."

In the guise of bolstering national unity, the president had put America's immigrants on notice. They would be to blame if the nation splintered into "camps of hostile opinion, hot against each other." The idea that the United States might be drawn into this great conflict was so alien that Wilson did not even mention it.

Andrew Christofferson had been on his Montana homestead for less than a year when the war began. It was plenty long enough for him to figure out that 160 acres of dry northern prairie was a tough place for a young man to live by himself, let alone scratch out a living. Andrew built a rough claim shack and set about breaking the tough brown sod and planting a crop, probably wheat, since it was too dry to grow much of anything else. When things got too lonesome, he walked south to Chinook to collect his letters from Norway and his copy of the *Decorah Posten*, the Norwegian-language newspaper published out of Decorah, Iowa, that was his sole source of news from the outside world.

By the terms of the Homestead Act, Andrew was required to farm his land for five years before taking legal possession—but his farming must have been pretty minimal during the 1914 growing season, because at the end of May, he filed notice of his intention to leave his claim for the duration of the summer. He likely took the train back to Larimore, North Dakota, to go back to work with his brother Tollef on one of the big farms in the Red River Valley. Andrew had already come to realize that the only hope of surviving on the grudging soil of eastern Montana was to secure a regular source of income from someplace else—a fact of life Montanans still reckon with.

Andrew was away from the homestead when the fighting broke out in Europe, but his mail and his copy of the *Decorah Posten* got forwarded from Chinook, so he was able to follow the momentous events with a few days' lag. "The insanity of war, like an epidemic, is again playing havoc with Old Europe, and Austria is the source of the infection," the *Posten* reported on July 28, 1914. Within weeks of the assassinations at Sarajevo, the *Posten* was publishing dire warnings of financial ruin and even famine in Norway.

What Andrew found most disturbing were the accounts of the German assault on Belgium—the "rape of Belgium," as it was soon being called. On August 18, the *Posten* reported that 30,000 Belgian civilians had fled before the German bombardment of Liège. "The town is a grim sight, rubble everywhere, bridges destroyed, walls shot to pieces, roofs collapsed. There is not a street that is not marked by the devastation. . . . Each morning the Germans collect all the bread from the bakeries, while the inhabitants themselves go hungry." August 21: "The air from Liège to the Dutch border is filled with the stench of rotting corpses." In three days starting August 25, German soldiers reduced the gemlike university town of Louvain, the so-called Oxford of Belgium, to rubble, killing 209 civilians outright and forcing 42,000 others to evacuate. The Great War, horrific as it proved to be, did not have a large number of civilian casualties—nothing like the firebombing of Dresden or the nuclear attacks on Hiroshima and Nagasaki or the systematic extermination of European Jewry: the rape of Belgium in the war's first weeks was the one glaring exception. Andrew, a newly converted Christian, read the news with outrage, sorrow, and amazement. How in the name of God could a nation as enlightened as Germany claim that it was justified in murdering civilians and burning ancient cities because of a handful of Belgian snipers? Andrew wondered about the Germans working beside him in the fields. How did they feel about what their fellow countrymen were doing in Belgium?

By the time he returned to Montana in October, there was a new word for German circulating in newspaper columns and railway cars and corner stores: Hun. "Why this injustice to the Huns?" one A. Acton demanded in a September 24 letter to the *New York Times* commenting on the coinage. "They [the real Huns] had not been blessed with opportunities to know better; and they made no pretentious boasts about their superior 'culture.'" Back on his farm, with another winter closing in, Andrew shook his head over the *Posten* and counted his blessings that he was a neutral Norwegian living in neutral America.

B utte, Montana, is 237 miles west of where Andrew Christofferson was homesteading—but in 1914 it might as well have been another country. Where Chinook was dry rolling prairie, monotonous, monochrome, sparse in vegetation and sparser still in people, Butte was a rough tough company town squatting on top of a copper-rich hill and packed tight with high-strung immigrants. For the better part of four decades now, Irish, Cornish, Welsh, Serb, Italian, Finnish, Croatian, and Mexican workers had been pouring in to work "the richest hill in the world," and in the process they turned Butte into a kind of Rocky Mountain Pittsburgh. By the turn of the century, Butte was dominated by one of the most powerful unions in the country—the Butte Miners Union—and one of the most ruthlessly successful companies in the world—the Anaconda Copper Mining Company. The place was a cauldron of rampant capitalism and radicalized labor.

In 1914, the year eighteen-year-old Peter Thompson arrived from Belfast with his blue eyes and shy irresistible grin, the cauldron exploded. Incensed at how weak and conservative the union had become under the iron rule of Anaconda, breakaway Finnish and Irish miners joined forces with the Wobblies to take matters into their own hands. On June 23, 1914, the radicals ransacked the Miner's Union Hall and then dynamited the building to matchsticks. Anaconda, which by this time controlled not only Butte's copper but Montana's politics and press, struck back swiftly and savagely. Martial law was declared by Montana's governor; the company, cracking down on Wobblies and "the avowed wreckers" of the mines, henceforth refused to recognize any unions; workers who refused to cooperate were locked out. The outbreak of war in Europe only made the situation more tense. When the fighting started in August, half of Butte's mines shut down because of the uncertainty of international trade. Peter Thompson had come to Butte innocently and heedlessly because he had family there and the family told him there were jobs; but now he found himself in a hornet's nest.

When they first laid eyes on him, Peter's relatives in Butte said he was too bright and promising to be a miner, but until he had some education under his belt there was nothing else for him in Butte, so to the mines he went. The company took him on as a nipper—his job was to get the miners their tools; later on, he worked as a motorman on the small train that carried ore through underground tunnels. Peter managed to hold on during the explosive summer, but when the war began and half the mines shut down, the company tossed him out. As German soldiers stormed through Belgium and Britain began mobilizing its vast army—including two of the Thompson brothers back in Ireland—Peter hung around his aunt and uncle's flat on Montana Street and listened to talk about the war. The kaiser was the one the family put their money on—anyone who could beat the hated English was fine by an Irishman. Peter kept his own counsel while the elders held forth. What happened in the old world really didn't concern him anymore. America, Montana, Butte—this was his home now. Hadn't he renounced his allegiance to the English sovereign when he signed his first papers declaring his intention to become a U.S. citizen? King George V was no longer his ruler; England's war wasn't his war. Irish he might be by heritage, looks, accent, and temperament—but America was his future.

Peter was not out of work for long. By September the West Colusa Mine took him back. He was making $3.50 a day now, and there was talk that the war would soon drive wages even higher.

Maximilian Cieminski in Bessemer, Michigan, was also hearing rumors that the war would boost wages for workers in the local mines. Rumors spread in Polish, Italian, German, Finnish,

Swedish—even English sometimes. Almost every European language, in fact, except Kaszubian, the language that Max had grown up speaking. Kaszubian, as Max's father never tired of pointing out, was like Polish only different—special, somehow both nobler and earthier. In Polonia, the central Wisconsin farming town where the Cieminskis lived, almost everyone was Kaszub, and from the time he was a small boy Max, the youngest of the nine surviving Cieminski children, had heard countless stories of the magical land of Kaszubia with its lakes and flat sandy fields and haunted forests. Like the Poles, the Kaszubs were Slavic and predominantly Catholic; but they had their own customs, folktales, churches, and peculiar beliefs. And of course they had their own homeland. Max's father, Paul, could pinpoint for the boy on a map of Europe exactly where their home village of Brusy was: right there in the western Prussian province of Pomerania, inland from the Baltic Sea, southwest of Gdansk, northeast of Berlin.

In the late summer of 1914, many other eyes were trained on maps of this area, because Kaszubia was now a highway for the German army massing against the Russian invaders on the Eastern Front. So when the guys at the Bessemer iron ore mine talked about war and wages, Max had other concerns. He wondered whether Brusy had been looted by German soldiers, which army his relatives back in Kaszubia would fight in, what would happen to his people after the war. It never occurred to this gentle, soft-spoken, blue-eyed man of twenty-two that one day he would be inserted into the same war. Of the twelve soldiers whose stories are told in this book, Max Cieminski was the eleventh.

Born in Polonia, Wisconsin, in 1891, Max was not technically an immigrant, but he might as well have been one. He had grown up speaking Kaszubian at home (along with Polish and German). The customs he had learned and the folktales he had heard had all been Kaszubian. He knew very little English and rarely encountered anyone who was not a Kaszub or a Pole. Officially, as the son of two naturalized Americans, Max was an American citizen, but in fact he had spent his childhood in a little American corner of Kaszubia. Were it not for the unmistakable Wisconsin landscape of fat cows and lush rolling farm fields, Polonia could have been a farm town in Pomerania. Even today, more than a century on, Polonia's Kaszub ties are unbroken.

A few incidents from Max's childhood have come down through the family. One day, no one remembers when or how it happened, Max got into an accident on the farm that resulted in the amputation of his right index finger—the trigger finger. The Cieminskis loved to hunt—the fields and woodlots around their farm were thick with game birds and deer—and after the accident Max had to learn to pull a trigger with his middle finger. He learned well. The boy grew up to be a crack shot. This, of course, would have a bearing on his future in the army.

Max's childhood ended in 1907, the year he turned sixteen. His father informed him that the time had come to go out and earn a living. But what was he supposed to do? The options were pretty much limited to hiring out to strangers as a farmhand or going to work in a mill or factory. For once Max caught a lucky break. His oldest sister, Mary—his favorite—was married to a Polish man of substance and property who lived in Bessemer, 150 miles due north of Polonia on Michigan's Upper Peninsula. Mary's husband, August Kondziela, owned a brewery—the Bessemer Brewing Company—widely considered the finest in the region, and he and Mary also ran the Michigan Hotel, a boardinghouse for workers at the local mine. Mary was only too happy to take in her little brother and put him to work at whatever odd jobs came along.

So at sixteen Max moved up to Bessemer, but life in Michigan did not work out as he had hoped. For a while, Max's brother-in-law put him in charge of the team of six horses that pulled the brewery wagon, but there just wasn't enough money to pay the kid a regular wage. So Max, like just about every other able-bodied male in Bessemer, was forced to go to work in the iron ore mine. All day, every day, he descended with a motley crew of immigrant miners into some of the deepest, richest iron deposits in the country. When he came out at night, his clothes were rust red from the mineral he'd been working and breathing all day. Max took his evening meal at the Michigan House and then, every night, he went over to the brewery to sleep beside his gun in an upstairs room. August had hired him to be the night watchman at his Bessemer Brewing Company out of fear of sabotage from the competition. In exchange he got meals and laundry at the boardinghouse that his sister ran. Not much of a life for a sweet-natured young man descended, as his father once told him, from Kaszubian nobility.

One of the only bright spots in Max's life was his little niece Marguerite, Mary's daughter. Marguerite was four years old in 1914, an adorable and adoring little girl who loved nothing better than to hang around with Uncle Max and jabber at him. Max knew by heart all the Kaszubian folktales he'd heard from his mother before she died, and on Sundays after church he would sit with his niece and tell her the stories his mother had once told him.

Marguerite loved Max's voice as much as the stories he told—a soft, musical voice that went with his easy grin and smooth, round, childlike face. Mary couldn't help melting a little when she saw her kid brother and her daughter together. Max was the last-born of the Cieminski kids, Mary the first—so she was old enough to be his mother. In fact, she had practically raised him once their mother died. It did her good to have Max under her roof, telling stories to Marguerite and keeping the memory of their mother alive.

Sweetness was scarce in Bessemer, Michigan, in 1914, and Mary and Max were grateful for what came their way through Marguerite. In Mary's eyes, it was almost as if Max and Marguerite were siblings— only they didn't quarrel and tease like brother and sister. They just loved each other in that deep, unfathomable way that sometimes binds two people together. Uncle Max and Marguerite were each other's favorites and would be as long as they lived.

Like Frank and Joe Chmielewski and Max Cieminski, Meyer Epstein had family in the war zone of the Eastern Front. His younger brother Zender, left behind when Meyer emigrated in 1913, was now seventeen, prime age to be snapped up into the ravenous Russian army and sent to fight the Germans. Meyer had wanted to take his brother with him when he sailed to America, but the aunt who raised the boy refused to let him go. Now it was all but impossible to get anybody out.

Meyer followed the war in Europe in the pages of *Forverts*, the popular socialist Yiddish daily published near his tenement building on the Lower East Side. In the late summer and early autumn of 1914, the news from the Eastern Front was increasingly alarming. At the end of August, as Belgium burned, Jews fled en masse before a massive Russian attack against Austrian forces in the southern Polish-Ukrainian province of Galicia. On September 1, *Forverts* reported that as Russian forces approached the heavily Jewish Galician capital of Lemberg (called Lviv by the Ukrainians and Lvov by the Russians), "200,000 inhabitants flee in terror. Whoever can is running wherever his eyes lead him." Two days later, the paper's front-page headline proclaimed in Yiddish, "Russia Boasting Great Victory Against Austrians near Lemberg; Russia reports to have buried 15,000 Austrian soldiers fallen on the battlefield." Meanwhile, the Germans had surged into western Poland, bombing and burning villages in their path. *Forverts* ran a harrowing account on September 6 of random violence against Jews in the ancient Polish city of Kalisz west of Warsaw: "Three elderly, frightened folk, in their bedclothes, were tied up and shot in front of everyone. The German soldiers looted and plundered. The city is now in ruins, partly bombed and partly burned. Corpses are strewn about as if after an earthquake." By the end of September, all of Galicia had fallen to the Russians and Jewish refugees numbered in the thousands.

The situation was even more desperate in other parts of the Pale. Stories trickled back to America of mass deportations of Polish Jews, synagogues looted and torched, summary executions of Jews suspected of collaborating with the enemy. Russian soldiers stationed around Minsk, where Meyer's brother Zender lived, were going door to door searching Jewish houses, interrogating and sometimes arresting Jewish residents; the same thing happened in shtetls throughout Belarus, Latvia, and Congress Poland. "Seven million Jews are involved in this war," reported *Tageblatt,* a Yiddish daily paper, on November 2, 1914. "Not only the 400,000, the flower of our European youth, in arms, but their parents, their wives, their children. . . . The great armies that have been swaying back and forth in Poland and in Galicia have fought every inch of ground in Jewish towns and villages."

Among the 400,000 Jewish youth in arms was Sam Goldberg's uncle—his mother's kid brother Duvid. At eighteen, only four years older than Sam, Duvid had been drafted into the Russian army when war broke out and sent to the Eastern Front to fight in the disastrous Battle of Tannenburg. Duvid was among the 95,000 Russian soldiers taken prisoner when the Germans routed the Russian army—the rout that precipitated the suicide of the Russian general, Samsonov.

With so many young Jewish soldiers fighting on both sides of the conflict, it was inevitable that Jews would be forced to fire on their fellow Jews. A story circulated from the shtetls of the Pale to the tenements of the Lower East Side of a battle in Galicia in which a Jewish soldier was about to make the fatal plunge with his bayonet when the enemy soldier cried out the Hebrew words "*Shema Yisrael*"—Hear, O Israel—the first words of the most fundamental Jewish prayer. The soldier dropped his rifle in horror. Jews on the Eastern Front, half the world's Jewish population, were "trapped between two infernos" as one reporter put it.

Meyer, only a year before, had plied his trade with a horse and cart through this very corner of eastern Europe. He never forgot to thank God for getting him out before it was too late. Life was not easy on the Lower East Side, but Meyer knew he had it good in America. Ever since Sam Minskoff hired him after that feat with the radiators up on Broadway, he had been holding down a steady job as a plumber and making decent money. He joined the plumbers' union and he was proud to work alongside other union men. He would never grow rich, but as long as he had his strength and his health and his union card, he would never go wanting in America. Freedom from want, freedom from fear, freedom from oppression—these were blessings more precious than ever after word reached Meyer that his brother Zender had been seized by the Russians and put in a prison camp in Poland.

Meyer, like thousands of other Jewish immigrants, didn't just get war news from *Forverts*; he also got a daily dose of insight into American life, manners, and mores. Only in America could there be a newspaper that covered strikes on one page (always on the side of the labor union) and gave guidance on coping with a vegetarian son on another (take him to a psychiatrist). From *Forverts*, Meyer learned how to dress and shave and shake hands like an American, how to deal with American bureaucrats and the officials who had the power to grant him citizenship. Without being overly polemical, *Forverts* pushed an agenda of moderate socialism, Jewish self-improvement, and accommodation to American ways that made good sense to Meyer. No *Forverts* reader could resist the Bintel Brief (bundle of letters) column, a kind of Dear Abby forerunner, with practical advice on everything from luring back runaway husbands to breaking in greenhorn relatives to locating Jewish aid societies.

It was only in its war coverage that *Forverts* veered, and urged its readers to veer, sharply away from the American mainstream.

Abraham Cahan, the paper's charismatic Lithuanian-Jewish editor, had fled the Pale in 1882 to avoid the mass roundups of Jews and revolutionaries following the assassination of Czar Alexander II. Like most Jewish intellectuals of the day, Cahan, "more exile than immigrant," in the words of one colleague, hated the Russian autocracy with a hatred so fierce and abiding that it became the cornerstone of his politics. What Russia did, Cahan opposed; what Russia stood for, Cahan despised; Russia's allies were Cahan's enemies, Russian's enemies his friends. And so, when Russia, as a member of the Triple Entente, went to war on the side of Britain and France, Cahan backed Germany and Austria. This was not simply a case of the enemy of my enemy is my friend. Germany and Austria, though no paradise of philo-Semitism in the early twentieth century, were much more tolerant of Jews than Russia. For years, Jews who had been barred from Russian universities pursued advanced degrees in Germany and Austria, and by 1914 there was a prominent and vocal group of German-educated Jews in the American immigrant community. To Jews, Russia was the land of the pogrom; Germany, by comparison, seemed enlightened.

As the American press grew increasingly pro-British in the aftermath of the rape of Belgium, *Forverts*, along with most of the Yiddish press, maintained a steadfast pro-German, anti-Russian line. "A German victory . . . will be a victory over Nicholas' pogromists," Meyer read in Cahan's paper; "every German and Austrian success in this struggle is a defense of the honor of Jewish women, a defense of the blood of innocent Jewish men."

In the autumn of 1914, the majority of the nation, its president included, publicly avowed American neutrality—but in private sides were being drawn, alliances and allegiances put to the test. Armed camps were forming, and increasingly the population was lining up on the basis of ethnic background and the degree of their "Americanness." The gravitational pull of blood and money, always powerful, becomes irresistible in time of war. And already, in September of 1914, those forces were pulling the country apart.

Whether or not Meyer agreed with the pro-German stance of the newspaper he read each day is impossible to say. What's clear is that he, like the Chmielewski brothers and Max Cieminski, had a different stake in the war from that of people born in the United States. What mattered most was family: if you had relatives fighting or living in the war zone, you weren't neutral, no matter what the president professed.

Epifanio Affatato was too busy trying to earn a living to pay much attention to the outbreak of war overseas. After discovering that Brooklyn's streets were not paved with gold, Epifanio had gone out to acquire gold the hard way. His father, Dominico, was a pick-and-shovel man—he labored shoulder to shoulder with hundreds of other southern Italian immigrants building the New York City subways—but that work did not appeal to Epifanio. Instead, he and his older brother Carmine decided to try their luck out west. The Affatato boys went to Des Moines and got a job with the railroad, repairing track and working a hand truck. But they found out soon enough that railroad work in the Midwest was not much better than subway work in the big city. The gangs were segregated by nationality—Italians together, Poles together, Slovaks together—and the Affatato brothers ended up with a bunch of *paesani* in a shack with no heat and no running water. Two, sometimes three men to a bed; lousy food; wages of a dollar a day—10 cents an hour for a ten-hour day. "No work could have been more perfectly calculated to disillusion young peasants who had romantic notions about America," wrote one Italian of life on a railroad gang. "Only those who were submissive and had the strength to endure, survived."

The Affatato boys had the strength, but they lacked the submissiveness to tolerate work on the railroad. After a stint in Des Moines, they worked their way back to Brooklyn and rejoined their father.

The Affatato men got their news from the popular Italian-language daily *Il Progresso Italo-Americano,* and all through the autumn of 1914, the front-page columns were full of accounts of the slaughter in Europe. "Eighth Day of the Battle of the Aisne," ran the headline on Monday, September 21, 1914. "Rheims, one of the most beautiful cities in France, with a population of more than 105,000, important juncture in the French defenses, is today in flames following three days of furious and continuous bombardment on the part of the Germans." In the following days, the paper printed a series of spellbinding dispatches from France by Italian war correspondent Luigi Barzini. In some of the first eyewitness accounts of the carnage to appear in the United States, Barzini wrote of the spectacle of "an enormous plain covered by hundreds and hundreds of corpses. . . . Almost all the dead lie face down or turned to the ground, having been hit in the head or the chest, and all hold the strange poses, inhuman, grotesque, sinister, of those who were condemned to death on the field of battle." By the middle of October, the paper was running daily reports of Belgian cities torched, Austria troops massed on the Italian border, *un' enorme, gigantesca battaglia* on the Eastern Front in which 3 million bayonets were poised to strike on a line running from Warsaw to Lvov in Galicia. Every week brought new horrors. In the second half of October, the Italian daily unfolded the details of what became known as the first battle of Ypres, which raged in Flanders for six weeks, from October 14 to November 22, in the course of which some 24,000 British soldiers and 50,000 Germans died. *Kindermord*, the Germans called this battle—the massacre of the innocents—because so many of their dead were young university students who had rushed to volunteer in the heady days of August.

It never occurred to Epifanio when he read Barzini's dispatches that a day was coming when the fields of death around Ypres would figure in his own innocent life.

Only the onset of winter and a shortage of artillery halted the killing temporarily—but the death toll all along the Western Front was already staggering: some 607,000 soldiers killed in four months of fighting. As winter closed in, two parallel lines of trenches ran 475 miles from the North Sea to the mountains of Switzerland. Western Europe was mired in a stalemate of mud.

It was a mystery to Epifanio how Italy had managed to remain neutral since, according to *Il Progresso Italo-Americano*, the country was technically a partner in a long-standing Triple Alliance with Germany and Austria-Hungry. By all rights, Italian boys should have been fighting and dying and shivering in trenches alongside their German and Austrian comrades. The paper was a little vague about why this hadn't come to pass. *Assai torbide, infine, e confuse*—extremely cloudy and confused—remains the outlook for how the war will affect Italy, one article began. So be it. To Epifanio's way of thinking, it was better for the family in Calabria, better for the family in Brooklyn, if they did their jobs and stayed out of this mess.

I n fact, Italy's role in the deepening conflict was not as *torbide e confuse* as it appeared. A month into the war, Italian diplomats in London had secretly agreed to withdraw Italy from the Triple Alliance. Then, on April 26, 1915, after months of hard negotiating, Italy entered into another, and far more significant, secret arrangement with England, France, and Russia. The Allies were desperate for manpower and eager to chip away at German-Austrian forces by opening a new front to the south; Italy was desperate for land. So the two parties struck a bargain that came to be known as the London Pact of 1915. Italy promised to "use her entire resources for the purpose of waging war jointly with France, Great Britain, and Russia against all their enemies." In exchange, the Allies promised, once they won the war, to let Italy expand its borders to the north and to the east around the Adriatic by acquiring Trieste, Trento, Gorizia, and Istria, among other territories.

For obvious reasons, the terms of the London Pact were to remain hidden from the public, and no report of this seismic shift appeared in *Il Progresso Italo-Americano*. Italy would soon enter the war, but still only a handful of diplomats were aware of it.

I n that same tender 1915 spring week of secret negotiations in London, the German army unveiled a new way of making people suffer and die. Chlorine gas, the first of three lethal gases introduced during the war, killed by overstimulating fluid production in the lungs. It was like drowning on dry land: the air spaces of the lungs filled with plasma, breathing became impossible, the brain blacked out from oxygen depletion, and you died—within minutes if you had inhaled enough of the stuff, agonizing hours later if the exposure was less intense. A new horror had been let loose on the world—death by toxic strangulation.

The spring of 1915 brought flowers to the fields of France and Belgium, but not hope. The world had been at war for nearly nine months on April 22, the day the first fumes of chlorine gas wafted over Ypres, and nobody was talking about ending the fighting. Both sides remained convinced they could humble and crush the other. The killing went on apace.

When Meyer Epstein picked up his copy of *Forverts* on Saturday, May 8, his eyes fell immediately on the word "LUSITANIA" in huge bold letters on the front page. Seeing the name of the fabled swanky steamer in print raised a spark of pride in Meyer, since this was the ship on which he had sailed to America nineteen months earlier. Then he took in the entire banner headline—"GERMANS SINK THE LUSITANIA NEAR THE COAST OF ENGLAND"—and the enormity of the event hit him. The article under the big headline was sketchy and vague—the paper had gone to press while the story was still unfolding across the Atlantic—and Meyer had to wait until Sunday, two days after the ship went down, to learn the full story. Early in the afternoon of May 7, while the cabin passengers were having lunch, a torpedo fired by a German U-boat ripped a hole in the starboard side of the *Lusitania* between the first and second funnels. Eighteen minutes later the great Cunard luxury liner disappeared into the sea. Some 1,198 people perished, including nearly 100 children. Of the 197 American citizens who had set sail on the *Lusitania* from

Manhattan's Pier 54 on May 1, 128 died when the ship sank. Among the American dead were Vanderbilt scion Alfred Gwynne Vanderbilt and New York theatrical producer and actor Charles Frohman.

Meyer wondered, everybody wondered: did this mean the United States would declare war on Germany? He puzzled over the editorial that *Forverts* ran that Sunday weighing this question: "American capitalists are waiting for the opportunity to throw themselves into the bloody mess. They are not even making believe that this loss is wrenching their patriotic hearts. In truth, they are overjoyed at the opportunity to fan the hellish fires of war. Their hatred for Germany is inciting them to use this as an opportunity to attack Germany. We socialists are forever against war and nationalistic fervor. . . . We believe we should remain neutral, not more belligerent toward Germany than toward her enemies. The horrible truth of the *Lusitania* is the result of the politics of war, not the fault of Germany."

A ndrew Christofferson got the news three days late from the *Decorah-Posten*. The paper's May 11 edition had the words "*Lusitania* Skudt Isænk" (*Lusitania* Sunk by Torpedo Fire)—blazoned across the front page. Beneath the headlines there was a vivid story based on the survivors' "grim descriptions" of the scene: "When the ship had vanished, the sea was filled with hundreds of persons struggling to stay afloat, and the air was filled with their cries for help and their calls to loved ones, bidding them farewell."

"The Affair Arouses Deep Indignation in All Countries," proclaimed one of the strings of front-page headlines in the *Decorah-Posten*, but indignation was by no means universal, even in the United States. The Cincinnati *Volksblatt*, one of the country's many flourishing German-language newspapers, hailed the torpedoing as "a shot straight into the heart of England," and other German-language papers rushed in to blame the tragedy on England and its American sympathizers. They insisted that the *Lusitania* was a fair target since it was

armed and transporting American-made weapons and ammunition
to the British; that passengers had in effect sacrificed their own lives
since the German government had issued advance warning not to sail
owing to the danger of German U-boats; that British incompetence
and ineptitude on the part of the Cunard line were the real reason so
many had died, and so on. A lot of this was bluster. German Americans were well aware that if the *Lusitania* crisis precipitated American
entry into the war, it wouldn't matter that the Cunard liner could be
technically classed an armed British auxiliary cruiser or that American
citizens had been advised to stay off the boat. War would be a disaster
for the German community in the United States.

Some 8 million individuals either born in Germany or with at least
one German parent, almost 8 percent of the total population, were
living in the United States. In the furor that followed the torpedoing, these German Americans were forced as never before to consider
the meaning and consequences of their dual allegiance. Where did
their loyalties lie? What would they do if they had to choose between
their old country and their new one? Just after the sinking, the editor
of Milwaukee's *Germania-Herold* framed this dilemma with eloquent
frankness that would soon become impossible: "We are Germans, of
course. I was an officer in the German Army, I have one hundred
and twenty-five relatives now fighting for Germany. When people
ask us, therefore, where we would stand in case of war between this
country and Germany, it is like asking a man where he would stand
in his own household as between his wife and his mother. However,
if war ever came between this country and Germany or any other
country, we would be American citizens, just as we were in the Civil
War." Not all German Americans, however, felt the same way. In the
early days of the war, the German embassy was flooded with requests
from German Americans to be shipped back "home" so they could
join the soldiers at the front, and ultimately upwards of half a million young German nationals returned to fight for the Fatherland.
Thousands of others rejected the armies of both wife and mother

on religious grounds. German-speaking Hutterites and Mennonites abided by a form of Christianity that opposed service in the military of any country. Most of them were American citizens in 1915, but if their country declared war on Germany, they would refuse to fight for the United States, just as they had refused to fight for Germany. They worried, with cause as it turned out, that such refusal would at best be deemed cowardice—but more likely treason.

In 1915, Teddy Roosevelt, the nation's foremost enemy of the hyphen, stood up in front of a gathering of Irish Catholics and thundered against the "moral treason" of immigrants who refused to sunder their ties to their country of origin. "The one absolutely certain way of bringing this nation to ruin, of preventing all possibility of its continuing as a nation at all, would be to permit it to become a tangle of squabbling nationalities. . . . There is no such thing as a hyphenated American who is a good American. The only man who is a good American is the man who is an American and nothing else." Before the war, such rage against the hyphen was part and parcel of Roosevelt's jingoistic rhetoric: you were either "100 percent American" or not American at all. But in 1915, with the European continent laced with trenches and Anglo-Saxons dying by the tens of thousands, it was a coded warning for immigrants, especially those who "at heart" felt any sympathy with Germany or its allies. If the United States entered the war, as Roosevelt fervently hoped it would, its strength in battle would crucially depend on the absolute loyalty of the immigrant population. It would not be enough for the foreign-born to sweat and bleed for America: to be real Americans they had to be "American and nothing else." They had to surrender their souls.

But would they really be better Americans, better soldiers, if they did? In the land of the free, it was still possible to ask such a question, as the editor of the Finnish American newspaper *Paivalehti* did in an impassioned editorial: "The man is good for nothing who at a moment's notice can forget his past, his kindred, his people and without a sigh change his views of life and ways of thinking even if he is aware of all the good this country offers it inhabitants and citizens. . . . [Immigrants] become Americanized gradually and surely; there is no cause to anxiety as to them and their loyalty is beyond any doubt."

The mood of a nation is always hard to gauge, but after the *Lusitania* went down there was an unmistakable shift in American attitudes toward immigrants. Roosevelt's rhetoric reflected a pervasive fear that the country was splintering along ethnic lines. And in fact, this fear was not ungrounded. By 1915, a shadow war of protest was erupting on the streets of America's immigrant cities. Irish Americans and German Americans had begun banding together to decry the United States' increasing moral and material support for Britain. The motives of German Americans were obvious; but for the Irish, the pro-German stance was a case of the enemy of my enemy is my friend. Irish Americans were desperate to keep their adopted country from joining the war on the side of the nation that had oppressed their people for hundreds of years, and they stood shoulder to shoulder with Germans at protest rallies and marches around the nation. In the highly charged atmosphere of 1915, the words *neutrality* and even *peace* came to be interpreted as support for the kaiser. At the end of June, a month and a half after the *Lusitania* sank, Irish and German protesters filled New York's Madison Square Garden and overflowed into the surrounding sidewalks in a massive peace demonstration. Irish firebrand Jeremiah O'Leary, president of a pro-neutrality group called the American Truth Society, traveled the country baiting Wilson, Roosevelt, and other critics of the "hyphenates."

"Each of the belligerent nations had children in the New World," write historians Meirion and Susie Harries, "and every outburst of passion in Europe was echoed in America." The streets of Chicago erupted in violent clashes between Slavs and Germans. Cincinnati Jews talked of raising a Jewish militia to join the German forces fighting the czar. Peace protesters were coming under suspicion as potential spies or even saboteurs. "None of us foresaw ... that in this broadcast sowing of hate [in Europe] so much of it should fall upon our shores, take root and grow," said a Jewish college professor who had emigrated from Austria-Hungary. Old loyalties of blood, custom, and faith were surfacing—and an increasingly venomous nativist backlash was rising in response. Even if the United States stayed out of the war, people were beginning to wonder how much longer peace could be preserved at home.

On May 23, 1915, banner headlines in *Il Progresso Italo-Americano* proclaimed: *La mobilitazione generale—Papa Benedetto XV richiama i catolici al dovere verso il tricolore! ... il primo sangue!* General mobilization—Pope Benedict XV recalls Catholics to the duty to the Italian flag! ... the first blood! Italy had entered the war on the side of the Allies. In the coming months, some ninety thousand Italian-born men would leave the United States and return to their homeland to join the Italian army. Epifanio and Carmine Affatato were not among them. The brothers stayed put in Brooklyn, worked as laborers for whoever would hire them, and sent what money they could afford back to Calabria. With Italy at war, the family still living over there would need American dollars more than ever. Epifanio concluded that he would be of more use working in Brooklyn than fighting in Italy. And anyway, if he went back it would be that much harder to become an American citizen. For Epifanio, at that point in his life, American citizenship was the number one goal, the key that would open every door. Two years earlier, when he turned eighteen, he had filed his first papers—his declaration of intention to become a U.S. citizen. He paid a dollar; renounced his allegiance to the Italian king, Vittorio Emanuele III; signed his name; and just like that his status changed. From alien to declarant. All he had to do now was to wait five years—until June 1918. Provided, that is, that he remained in the United States the entire time. Only five years of *continuous* residence qualified you for citizenship. So Epifanio did not even consider going back to fight for Italy. For what? He joined a club that called itself La Società—the society—where Brooklyn guys from Calabria could get together, talk, socialize, shake their heads over the world situation, trade tips on jobs. Men only, of course. Let other *paesani* go home and shoot at the Austrians. His life was here.

Tony Pierro up in Swampscott felt the same way. Like the Affatatos, the Pierro men followed the war in the pages of *Il Progresso Italo-Americano*—but despite the ominous headlines, the daily reports of carnage, the feverish flag-waving over the Italian mobilization, Tony and his brothers and father had little connection to the war. Basilicata was at the opposite end of the country from the Austrian front. And anyway, all the men in the immediate family of draft age were already in the United States—the three brothers, Daniele, Nicola, and Vito, who remained in Italy were just kids. Italy may have been at war with Austria-Hungary—but what was Italy to Tony? Italy meant taxes and conscription—that was why they had left. Why go back now when there was so much work here? Swampscott had undertaken a massive public works project to widen its main street—which meant months of employment for the Italian community. Tony kept his job on Mr. Clapp's estate and looked around for something better—better than digging, better than widening a street. What he wanted was a nice clean job and, someday, a wife and a car. A soldier's life, from what he read in the paper, was a lot like a laborer's—only you slept in the mud and got shot or stabbed with a bayonet or blown up by an exploding shell. Not for him.

So far, the war in Europe had been good to Peter Thompson. With wartime demand pushing the price of copper sky high, Butte's mines were running round the clock, money was pouring in (most of it landing in the coffers of the Anaconda Copper Mining Company), and wages were rising steadily for workers willing to play by the company's rules. Peter was willing. By 1915, he was pulling in $3.83 a day, up from $3.50 the year before—and the next year his pay jumped to $4.40. A year after arriving in America, Peter had put enough aside to bring his father over. In July 1915, Sam Thompson took his son's wages and bought a ticket aboard the White Star line *Arabic*, braved the German U-boats that infested the North Atlantic, and joined the growing Thompson clan in Montana.

For Peter, his father's arrival meant an immediate uptick in the anti-English rhetoric noised about the house. "I don't care who wins the war as long as England gets beat!" Sam Thompson was fond of saying, and if he said it once he said it a hundred times. He had plenty of fellow countrymen in Butte who felt the same way. In those days, the town was home to radical Irish nationalist fraternal groups—the Ancient Order of Hibernians and the Robert Emmet Literary Association (the Butte wing of Clan na Gael)—both of which were outspoken in their hatred for England and their fervent desire to keep the United States from joining the war on the side of the Allies. Butte's hardworking wild-eyed Wobblies were in the same camp. Anti-British and pro-neutrality sentiments did not always spring from the same source—but they overlapped and merged in working-class immigrant bastions like Butte. Local socialists never tired of pointing out that even though wages were climbing, the cost of living was going up even faster. Inflation was becoming a problem nationwide, but in Butte because of its isolation, its dependence on a single industry, and the large transient foreign-born population, the problem was out of control. As one historian pointed out, "Miners with families did not make enough to maintain a 'minimum comfort level' at any time during the war period." To make matters worse, mining accidents had become rampant with stepped-up wartime production and new technology. Innovations like steam-powered hoists and two-man handheld air drills were killing more miners in accidents and putting more dust in the shafts, which resulted in higher rates of silicosis.

Peter listened to his elders rail against the company, the war, the filthy English king. He listened but he didn't join in. Peter had his mind on making as much money as he could, and his eye on a pretty girl. Pushing twenty, he had already been working for the last eight years between Belfast and Butte. Let his father and his uncle talk. Peter was happy to ride the wartime boom as long as it lasted.

Matej Kocak had been a marine for eight years when he finally got a medal—the navy's good conduct medal awarded on December 11, 1915, at the end of his second enlistment. Private Kocak was discharged from the marines two days later with an "excellent" character, having received impeccable marks on his "professional and conduct record." A couple of weeks shy of his thirty-third birthday, Kocak was now a seasoned marine in the prime of his life. A trained warrior without a war.

After his discharge, Kocak likely spent some time in Binghamton with his brother Paul and Paul's wife, Julia. The Kocak clan was adding its numbers to Binghamton's already considerable Slovak community. A cousin also named Paul, ten years younger than Matej, had come over from their village in Austria-Hungary and gotten a job in town as a shoemaker. Good Catholics, the Kocaks attended Mass at Ss. Cyril and Methodius Church, the Slovak church that had been built in 1904 to serve the local immigrant community. Matej went to church with the rest of the family when he visited Binghamton, though in general he preferred taverns to churches. The word in the family was that Matej liked to have a good time and could more than hold his own when it came to drinking—no mean feat among fellow Slovaks. Something else he had picked up during his years with the marines.

Whether he was in the tavern or the church, Matej heard talk about the war from his fellow Slovaks. Though winter had once again brought the fighting in Europe to a sullen halt, the Slovak community was still burning with excitement and hope. Every Slovak church in

America prayed that the war would turn out to be the hammer that shattered the Austro-Hungarian Empire and delivered their people from decades of Magyar oppression. If only the United States would add its muscle and influence to the Allied cause. What was Wilson waiting for? Fiercely loyal to their new country but still deeply attached to the old one, Slovak Americans had to live with the bitterness of seeing relatives in Europe drafted to fight and die in the Austrian army, while the United States insisted it was neutral and did nothing. Such was the lot of a people without a country of their own.

Matej didn't spent a lot of time in Binghamton. On December 29, 1915, two days before his birthday, he went to the New York City recruiting office and signed on for another four years. By the start of the new year, Matej was on his way to the marine barracks down in New Orleans.

The winter lull in the fighting was short-lived in Europe. On February 21, 1916, the German army launched a major offensive aimed at the ancient fortified city of Verdun in northeastern France. German general Erich von Falkenhayn declared his intention to "bleed white" the French army—but the bleeding proved to be universal. Generals on both sides ordered their forces to keep pounding away at whatever cost in human life, and the cost mounted daily. One historian wrote that by June, "About twenty million shells had been fired into the battle zone since 21 February, the shape of the landscape had been permanently altered, forests had been reduced to splinters, villages had disappeared, the surface of the ground had been so pockmarked by explosion that shell hole overlapped shell hole and had been overlapped again. . . . To both armies Verdun had become a place of terror and death that could not yield victory." The slaughter ultimately dragged on for ten months, the longest battle in the war, at the end of which the French and German positions were virtually unchanged. Casualties were appalling even by the standards of this appalling war—nearly 400,000 French soldiers killed, some 340,000 Germans dead or missing.

A month after the Germans mounted their assault on Verdun, the United States embarked on a military engagement of its own against Mexican general Pancho Villa. It did not turn out to be one of the finer moments in the nation's military history. For the past six years, Washington had been following the revolution in Mexico with mounting concern that the chaos would spill over the border. The marine occupation of Veracruz that President Wilson authorized in the spring of 1914 did little except to inflame anti-American senti-ment. After war broke out in Europe, Mexican strongmen and gener-als of various ideological stripes continued to wrestle for control, while American politicians and businessmen puffed, postured, worried about a secret German-Mexican alliance, and sold arms and ammunition to whoever had the cash to pay for them. Pancho Villa was among the Mexican generals who relied on American weapons, and for a couple of years Sam Dreben, his trusted agent in El Paso, ran a tidy covert operation keeping the general supplied with rifles and bullets.

Then President Wilson threw his support behind the government of reformer Venustiano Carranza, and Villa found himself out in the cold. Stung by the Yankee betrayal and by the drying up of American arms shipments, Villa decided to retaliate. In January 1916, he hit a train near Santa Isabel that was carrying employees of the copper and smelting company ASARCO, owned by the Guggenheim family. Eighteen American ASARCO employees died at the hands of Villa's men. Then on March 9, Villa made an even bolder move and took the fight onto American soil. In the early morning hours, 485 Vil-lista raiders crossed the border and invaded the town of Columbus, New Mexico. Villa's men opened fire on a detachment of the 13th U.S. Cavalry, killing ten soldiers and eight civilians. With the town ablaze behind them, the raiders rode back to Mexico with a string of American horses and mules and a cache of American weapons and ammunition. Not all of Villa's men, however, made it back. In the course of the raid, an American machine gunner opened fire and cut down eighty of the Villistas.

A howl for revenge went up from "the one great nation at peace." American soil had been invaded; American lives and American property had been destroyed; the U.S. military had been attacked. President Wilson's response was swift. On March 19, ten days after the raid, Wilson summoned General John J. "Black Jack" Pershing, a former West Point instructor and career officer who had served in the Philippine-American War, to mount a "Punitive Expedition" against Villa and his forces in Mexico. Six thousand army troops were dispatched south with orders to track down the Mexican bandit-general. Army and National Guard divisions were mobilized to patrol the U.S.-Mexican border. Eight planes were called up to fly reconnaissance missions over the dry hilly terrain where Villa and his men might be lurking—the nation's first airborne military action. Pershing, headquartered in the town of Colonia Dublán near the Casas Grandes River in the Mexican state of Chihuahua, put out a call for volunteers from Texas. Among those who responded was Sam Dreben.

Despite his personal and business ties to Villa, Dreben, now pushing forty, jumped at the chance to serve once more with the U.S. Armed Forces. "Love of his adopted country and his country's flag was his religion," journalist Damon Runyon, Dreben's great champion and friend, once wrote of him, and war was his avocation. The Fighting Jew was back. Pershing was quick to recognize the value of Dreben's intimate knowledge of Villa's character and tactics, and he made Dreben his chief scout. Disguised as an Arab peddler, the Fighting Jew infiltrated Mexican villages ahead of the American troops and gathered what intelligence he could. As usual, he had his share of colorful adventures, lightning escapes, and ruinous gambling parties. Though his efforts as a scout and spy ultimately proved fruitless, Dreben won

Pershing's respect and friendship. "He never failed to volunteer for the most dangerous tasks," wrote the general at the end of the punitive expedition. "I never knew him to make a false or misleading report."

Pershing acquitted himself rather less favorably in the eyes of his own superiors. Not only did his forces fail to track down Villa, but Villista raids across the border continued unabated despite the presence of sizable garrisons from both the National Guard and the Texas Rangers. In late June, Pershing's forces, including the African American 10th Cavalry Regiment, suffered humiliating losses at what became known as the Battle of Carrizal. Pershing had moved into the territory on a tip that Villa was camped there; but instead his cavalry found itself taking the field against a contingent of soldiers from the Mexican National Army. Seven men were killed, many more wounded, and twenty-three taken prisoner. Pershing received orders from Washington to desist from dispatching long-range patrols that might meet a similar fate.

Pershing's Mexican campaign limped along until the close of 1916, fought during the same months as the battle of Verdun and with just as little to show in the end. Despite Dreben's best efforts, Pershing never came close to capturing or killing Pancho Villa, though his men did eventually take out two Villista generals. American forces, by all accounts, spent most of their time in Mexico gambling, drinking, and whoring. When Pershing pulled out in February 1917, he admitted that the Punitive Expedition was a failure. Given the state of the world, it was a disturbingly unimpressive show of American military might.

The Punitive Expedition, lame as it was, did have at least one positive outcome: it inspired a future hero to enlist in the U.S. Armed Forces. Michele Valente had come to America three years earlier with the flood of prewar immigration from the south of Italy. Eighteen years old, strapping, tall, ruddy-skinned and blue-eyed, a son of the village of Sant'Apollinare near the ancient monastery of Monte Cassino, Michele stood out from his compatriots not only in his appearance but because of his easy, confident manner. Though just a kid, he already carried himself like a man. Michele landed in New York City with $30 in his pocket and the address of his uncle Paul in a town called Ogdensburg, which turned out to be a long, long train ride away, up in the northern edge of New York State. So far north that you could see Canada across the St. Lawrence River. Michele moved in with his uncle and got a job as an orderly at St. Lawrence State Hospital, the local mental hospital. Greenhorn though he was, Michele knew America had something better to offer than that—something more manly, better paying, more exciting. When he heard about the Pancho Villa raid and the military campaign Pershing was leading down in Mexico, Michele found it. He decided to sign up with the 1st Regiment of the New York National Guard. He'd been in the United States for only three years, but that was good enough for the officers who were activating this upstate National Guard regiment. Michele Valente—now known as Mike—was tall, fit, muscular, courageous, just what the Guard was looking for to chase Mexicans in 1916. If it weren't for the accent, nobody would even know he was Italian.

In the event, the Punitive Expedition ended before the 1st Regiment could be dispatched to the Mexican border. But no matter. For this eighteen-year-old immigrant hospital worker, it was the start of a military career. When the country entered the war, Mike would have ample chance to show what he could do under arms.

Mike Valente is the twelfth of the dozen immigrant soldiers whose stories come together in these pages.

In his ninth year with the marines, Matej Kocak fired a gun in battle for the first time. In the middle of May, while Pershing was futilely chasing Pancho Villa around Chihuahua, a contingent of marines was dispatched to the Caribbean Island of Hispaniola to quell political turmoil in the Dominican Republic, then more or less an American client state. Kocak arrived off the coast of the capital city of Santo Domingo on board the *Kentucky* on May 27, and he went ashore on June 11 with the 14th Marine Company to set about restoring American-style order. At Las Canitas in Azua Province, Kocak's company engaged in a series of skirmishes with Dominican forces alternately described as "native bandits" or "revolutionaries." Though seriously outnumbered, the 14th Marine Company prevailed, at least for a while. As time passed, the local population came to hate living under American occupation and their hatred emboldened guerrillas to keep fighting in the eastern part of the island. As long as there was resistance, the marines stayed put—trapped in the miserable role of peacekeeper in a country that wanted them out.

Kocak sweltered through a Caribbean summer and the hurricane season that followed. Thankless and miserable though the mission may have been, it was not a total loss for him. In the rugged hills outside Santo Domingo, the Slovak private found out that he could remain cool under fire. And before his tour in the Dominican Republic was over, he got promoted to the rank of corporal.

N ineteen hundred and sixteen was an election year in the United States, and the two parties held their conventions within a few days of each other in June. The Republicans went first, nominating the Supreme Court Justice Charles Evans Hughes in Chicago on June 10. The fact that Hughes had been on the court since 1910 and thus largely out of the public eye was considered an advantage: his views on the war, the increasingly divided immigrant communities, and the Mexican situation were largely unknown, and that's the way the party wanted it. The Republicans, well aware of the bad blood between Wilson and the nation's large German and Irish communities, hoped to send their candidate to the White House on a rising tide of German and Irish support.

The Democrats, convening in the heavily German city of Saint Louis, Missouri, from June 14 to June 16, overwhelmingly renominated Wilson. In the rancorous campaign that followed, the war, loyalty to America, preparedness, 100 percent Americanism, and the conflict between labor and business were the issues that dominated—and each of these issues implicated the nation's large restive immigrant populations. Just days before Wilson was renominated, huge numbers of preparedness supporters marched down New York's Fifth Avenue under an electric sign that pulsed the words ABSOLUTE AND UNQUALIFIED LOYALTY TO OUR COUNTRY. The coded message was lost on no one: immigrants and "hyphenates" beware.

Wilson, harping on the absolute-loyalty theme, declared in a much publicized run-in with Irish agitator Jeremiah O'Leary that he would be "deeply mortified" to have the votes of "disloyal Americans" like O'Leary and his followers. The German press was quick to point out that whenever Wilson used the terms "hyphenated-American" or "individual passions" or "national interests," he was in fact attacking the German American community. Now a practiced politician, Wilson shrewdly adapted his message to the views and fears of his audiences. Somehow the president managed to attack his Republican opponent as both a pro-German weakling and a puppet in the pocket of the fanatically bellicose, pro-British Teddy Roosevelt. In politics, as usual, defamation trumped truth.

But in the end, what voters responded to most powerfully was the Democratic slogan "He Kept Us Out of War." For all the heated rhetoric of the immigrant press and the noisy partisan demonstrations, the majority of Americans in 1916 had no appetite for war. The news from Europe that summer was just too grim.

Grimmest of all were the reports that came back from the first day of the battle of the Somme on July 1, 1916. Douglas Haig, the British general behind this new Allied push, had worked up a plan to open the campaign with a week of ferocious shelling—the figure was set at a million shells—intended to stupefy the German infantry and shred the barbed wire strung in front of the German trenches.

Then nineteen British divisions and three French divisions were to stream onto no-man's-land, race through gaps in the shell-blasted wire, and rapidly push the Germans east. Haig, in short, was convinced he had the formula for the long-awaited breakthrough.

Envisioned as a rout, the Somme instead became a bloodbath. The million British shells failed to make much of a dent in the deeply en-trenched German line, and the vicious German barbed wire, far from being destroyed, became twisted into even more impenetrable tangles. When the Tommies went over the top, the Germans let them advance across no-man's-land to within easy firing range and then mowed them down as they got caught up at the wire. Staggering under their 60-pound packs, zealous and confident of sure success, the British soldiers didn't stand a chance.

Stories came back from the Somme of grinning raw recruits kick-ing a soccer ball before them as they trotted out into no-man's-land. When German machine guns, rifles, bayonets, and entrenching tools had finished their work that day, 19,240 of the 100,000 British ad-vance force had died and another 36,000 had been wounded: one out of every five men cut down on the battle's first day. Never, before or since, had so many British soldiers perished in a single day of war.

The British Expeditionary Force buried its dead—at least those they could get to—and pressed on at the Somme for five more months. In August, Peter Thompson's brothers Denis and John, both fighting with the 8th Royal Irish Fusiliers, were rotated to the front. The Thompson boys were lucky. They survived. Denis even got a break when he was sent home to Belfast to recover from a bout of trench fever.

In the early hours of July 30, 1916, a month into the carnage at the Somme, a massive explosion rocked New York and New Jersey. Black Tom Island in New York Harbor, site of the largest munitions dump in the nation, had blown up—or rather had *been* blown up—setting off an earthquake of detonating shells and dynamite that shook the ground as far away as Philadelphia. Lower Manhattan would not see the like again until September 11, 2001.

The explosions began at 2:08 A.M. when 100,000 tons of TNT stashed in a barge tied up off Black Tom Island were ignited. By morning, after more than 2 million pounds of shells, fuses, and nitrocellulose detonated, not a single pane of glass survived in the shops lining Jersey City's main street. Lower Manhattan skyscrapers were battered, and windows lay shattered in Times Square. Shell fragments had ripped through the flowing drapery that cloaked the Statue of Liberty's breast and damaged her torch. On Ellis Island, five hundred newly arrived immigrants had to be evacuated after the ceiling of the hospital building caved in. Rescue workers trying to approach the blast site were driven back by swarms of bullets that kept spitting from the incinerated ruins. Astonishingly, only five people, including a ten-week-old infant, were reported killed, though New York police believed many more perished on barges that were being used as illegal houseboats near Black Tom Island.

Rumors had been swirling for months of German spy rings, schemes to sabotage American food supplies and smuggle in chemical and biological weapons and kidnap leaders. Supposedly secret fortifications were in place on the New York City shoreline in the event of a German invasion. In this paranoid climate, one would think that the destruction of the nation's largest munitions dump, occurring at the height of the battle of the Somme when British forces were desperate for ammunition, would unleash a massive hunt for German saboteurs. Yet American officials were curiously numb. President Wilson dismissed the explosion as "a regrettable incident at a private railroad terminal," and initial investigations concluded that "the fire and subsequent explosions cannot be charged to the account of alien plotters." This conclusion turned out to be dead wrong—it later emerged that a secret ring of German saboteurs had carefully plotted and carried out the explosion with the aid of German diplomats, sympathizers, and all manner of shady Teutonic characters.

Perhaps the American public in 1916 was not ready to face the truth. The world had been at war for two years now. Like it or not, the epidemic of violence was spreading.

Joe Chmielewski lasted three years at the Argyle Coal Company, and then, when he was eighteen, he quit to look for something better. Different, anyway. For once, luck was with him. An uneducated unskilled immigrant couldn't have picked a finer time than 1916 to look for a new job. Thanks to the war, demand for American raw materials and manufactured goods was skyrocketing at the same time that the supply of cheap immigrant labor was drying up—so the mine and mill owners were practically rounding up able-bodied workers on the street. Not that wages went up much, or working conditions improved any. In fact, in the Pennsylvania coal mines as in the Butte copper mines, stepped-up production brought a spike in accidents: explosions, cave-ins, floods, cages plummeting down shafts, underground fires, pockets of gas or smoke. Just that past January, Fifficktown and South Fork were rocked by an explosion when the powder house of Argyle's Number 1 mine blew up at dawn. Three large boardinghouses inhabited by immigrant mine workers burned to the ground. Anyway, Joe had had enough of coal. His brother Frank had gotten used to crawling on his belly all day and coming home at night with black knees and a black back, or so he said; but Joe was still young and he needed a change. In 1916 he signed up with the Carnegie Steel Company, a short train ride away in Johnstown.

The work was different, but otherwise Joe's life went on pretty much as it had before. He still boarded with the Yablonskys; he still went to St. Anthony's Church on Sunday morning; he still had Sunday dinner with Frank and Mary and the kids. If there was a girl in Fifficktown who caught Joe's eye in church or at one of the dances, no one heard about it. What schooling Joe had had in Poland would have to serve him for the rest of his life—in America there was no time left for education. Maybe once he became a citizen, Joe could get out of Fifficktown, quit the steel mill and find another job—something cleaner. But first he'd have to learn some English.

S am Goldberg left home for good that fall. The situation with his father had finally reached the breaking point. First Asriel had banished his surly, smart-mouthed sixteen-year-old son to the attic. Then he'd quit feeding the boy. Then there was a terrific row in which Sam had threatened to kill the old guy if he ever raised a hand to his mother. After that, Asriel threw him out of the apartment in Newark, and that was the end of Sam's so-called childhood.

Sam took the train from Newark up to Hartford, where he had a distant relative who was living in a boardinghouse. Sam got a room there—actually more like a luggage storage closet than a bedroom. No heat, no window. When the weather turned cold, Sam got sick. Trembling uncontrollably with fever, he somehow got himself to a hospital, where the nurse on duty recorded his temperature at 104 degrees. In his delirium, Sam thought he was back in Russia visiting his grandfather and telling the old man about his father: "That bastard of a father is the worst father who ever lived. He takes pleasure in physically hurting people."

When he recovered and put some flesh back on his bones, Sam decided he'd had enough of Hartford, so he headed south to try his luck in Atlanta. He landed an accounting job in an automobile company. To test him out on the first day, the boss gave him a math problem: "Okay, Goldberg, what's 60 percent of $8.50?" And just like that, Sam replied, "$5.10." The boss figured it was some kind of trick, so he left him with a bunch of problems and came back half an hour later to check up. "Where's the scratch paper where you did the figuring?" the boss wanted to know. Sam pointed to his forehead. "I don't need paper. I do it all in my head."

He was still lonely as hell and on fire with anger—but as long as he had his brain he knew he could always get a job.

Woodrow Wilson was reelected on November 7, but barely. In the end, the results hung on four thousand Democratic votes that carried California: that was all that kept Charles Evans Hughes out of the White House. The ethnic vote, courted avidly by both parties, turned out to be too divided to be decisive. Irish Americans, incensed by Wilson's attack on O'Leary, bolted in disgust from the Democratic Party; but Poles, who had resoundingly opposed Wilson four years earlier, strongly endorsed him this time because of his backing for an independent Poland. German-language newspapers in New York, Cincinnati, St. Louis, Milwaukee, and Sioux Falls hammered away at Wilson's pro-British policies, but on election day German American voters did not cast their ballots with sufficient solidarity to determine a single electoral vote.

Wilson slipped back into the White House under the promise to keep Americans out of the war, but even before his second inauguration on March 5, 1917, that promise was crumbling to ash. On February 1, 1917, the Germans resumed unrestricted submarine attacks on all merchant ships bound for Britain and France, including those belonging to neutral powers, and once again there were headlines about American vessels sunk by German torpedoes. More bad news came from Mexico at the beginning of March. Playing off the anti-Americanism rampant after Wilson's occupation of Veracruz and the bungled Villa campaign, Germany entered into a secret alliance with the Mexicans. On January 19, 1917, Germany's foreign secretary, Arthur Zimmermann, sent a telegram to the German ambassador in Mexico with instructions to enlist Mexico in a war against the United States: "We propose an alliance on the following basis with Mexico:

That we shall make war together and together make peace. We shall give general financial support and it is understood that Mexico is to reconquer the lost territory in Texas, New Mexico, and Arizona." When the so-called Zimmermann telegram was intercepted and published in American newspapers that March, it raised a terrific public outcry. For Wilson, this was the last straw. On March 20, the president assembled his cabinet secretaries and sought their advice on the situation with Germany: the vote was unanimous to go to war.

The same day that the president secured his cabinet's approval of war, twenty thousand American Jews packed New York's Madison Square Garden, shouting and dancing in the aisles. They were celebrating the fall of the czar and the birth of a democratic government in Russia. Russian workers had been rioting for weeks in the capital Petrograd, and when a general strike shut down the city on March 8, the war-battered Russian army refused to quell the insurrection. In desperation, Czar Nicholas II had abdicated the throne on March 15 and a liberal provisional government took control. For the millions of Jews who had emigrated from the Pale, the start of the Russian Revolution was the answer to their prayers—and it occasioned a sudden and radical shift in Jewish allegiance in the war. "As if by magic, the debates and discussions on the Jewish streets have disappeared," *Forverts* reported. "There is nothing more to discuss. . . . Feelings dictate, reason dictates, that a victory for present-day Germany would be a threat to the Russian Revolution and dangerous for democracy in Europe." With the United States on the brink of war with Germany, America's Jews had providentially fallen into line as staunch, indeed ecstatic, supporters of the Allied cause.

T he world must be made safe for democracy," President Wilson announced when he went to Congress on the night of April 2, 1917, to ask for a declaration of war against Germany. The president acknowledged that he was urging the nation to join "the most terrible and disastrous of all wars," but he insisted it was worth it because "the right is more precious than peace." The applause was long and loud when the president stopped speaking, and when the assembled lawmakers weren't clapping, most of them waved little American flags that they had tucked into their pockets. But regardless of the roar on Capitol Hill, the country was by no means united in its support or appetite for war—and the debate that followed in Congress reflected this. "We are going into war upon the command of gold," Nebraska senator George W. Norris thundered on the Senate floor. "I feel that we are about to put the dollar sign on the American flag." Shouts of "Treason!" erupted. There was much soul-searching on the question of the nation's huge divided immigrant population, particularly the German Americans. Wilson had been careful to extend a hand of friendship to the "millions of men and women of German birth and native sympathy who live among us and share our life," but the handshake was quickly followed by a fist brandished at those who fell short of the absolute standard of "true and loyal Americans." "If there should be disloyalty," Wilson warned, "it will be dealt with a firm hand of stern repression." Prophetic words. On the last day of the debate, one congressman, contemplating the inevitability of a draft, predicted that compulsory military service would create a true melting pot "which will . . . break down distinctions of race and class and mold us into a new nation and bring forth the new Americans." But how many of these "new Americans" realized that they would soon be called on to fight? And who in America, whether native or foreign-born, truly understood what "a firm hand of stern repression" would mean for the country in the coming months? There was no time for these fine points. For America, the "most terrible and disastrous of all wars" was at hand.

The nation held its breath while Congress debated. The Senate reached its decision on the night of April 4—of the eighty-eight senators present, only six voted against the war. The following day, the

House took up the measure in a marathon seventeen-hour debate that lasted into the wee hours of April 6, which happened to be Good Friday. At 3:15 A.M., the vote was finally tallied: 373 members of Congress in favor, 50 opposed. The United States had entered the Great War.

CHAPTER 6

THE ARMY OF FORTY-THREE LANGUAGES

Joe Chmielewski didn't regret leaving Poland—especially the way Poland became once the war started—but he had never really taken hold in America like his brother Frank. In the spring of 1917, Joe was twenty and restless. Three years at the Argyle mine, then starting the new job at the steel mill; the grimy train to Johnstown every morning and back to the smoke and soot of South Fork every night; the humid slag-ridden hills and the little houses that climbed the hill toward the cemetery; the taverns, the church. It was a tough life for anyone, but especially hard on a quiet loner like Joe. He was too reserved to cut loose with the other guys in the tavern; too withdrawn to attract much attention outside of the family; never messed around with women as far as anyone knew. *Mild* was the word Joe's brother Frank used. Mild-mannered, mild-tempered—but not so mild that Joe didn't yearn for something better than Fifficktown, Pennsylvania.

Before he made a move, Joe was in the habit of going to Frank for advice. All the Polish guys in town did the same. Frank was the one who took them to the county courthouse in Ebensburg to show them how to take out their first citizenship papers. Frank wrote their letters home to families in Poland, translated their legal documents, made sure their payments were up to date with the Polish Roman Catholic Union of America, from which they all bought life insurance and annuities. So when the time came to register for the draft on June 5, 1917, Joe sat down with Frank to ask him what it was all about—and from Frank he learned that he didn't have to worry. By the terms of the Selective Service Act of 1917 that Congress had enacted three weeks earlier, reinstating the draft for the first time since the Civil War, only men between the ages of twenty-one and thirty had to register. Joe, who had turned twenty in May, was too young and Frank, at thirty-two, too old. America was going to war—but it looked like it didn't need the Chmielewski brothers. At least not yet.

But there were plenty of other draft-age Polish guys, along with Slovaks and Lithuanians, who crossed the river from Fifficktown into South Fork that June morning, walked past the hotels and banks and taverns bedecked with American flags, and lined up at the registration station to fill out and sign their draft cards. For once, it didn't matter where you were born or how you earned your living or whether your father owned the mine or died in it. As one newspaper informed its readers, you had to register even if "you are blind, deaf, dumb, legless and armless; if you are in jail or a minister of the Gospel; if you are white, black or yellow; if you have conscientious scruples against the war, or if you are a citizen of Germany, Senegambia or other land." On June 5, if your birthday fell between 1887 and 1896, you either filled out a card or faced arrest.

Though the draft was unpopular with people and politicians alike—Wilson had to wage a fierce battle with his own party to push it through—it became clear in the weeks after the nation declared war that a draft was the only way to raise an army sufficient to satisfy the ravenous appetite of this Great War. For all the bluster about preparedness, the U.S. military was more or less a phantom in 1917. With a scant 210,000 men in arms between the Regular Army and the National Guard units maintained by individual states, the nation's armed forces ranked seventeenth in the world. In strength and training, the American army was "laughable compared to that of European armies," in the words of military historians Douglas V. Johnson II and Rolfe L. Hillman Jr., "and its organization was that of a constabulary force fit only to chase *insurrectos* and *bandidos*." Considering that the British had suffered more than 55,000 casualties (dead and wounded) on the *first day* of the battle of the Somme, it was obvious that the country's manpower shortage was acute. Equally obvious was the fact that Americans, foreign-born and native-born alike, had little appetite for making up this shortage voluntarily. In the two months since the country had declared war on Germany, only 32,000 men had stepped up to make the world safe for democracy. American workers, many of them immigrants, were taking to the streets that spring and summer not to enlist in the army but to go on strike to protest working conditions, low wages, and the spiraling cost of living. A draft was inevitable, and on June 5, aliens were caught up along with everyone else in the mandatory fervor of National Registration Day.

Y ou will have the streets of American cities running red with blood," Missouri Senator James Reed warned the secretary of war during the congressional debate over reinstating the draft. The senator was not alone in fearing a mass uprising, especially in cities with large immigrant populations. When a draft was imposed during the Civil War, over a hundred people died in New York City during five days of rioting—an event that the *New York Tribune* chose to commemorate in a full-page story right before National Registration Day, lest anyone forget. Now, gearing up to fight a world war, the nation seemed even more combustible. One-third of the population was either foreign-born or the child of a parent born overseas; more than one hundred languages and dialects were spoken in the forty-eight states; fifteen out of every hundred Americans traced their ancestry to the powers we were fighting. How would this divided polyglot population react to an enforced call to arms? The government braced for riots, mass protest, civil disobedience, fraud, or at the very least widespread incomprehension of the terms of the Selective Service Act.

Yet on the whole, Registration Day proceeded peacefully and the foreign population acquitted itself admirably. In New York City, at the dot of 7:00 A.M., Meyer Epstein and Epifanio Affatato heard church bells ringing, horns blasting from the harbor, and factory whistles going off all over town, as if to announce the start of some festive public holiday. The National Guard and the New York City police were out in force to break up protests threatened by anarchists and socialists—the city was "prepared for anything short of an invasion" reported the New York *World*—but the only incident of note was the arrest of two Italian men from the Bronx who tried to push their way to the head of the line so they could be first to register. Meyer had no trouble finding a place to sign up—almost every block in Manhattan had a booth set up in a storefront, school, or barbershop. Even funeral parlors were pressed into service. Some eight hundred interpreters were on hand to help foreign-born registrants. As the *New York Times* reported in a rare show of sympathy, "The foreigner who too often had been snarled at or ignored when looking for a job found an interpreter for use in case of need, an official who was willing to explain everything to him and to listen patiently to his own statements."

In Swampscott, Tony Pierro affixed his neat signature at the bottom of the card affirming that he was an unmarried alien laborer in the employ of E. H. Clapp. At twenty-one years and four months, Tony had just made the cutoff. His brother Michele, born in 1891, also had to register. Tommaso Ottaviano, who had celebrated his twenty-first birthday only two weeks earlier, registered near his home in North Providence, Rhode Island, duly noting that he worked at Esmond Mills and was the sole support of his mother, Antonia. In Cambria County, where the Chmielewskis lived, so many guys turned out that the blank registration cards ran out before day's end; in Johnstown, the big burg near South Fork, there were long delays because the men who were registering spoke so many different languages.

Mass protest largely failed to materialize. Even the staunch social-ist Abraham Cahan, who in 1914 had championed German victory in the pages of the *Forverts*, fell into line behind Uncle Sam: "The paper I represent preaches faithful and loyal citizenship," wrote Cahan; "every man between the ages of twenty-one and thirty must do his duty."

The overwhelming majority, foreign and native-born alike, did. When the numbers were tallied, the War Department found that some 10 million men, 610,000 of them from New York City alone, had signed draft cards that day. The nation now had the raw material for a modern army.

The only immigrant groups to stir up trouble on June 5 were the Irish and Finnish copper miners of Butte, Montana. Butte had been a tinderbox since the labor unrest of 1914, and tensions continued to mount in the run-up to the U.S. declaration of war on Germany. Local Irish nationalists had hoped to the bitter end that some miracle would keep their adopted country from joining the hated English imperialists—and when that hope died they were fierce in their dis-appointment. "There is not one Irishman in the United States that would fail Old Glory in time of need, but I hope that everyone that goes across the water to the trenches will never get back," one Irish radical declaimed bitterly at a meeting of the Butte Hibernians. Come registration day, fifty federal troops were dispatched to the mining city to bolster the National Guard regiment. Hoping to head off trouble, the mayor declared National Registration Day a public holiday. The mines were closed, but so were local saloons, as Irish miners were none too pleased to discover. Throughout the day, Irish radicals distributed handbills blasting England for putting "chains of slavery around Ire-land" and advising workers to stay out of the war. That evening, de-spite a warning issued by the National Guard that protesters would be arrested, Irish and Finnish miners joined forces with Wobblies to stage an antidraft march. The accounts of what ensued varied wildly. Some newspapers reported six hundred Irish miners facing down federal authorities, martial law declared, protesters clubbed, soldiers massed with bayonets at the ready; others insisted that registration had proceeded without incident. Twenty-two-year-old Peter Thompson, Belfast-born though he was, ignored the hot anti-English talk at home and the angry protests in the streets and dutifully signed his name to his registration card.

Three days later, Butte's antiwar protests were overshadowed by one of the worst mining disasters in the town's history. Just before midnight, a fire was accidentally ignited in the shaft of the Granite Mountain Mine—then going round-the-clock because of wartime demand for copper—and spread rapidly through thousands of feet of wooden shafts. A crowd gathered to stare in horror at the "mighty geyser" of flame that shot into the summer sky. By Sunday, the corpses of 168 miners had been removed from the bowels of the Montana hillside. The next day, as Butte's miners began walking off their jobs, management blasted the strikers as "unpatriotic and seditious persons." It was a cheap shot aimed at foreign-born workers—and the harbinger of many more shots to come.

The unrest in Butte was the exception. On the whole, National Registration Day had been a stunning success, particularly among the foreign-born population. "The most remarkable thing is how well and willingly the foreign element has responded," a New York registry board head told a reporter for the *New York Times*. "These people are on the 'job.' They seem anxious to serve the country of their adoption." Most anxious of all the "foreign element" were the Poles. In anticipation of registration day, every Polish-language newspaper in the country ran a message entreating "the Poles of America" to "give new evidence of their spirit of loyalty" and "to set a high standard of enthusiasm." The Sunday before, priests in Polish American churches across the country agreed to preach this same message of enthusiastic loyalty to their congregations. But the Polish priests were preaching to the choir: Polonia had been behind the Allied war effort ever since President Wilson announced his support for a "united, independent, and autonomous Poland" in his "Peace Without Victory" speech in January 1917. In the weeks after the United States declared war on Germany, young Polish men had rushed out to enlist. By some estimates, as many as 40 percent of the first hundred thousand men who answered Wilson's call for volunteers were Polish, even though Poles made up only 4 percent of the U.S. population. When registration day came, Polish communities in Chicago, Buffalo, and Milwaukee declared a "patriotic holiday" for young draft-age Polish American men. Milwaukee went a step further: its Polish community raised an entire infantry battalion of Polish Americans under the command of Lieutenant Colonel Peter F. Piasecki. "Even though glory consisted of getting maimed or killed (and having your name misspelled in the newspapers)," wrote one Polish historian, "more than 300,000 Polish immigrants and Americans of Polish descent served before the conflict terminated. . . . The Polish boys were the first and most numerous to respond when the call to arms was sounded."

In the surge of patriotism that swept through Polonia, the ambiguous status of Polish immigrants was a complicating factor that many overlooked or failed to understand. The core issue was the difficulty in determining Polish exemptions. The Selective Service Act of 1917 specified that men born overseas who had not taken out their first citizenship papers were exempt from the draft, as were enemy aliens, that is, noncitizens who had been born in Germany. But what if you were an ethnic Pole born in the Prussian partition of Poland? Technically you'd be considered an enemy alien even though in reality you'd be likely to want to fight for the liberation of your homeland. When the United States declared war on Austria-Hungary in December 1917, this problem spread to the host of ethnic groups who had been born in the Hapsburg empire—not only Poles but Czechs, Slovaks, Ruthenians, Hungarians, and Jews. The status of nonnaturalized men born in neutral countries or countries allied with the United States was also complex. The first iteration of the Selective Service Act made no special provision for these men, so declarant immigrants from Italy, Britain, France, and Greece were subject to the draft along with everyone else. The problem was, Allied nations as well as some neutral countries had preexisting treaties and conventions that expressly exempted their citizens from being drafted into a foreign army. The emergence of a semiautonomous Polish Army organized by a group known as the Polish Falcons and made up of Polish immigrants presented other problems. In time, the Polish Army—also known as Haller's Army after its leader General Jozef Haller, as the Blue Army for the uniforms they wore, and as the Polish Army in France—would be recognized by the U.S. government. But in the first flush of recruitment, draft boards had no idea how to handle such fine points of national identity and allegiance.

Compounding the official muddle was the immigrant grapevine of rumor, misinformation, and wishful thinking. After the United States entered the war, word spread through cities and mining towns that if you signed up to serve in the army, you automatically became an American citizen. Even though this wasn't true in the summer of 1917 (an amendment to the immigration act granting immediate citizenship to soldiers would be enacted the following May), a lot of foreign-born guys believed it nonetheless and signed up for that express reason.

One of them was Joe Chmielewski. After five years in this country, Joe still barely spoke English and had seen precious little of the United States outside of the coal mine and steel mill. But he understood enough to know that he wanted to stay here and become a citizen. So on June 17, twelve days after registration day, Joe presented himself at Columbus Barracks in Ohio's capital city and enlisted in the U.S. Army. The Chmielewskis can't explain why Joe went to Ohio to sign up when his legal residence was still Oak Street in Ffificktown. But they're certain that his desire to become a citizen was his motive for enlisting. For this husky, self-effacing, twenty-year-old Polish Catholic laborer, that was reason enough to go to war.

A month later, Mike Valente reported for duty with the New York National Guard. Having enlisted in the Guard the previous year during the Villa campaign, Mike was already a member in good standing of Company D, 1st Regiment. In the course of the Punitive Expedition, some 110,000 National Guardsmen had been posted to patrol the Mexican border, but Company D was not among them. Mike's unit had never even left the state. But this time it was going to be different. Guys from every town and city north of Albany were pouring into the regiment, all of them raring to get kitted out, trained up, and shipped off to France.

The nation shambled into war, slowly and awkwardly. Eight weeks elapsed between the declaration and draft registration day, and then another six weeks ticked by before the first draft lottery drawing was held in Washington, D.C. While mutiny spread through the war-weary French army and British Tommies mounted yet another costly offensive at Ypres, American families with draft-age sons held their breath and waited. On July 6, two weeks before the lottery, General Pershing, who despite his failure to capture Villa had been promoted to full general and appointed commander of the U.S. forces in Europe, sent word from the battlefields of France that he would need an army of a million American men to win the war. Five days later the commander tripled that figure.

The massive campaign to find and cull these men from the population at large triggered another wave of anti-immigrant anger—and immigrant protest—that summer. The War Department had assigned each state and territory a fixed quota of recruits based on population, and it gave the nation's 4,500 civilian-run draft boards the task of filling those quotas. Problems quickly arose in districts with large immigrant populations. Since so many of the men in these districts were exempt, either because they had yet to take out their first papers or because they were classified as enemy aliens, draft boards had to dig deeper into the pool of native-born men to fill their quotas. Even before the July 20 lottery, nearly half of the 610,000 men who had registered in New York City had dropped out—and most of the dropouts were foreign-born.

A howl went up against the nation's 4 million nondeclarant "alien slackers" who enjoyed all of the benefits of living in the land of the free but shouldered none of the burdens of defending it. Examiners at one Brooklyn board cabled President Wilson to complain of a large number of Russian Jews seeking exemptions: "The flower of our neighborhood is being torn from homes and loved ones to fight for these miserable specimens of humanity, who under the law may remain smugly at home and reap the benefit of the life work of our young citizens." Jews on the Lower East Side provoked even more resentment when they staged protest rallies against the draft. All that summer, newspapers printed increasingly nasty slurs against foreigners, slackers, union organizers, and Bolsheviks—all fast becoming synonymous in the public arena. Fingers were also pointed at aliens who claimed exemptions for religious reasons, especially German-speaking Mennonites and Hutterites whose faith forbade military service. The Wilson administration, having yet to determine how to deal with conscientious objectors, was proffering vague promises to Hutterite and Mennonite elders that their sons would be assigned to some sort of noncombatant service. Some young Mennonite and Hutterite men fled to Canada rather than risk being drafted; others, on the advice of their pastors, reported to induction centers but refused to follow orders once there. The U.S. Armed Forces still had not fired a shot overseas, but on the home front it was already dangerous to be a German-speaking pacifist. The failure to resolve the status of conscientious objectors would haunt the nation in the months to come.

The War Department's goal was to have 687,000 new recruits in uniform by the autumn, and call-up papers going out at the end of August. Tony Pierro was one of the lucky men to be drafted in this first round. A notice arrived in the mail ordering him to report to the Swampscott draft board and warning him that from now on he was "in the military service of the United States and subject to military law. Willful failure to present yourself at the precise hour specified constitutes desertion and is a capital offense in time of war." When he duly presented himself, Tony was subjected to a thorough physical examination. First his vision and hearing were tested, then his scalp, face, mouth, throat, teeth, and nose examined. Everything from the neck up having checked out okay, Tony was ordered to remove all his clothing and stand up straight with his arms out to his sides. Every square inch of Tony's skin was inspected, his chest poked, his abdomen scrutinized for the tell-tale bulging of hernia, his testicles squeezed, his anus peered at while he bent forward at the waist with his hands pulling his buttocks apart. Still naked, he was made to bend, flap, and vigorously jiggle every joint in his arms, legs, and neck. He was measured at five feet, four and three-quarters inches. Dark hair. Medium complexion. Neither flat-footed, alcoholic, myopic, nor sexually degenerate, Tony was in the army now.

Decades after the war, his family remembered a story that Tony used to tell about his first days in the army. One of the officers called Tony out and told him to line up with the other Italian immigrants. The officer informed the men that before things went any further they had a choice: either they could serve with Italy or they could stay and serve with the American Expeditionary Forces. The officer stood before Tony and demanded, "Which will it be, private—Italy or America?" "If I can understand your orders I'll stay and serve with the U.S.," Tony replied. "There shouldn't be any problem," the officer told him, and that was it. Tony was sent by train to the sprawling Camp Gordon training facility near Atlanta.

Like many an Irishman, Peter Thompson relished a good fight and he figured he'd be fighting faster if he signed up than if he waited to be drafted. And besides, the army officer at the recruiting office in Butte told him that if he enlisted he'd get his citizenship papers right away. Twenty-two years old and fitfully employed in the Butte copper mines, Peter was hell-bent on becoming a U.S. citizen—and putting as much distance as possible between himself and Ireland. So he volunteered.

Peter Thompson had a lot of company that summer—not only in Butte but all over the Treasure State. Montana men, foreign- and native-born alike, were lining up to volunteer at recruiting offices— almost twelve thousand in all, giving Montana bragging rights as the state with the highest number of volunteers per capita. By the end of the war, some 8 percent of the state's population would be in uniform. When the bodies were counted, Montana had the sad distinction of claiming a greater number of the dead as a percentage of men mobilized than any other state.

But no one was thinking about the body count in the summer of 1917. Right before summer's end, Peter and a couple of his Irish buddies left their jobs at the copper mine, kissed sweethearts or doting aunts good-bye, and departed for Camp Lewis near Seattle. Local families, immigrants most of them, turned out in force to see the boys off. Serbs, Croats, Greeks, Irish, Boy Scouts, and Spanish American war veterans cheered and wept while the young recruits paraded to the Northern Pacific train depot. Peter smiled and waved with the rest and accepted his Red Cross comfort kit. Off to the war! Actually, off to the fir- and alder-fringed flatlands of Puget Sound to learn to march, salute, load and shoot a rifle, box, and sing with their comrades—and then off to the war.

A story circulated about an incident at Maryland's Camp Meade, one of the thirty-two boot camps being hastily knocked together that summer. On day one, an officer called the roll for a bunch of new recruits, and not one man recognized his own name; then the officer sneezed, and ten men came forward. The story may have been apocryphal—or at the very least embellished—but it was a sign of how ill-prepared the country was for the ethnic and linguistic diversity of recruits pouring in from all over the country. Fully 20 percent—one in five—of these recruits were foreign-born. The commander of the 77th Division—the so-called Melting Pot or Times Square Division, assembled largely from the back streets and tenements of New York City—reported that his men spoke forty-three languages and held "quite as many shades of religious belief and disbelief." One 77th Division officer wrote of "the difficulties of teaching the rudiments of military art to men, however willing, who couldn't understand. Officers have had sometimes to get right down on their hands and knees to show by actual physical persuasion how to 'advance and plant the left foot.'" This was extreme, but not by much. Like it or not, America was going into the Great War with a polyglot, multiethnic, internally dissonant army.

"The military tent where they all sleep side by side will rank next to the public schools among the great agents of democratization" predicted Teddy Roosevelt in the run-up to war. But the men arriving at the chaotic, half-constructed cantonments in Georgia, North and South Carolina, Maryland, Texas, and New Jersey had other ideas. "Never in my wildest flights of fancy can I picture some of these men as soldiers," a drafted newspaper reporter named Irving Crump (who went on to become a popular children's book author) confided to his diary on his first day at Camp Upton near Yaphank, Long Island. "Slavs, Poles, Italians, Greeks, a sprinkling of Chinese and Japs—Jews with expressionless faces, and what not, are all about me. I'm in a barracks with 270 of them, and so far I've found a half dozen men who could speak English without an accent. Is it possible to make soldiers of these fellows?"

Crump was not the only one wondering. New York newspapers had a field day making mock of the "Izzies, Witzers, Johnnies, Mikes and Tonies" who boarded the train in Brooklyn and Manhattan and stepped off at Yaphank into the scrub oak and poison ivy of central Long Island. Native-born officers and soldiers were appalled at the immigrant recruits they were being forced to live with and train. One Camp Upton officer shuddered at finding himself burdened with a "chap" named Herschkowitz "who seemed the worst possible material from which to make soldier-stuff. He was thick-set, stupid looking, extremely foreign, thoroughly East Side." Sizing up the abilities and prospects of the huge pool of foreign-born recruits, Captain Edward R. Padgett of the General Staff concluded that "not more than one in a hundred knew the English language well enough to understand the instructions necessary to make them first-class fighting men." *Was it possible to make soldiers of these fellows?*

The strangeness any person feels at being yanked from ordinary life and sent to war was compounded by the strangeness of being tossed in with guys from such different backgrounds, different parts of the country, different stages of assimilation. "Many, many" of the men arriving at Camp Meade "were sullen, subdued, sad," noted one officer; "and I saw a good many of them who were having a hard time (and not always successfully) in keeping the tears back." Alexander Raskin, who had emigrated as a boy with his family from the Russian Pale and grown up in a nice Jewish neighborhood in Jersey City, New Jersey, was floored when the Gentile recruits on the train to boot camp got drunk and rioted. He described the "events I shall never forget in my lifetime" in a letter to his girlfriend Gus (Augusta):

> *Soon the "booze" began to work on a bunch of the fellows and the fun*
> *started. One half drunk jumped from the train—luckily the train was not*

going fast at that time. He sprawled out and was picked up by a cop and put back aboard. The train started up again and didn't stop until Perth Amboy. Then someone in the crowd noticed a boxcar with barrels and boxes of beer. In a minute about a dozen barrels and hundreds of bottles were on the train, and the car in which I was—or rather on the platform on which I stood—I stood all the way, resembled a third rate Polish saloon—more beer on the floor than anywhere else. By this time half the crowd was "beautifully drunk."

The drunk recruits took to smashing windows and tearing out train seats. Order was only restored when a dozen armed military police charged on board at Bordentown. Raskin, accustomed to sober, hard-working, law-abiding Jewish immigrants, had never seen anything like it. "I felt as I never before felt in my life," he confided to "Dear Gus," "and I have [had] some different 'feelings' in my young life."

Others, foreign- as well as native-born, complained of the "hard boiled language" and humiliating orders and curses hurled at them by gruff drill sergeants.

Tony Pierro arrived at Camp Gordon in the sandy flats outside of Atlanta around the same time as a devout twenty-nine-year-old Evangelical Christian from the Cumberland Mountains of Tennessee by the name of Alvin Cullum York. A born-again elder in the fundamentalist Church of Christ in Christian Union, York had initially tried to get out of military service on religious grounds. But he was a crack shot ("my daddy threatened to muss me up right smart if I failed to bring a squirrel down with the first shot or hit a turkey in the body instead of taking its head off," he wrote in a diary published, and perhaps somewhat fabricated, after the war), and eventually he was prevailed upon to do his duty. York had never laid eyes on a Jew, an Italian, a Greek or a Pole, and he went into shock at Camp Gordon when the lieutenant in charge "throwed" him in with a bunch of tough foul-mouthed street-smart bartenders, bouncers, ice men, coal miners, and mill hands named Muzzi, Dymowski, Sok, Konotski, and Saccina. "A right-smart number of them couldn't speak or understand the American language," York wrote in his diary. "And a whole heap couldn't read or write or even sign their own name. . . . I had never had nothing to do nohow with foreigners before. When I first heard them talk I kinder thought they were angry with each other; they seemed to talk so fast and loud." No matter what language the men spoke, all of them were issued the standard army six-piece mess kit, a blanket, and a mattress to stuff with straw; then they were ushered into barracks filled with rows of iron bunks. The first night, York was assigned a cot next to some Greeks and Italians. "Well, I couldn't understand them and they couldn't understand me, and I was the homesickess boy you ever seen."

One of the Italian boys York was throwed in with was Tony Pierro. What Tony remembered most about boot camp was not the homesickness or the Babel of different languages or the trouble he had loading and shooting a gun for the first time, but the indignity of being called a wop: "One time I lost a piece of steak on my dish—you know, you line up on the chow line, and they'd give you so much. So I said to the man that was giving the bread, I said, 'Put in two slices, please.' 'Oh, come on, Wop,' he answered. Oh, I took that ba-boom. I lost a piece of meat. I lost my appetite, I couldn't eat any more. I hated that name: Wop, Dago, Guinea. I hated all those names."

M atej Kocak got a furlough from the marines on June 12—ten days to visit with his family in Binghamton before going to war. It must have been an emotional reunion. Things had not been going well for the Kocaks. Matej's older brother Paul had married and fathered a son; but soon after the baby was born, Paul had gone insane and he was now a patient in the Binghamton State Hospital. Paul's wife, Julia, a widow in all but name, was left to fend for herself and her son John, now six years old, in a new world wild for war.

Matej had gotten used to the cockiness of his fellow marines chafing for a good fight "over there"—but the war frenzy he encountered in Binghamton's Slovak community was something different: there was an almost spiritual fervor feeding the flames. American Slovaks had been praying for months that the United States would enter the battle against Germany and the hated Austrian Empire, and after the declaration in April, support for the war became practically an article of faith. Slovak priests and community leaders stood shoulder to shoulder exhorting local boys to do their duty to their homeland and their adopted country, now joined in a single glorious cause. The Binghamton Slovak paper proudly printed the names of the twenty local Slovak boys who had volunteered to serve with the U.S. Army. At dances and dinners at Slovak Hall on Starr Avenue there was excited talk about the Czechoslovak Legion, a national liberation army being raised and funded by the immigrant communities. Matej, a career marine with a decade of military service under his belt, would have been welcomed back as a shining exemplar of the Slovak American warrior. The uniform alone was enough to secure him a round of drinks, or more likely several rounds, at the local tavern.

The Kocak family was tight-knit and lived close together in the same neighborhood, so it's likely that Matej spent some time with his cousin Paul in the course of his furlough. Though Paul, at twenty-five, was ten years younger than Matej, the family resemblance was unmistakable—both men were solid, stocky, open-faced, broad-shouldered, with piercing eyes and a shock of unruly hair. Paul, who had a good steady job stitching shoes for a big local shoe manufacturer, was not among the Binghamton Slovaks who had volunteered for the army; but as a declarant he was eligible for the draft, and he had duly registered on June 5.

There was, however, one cloud hanging over Paul's military future: his father was fighting in the Austrian army. If he was drafted and sent to France, Paul might one day have to fire on his own father. For Matej, this was a more pressing concern since it wasn't a question of *whether* he would fight overseas but *when*.

At the end of his furlough, Corporal Kocak was ordered up to Utica, New York, ninety miles north of Binghamton, for a two-week intensive course in machine guns at the Savage Arms Company. Named not for the ferocity of its products but for Arthur Savage, the man who founded it in 1894, the firm was a leading manufacturer of the lightweight Lewis machine gun, the British army's automatic weapon of choice and the U.S. Army's standard machine gun until it was replaced by the Browning automatic rifle. By the end of his two weeks at Savage, Matej was intimately familiar with the Lewis gun's sixty-two component parts, its stout black two-foot-long telescopic barrel, and its lethal capacity for spitting out 550 rounds a minute. On the completion of his course on August 8, Matej was assigned to the Marine's 92nd Company artillery regiment and ordered to Quantico, Virginia, for additional training.

His cousin Paul was drafted the following month. On September 15, Binghamton's Slovak families, Kocaks among them, turned out en masse at Slovak Hall for a farewell banquet honoring the boys going to war. "Slovak men!" local luminary Imrich Mažár addressed the assembled. "Please keep in mind that you are sons of the great and oppressed Slovak nation. . . . We were not even allowed to say that we were oppressed in the old country, in the Austro-Hungarian Empire. It was only when we arrived in our new home, the free America which accepted us as its sons, that we started to yell aloud for the whole world to hear about our persecution." Mažár went on to remind the men that they would be fighting not only for their "new country, which accepted you as its sons" but also for the liberation and rights of "the Czecho-Slovak nation" that would emerge from the ashes of war. A Catholic priest blessed the new soldiers; their mothers wept; their fathers beamed with pride.

At the end of the month, Private Paul Kocak got on a train heading south and reported for training at Camp Gordon. Whether it was the language barrier, the immigrant's instinctive reticence, or that the subject simply never came up, Paul kept the fact that his father was fighting in the Austrian army to himself.

Tony Pierro, who was already at Camp Gordon when Paul Kocak arrived, was in theory deep into training to be a soldier—but in fact a lot of the foreign-born guys at the camp had been shunted into "pick and shovel" work by officers who couldn't understand them and didn't have the time or inclination to try. Three-quarters of the recruits who showed up at Camp Gordon could not speak English; nearly half of them had only recently arrived from Europe. All that fall and winter, morale deteriorated. One officer reported that men from different ethnic groups were getting into scuffles with each other: "Old scores from the pages of history were reopened, [and] the scant harmony that did exist" in the camp was frequently disrupted. Every immigrant group felt it was being singled out for mistreatment. Soldiers of German and Austrian ancestry passed around stories that their families in the old country would be found and killed once word got back that they were serving in the American army (supposedly German agents had initiated the rumor as a way of undermining morale). Polish and Italian and Slovak Catholics didn't believe that the Sunday mass celebrated in the Knights of Columbus huts was really Catholic since no one spoke their language—so they refused to go to mass or confession. The prospect of being shipped overseas and dying in a state of sin weighed heavily on their souls. Far from learning to be American soldiers, foreign-born guys at Camp Gordon and other boot camps around the country were learning to be scared, bitter, and resentful at having to serve. When the army sent a multilingual foreign-born officer to investigate conditions among immigrant recruits at a New England boot camp that winter, Polish recruits told him with tears in their eyes that their officers neither understood their language nor cared about their plight. "They have had no chance to speak, to tell their troubles since the time they were put in the rank," the officer reported. "After three months' training, wasting time and energy, most of them had learned absolutely nothing." At Camp Gordon, not a single foreign-born recruit came forward when an officer asked which of them would be willing to go overseas.

To its credit, the army recognized the problem and moved quickly to address it. Under the aegis of the Foreign-speaking Soldier Subsection (FSS), two multilingual officers went down to Camp Gordon to take matters in hand. After canvassing the men to determine what languages they spoke, what their gripes were, and what abilities they had, Lieutenants Stanislaw A. Gutowski and Eugene C. Weisz divided the immigrant soldiers into three groups—noncombatant battalions made up of those who had some skills or trade but were not physically fit for combat; labor battalions for enemy aliens and those deemed seriously disloyal; and development battalions for all the rest. Camp Gordon's development battalion started off with two companies—a Slavic company for Poles and Russians and an all-Italian company. Polish, Russian, and Italian officers were found to lead the companies, while the men were given intensive courses in English (along with a heavy dose of civics, U.S. history, and kaiser-stomping patriotism). Eventually, Russian, Greek, and Jewish companies were added and staffed by officers who spoke the language, shared the culture, and understood the concerns of their men. The army also provided for the religious needs of immigrants and even ordered cooks to prepare ethnically appropriate food. The program worked so well at Camp Gordon that the FSS was commissioned to implement the "Camp Gordon plan" at fifteen army camps around the country. In time, a quarter of a million immigrants would pass through development battalions, more than half of them eventually moving on to regular army divisions. (It's worth noting that the Camp Gordon plan, though it benefited immigrants, grew out of a massive, coordinated endeavor to monitor the political views and loyalty of foreign-born soldiers: the FSS was originally organized under the Military Intelligence Division, whose first wartime director, Major

General Ralph Van Deman, was obsessed with possible plots hatched by subversive immigrants and German spy rings. It was Van Deman who ordered intelligence officers to set up surveillance networks to spy on immigrant recruits. To the general's surprise, America's foreign-born soldiers were found to be overwhelmingly loyal. After the war, in the Red Scare era, Van Deman went on to promote an insidious campaign linking Jews and Bolsheviks.)

Camp Gordon's intelligence officer, Captain Eugene C. Bryan, reported a dramatic improvement in the morale of immigrant soldiers after the plan went into effect. In fact, when foreign-born recruits were again asked about serving overseas, 85 percent now responded "in no uncertain terms" that they were willing to fight in France. (Bryan later revised that figure up to 92 percent, noting that the only holdouts were German Americans reluctant to fire on their relatives in the German army.)

There were so many men being recruited and processed in the last months of 1917 that the War Department created a new wing of the army to supplement the already existing branches of the Regular Army and the state-based National Guard. Dubbed the National Army, this new wing was composed entirely of draftees. One National Army division had the distinction of being manned by draftees drawn from all forty-eight states—twenty-eight thousand men organized into a "square division" (twice the size of the divisions deployed by the French and English) of two infantry brigades, an artillery brigade, a machine-gun company, and support troops. This was the 82nd Division, which adopted "All-American" as its nickname. The men of the All-American Division, wrote Teddy Roosevelt's son Theodore, were "of every racial stock that goes to make up our nation, from the descendants of colonial English to the children of lately arrived Italian immigrants." Among the latter was Private Antonio Pierro, assigned to the All-American's Battery E, 320th Field Artillery. Among the former was the born-again Tennessee marksman Private Alvin York, soon to distinguish himself as the best shot in Company G, 328th Infantry.

Despite the army's efforts to accommodate draftees who did not speak English, inevitably there were men who were ignored, misunderstood, or denied exemptions they were entitled to or who simply fell through the cracks. One of these was Max Cieminski. Even though Max was an American citizen, born in Wisconsin, he neither spoke nor understood English. He'd been raised in a household where Kaszubian was the first language, Polish the second, German the third—and when he moved to Bessemer, Michigan, to join his sister Mary, the only thing that changed was the order in which these languages were used. Since Mary's husband was a Prussian Pole, in their household Polish was spoken first, German second, Kaszubian last. For some reason Max could never understand, Mary told her children never to speak Kaszubian outside of the house, as if she were ashamed of it. At the iron ore mine where he worked, Max picked up the few English words he needed to get by and get paid. If any of the Cieminskis felt it was strange not to know the language of their native land, they never mentioned it. The way they lived, the need just never arose. In any case, their true native country was Kaszubia, no matter what it said on their birth certificates.

Max's induction notice arrived on November 19, 1917, a month after he turned twenty-six, and he duly reported to Camp Custer outside of Battle Creek, a training camp packed with Polish guys from Detroit and its heavily Polish satellite of Hamtramck. Max lined up with the other men for his physical exam and went through the whole humiliating business of being stripped, weighed, measured, poked, and peered at. The examiner shook his head when he took a look at Max's right hand—no trigger finger, no soldier. Max was told to put his clothes on, gather his stuff, collect a form, and go. But Max didn't understand what the examiner was saying to him and evidently none of the guys in line with him did either—or if they did they couldn't explain it. Officially, Max was exempt because of the farm accident in childhood that resulted in the amputation of his trigger finger—but since he didn't speak English, army rules didn't matter. Max stayed put; the officers in charge figured he was waiving his exemption; and he was processed along with all the rest.

At Camp Custer, Max got assigned to Company B, 337th Infantry, 85th Division—but for reasons known only to the army, in less than a month he was shuffled out and transferred down to Camp Pike near Little Rock, Arkansas. Max did his basic training that winter with an infantry regiment of the 87th Division largely made up of recruits from Arkansas, Mississippi, and Louisiana. His first protracted exposure to English was the drawl of farm boys from the Deep South serving in Company B of the 345th Infantry. Such was the army in 1917.

Max's beloved niece Marguerite, now pushing eight years old, was devastated when her favorite uncle went away to war—and she hung on every letter he wrote to her mother Mary. In one letter, Max complained bitterly about how abusive the German American drill sergeant was to the Polish recruits. Since Max's German was better than his English, he asked the sergeant in German for help with something. "If you don't shut up," the sergeant barked in reply, "you're going to get shipped back to Prussia to shovel coal for the Kaiser." Max was baffled. After all, he wrote Mary, weren't they all American soldiers together? Just because he spoke German didn't mean he was disloyal. In another letter, Max wrote at length about the War Risk Insurance program he had just subscribed to. He told Mary that he was making her the beneficiary of his army life insurance—his way of thanking her for raising him after their mother died and for looking after him during his years in Bessemer. If he died while he was in the army— whether from a German bullet or a stupid accident, it didn't matter how as long as it occurred during his time in the army—Mary would get a check from the government every month. Max's English may have been rudimentary, but he knew enough to understand a deal. A few dollars out of his pay every month, and Mary would be fixed for life if anything happened to him. Marguerite too.

But of course nothing was going to happen. The Cieminskis were all devout Catholics, and Mary had faith that God would look out for her baby brother Max.

While York and Pierro drilled with the All-Americans outside Atlanta that winter and while Max Cieminski drilled near Little Rock with the nasty German sergeant of the 87th Division, politicians in Washington, D.C., were lashing out at the Wilson administration for its miserable ineptitude in training and equipping an army. York, Pierro, and Paul Kocak were lucky to be in Georgia: in barracks in the northern states, hit by one of the coldest winters on record, poorly clad soldiers were dying by the score of influenza and pneumonia. Eight soldiers out of ten had been issued shoes that were too small and failed to keep their feet dry. Former army chief of staff Major General Leonard Wood raged: "Never in the history of any country has there been more incompetence than in our preparation for this war." Hundreds of thousands of recruits were drilling with wooden sticks on their shoulders because of foul-ups in the ordering and manufacture of rifles. Without uniforms or weapons, the enlisted men gathered in drafty mess halls to be lectured by wounded British veterans on the realities of trench warfare. One instructor vividly described the nightmare that awaited them in France: "We made an attack one day. As our first wave carried the enemy trench, they heard shouts from a dugout: 'Kamerad!' The Germans surrendered. The first wave rushed on, leaving it to the second wave to take the prisoners. As soon as the first wave had passed, the Germans emerged from their dugout with a hidden machine gun and broke it out on the backs of the men who had been white enough not to give them the cold steel. So now, men, when we hear 'Kamerad' coming from the depth of a dugout in a captured trench we call down: 'How many?' If the answer comes back 'Six,' we decide that one hand grenade ought to be enough to take care of six and toss it in."

The instructor's casual brutality was a sign of how desperate and exhausted Europe's armies had become. For the Allies, all the news that fall and winter was bad. The season began with a catastrophic blow to the Italian army. At the end of October, a combined Austrian and German force broke through the Italian line at Caporetto and, in the most dramatic advance of the war, swiftly pushed the Italians fifty miles south to the Piave River. Italy's losses of men and material were staggering: nearly 300,000 soldiers were taken prisoner; 11,000 Italians died and 20,000 were wounded; one-third of the nation's wartime supplies, including almost the entire artillery, fell to the enemy. By the middle of November, German and Austrian divisions were dug in some thirty miles from Venice. On November 7, while the rout at Caporetto was in progress, Bolshevik workers, peasants, and soldiers led by Lenin toppled Kerensky's provisional government in Petrograd, plunging the future of the vast beleaguered Russian army into uncertainty. That uncertainty ended in mid-December when Lenin declared a provisional armistice with Germany that in effect removed Russia from the war. "Russia," in the words of one historian, "had gone effectively out of the war faster than the United States came in." Even before the Bolshevik revolution, the battered Russian army had been "melting away" as deserting soldiers went back to their villages in the hopes of profiting from the political chaos. But Lenin's separate peace with Germany made it official: the Allies had lost their ally to the east. Inconclusive negotiations for a permanent armistice between Russia, Germany, Austria, Turkey, and Bulgaria dragged on until the spring, but Russian soldiers had already "voted for peace with their feet," as Lenin declared. With the Eastern Front shut down, scores of German divisions were now free to join their comrades in France and Belgium. Already German generals were planning to deploy this new infusion of manpower in a series of springtime assaults intended to win the war.

Dark days for the Allies—and where were the Americans? Nine months after Wilson had declared war on Germany, only four American combat divisions were on French soil, an "expeditionary force" totaling 165,080 men and 9,804 officers out of the 3 million men promised by Pershing. Parisians had cheered on July Fourth when the first fresh-scrubbed Americans with the 1st Division paraded through the capital—but now in the depths of winter the French were starting to wonder when, if ever, their new allies would be ready to fight. While the armies of Europe entered the third year of mass slaughter, the Americans had scarcely shed or spilled a drop of blood in battle. As one Frenchman remarked in disbelief, "We expected to see two million cowboys throw themselves upon the Boches and we see only a few thousand workers building warehouses."

Matej Kocak was among these few thousand. Matej shipped out on December 8, 1917, aboard the USS *DeKalb*—a German passenger liner that had been interned in the United States in 1914 and then seized and converted into a troop transport ship after the United States entered the conflict. Except for one U-boat attack that the *DeKalb* dodged in the mid-Atlantic, the crossing was uneventful, so Matej had a lot of time to mull over a piece of news that had a direct bearing on his family. The day before his departure, the United States had declared war on the Austro-Hungarian Empire in the wake of the Italian disaster at Caporetto, with the immediate consequence that nondeclarant immigrants from the Hapsburg empire—Slovaks among them—were now reclassified enemy aliens. Matej, a declarant, was not subject to this change of status—but what about his cousin Paul and his sister-in-law Julia? With an ocean and a war between them, there wasn't much Matej could do for his family. Paul would have to work this out for himself with the officers at Camp Gordon.

Corporal Matej Kocak, 66th (C) Company, 5th Marine Regiment, began his service in France on the day he turned thirty-five—December 31, 1917. On disembarking, he was jammed with a bunch of other marines into one of the French boxcars that passed for troop transports—the dun-colored rail cars labeled on the outside with their carrying capacity, *40 hommes, 8 chevaux,* 40 men, 8 horses, that would soon become familiar to hundreds of thousands of Doughboys—and taken east toward the front. The 1st Division had set up a training base they called Washington Center outside the village of Gondrecourt, an easy march to the line at the St. Mihiel salient south of Verdun, and it was here that Matej and his fellow marines spent the winter learning how to fight a trench war. Seasoned French warriors barked out commands and the Americans got used to tossing hand grenades, taking out pretend machine-gun nests, and going over the top in mock assaults from the training trenches. In the off hours, Matej picked up enough French to converse with the local girls and order coffee spiked with cognac at the local estaminet. A few weeks into Matej's deployment in France, the American command decided to reshuffle the troops it had on the ground, and his unit got folded in with the 6th Marines and a couple of Regular Army infantry and artillery regiments to form the 2nd Division. In the course of the winter, the 2nd Division was rotated out of Washington Center and into an active sector of the front line west of the St. Mihiel salient. Matej and his comrades were still technically being trained, but now the training came in the form of genuine shells, deadly gas canisters, and real raids launched by the "Verdun Boche," as the French referred to the battle-scarred veterans on the other side of no-man's-land. The Americans were too green to mount their own attack, but they were nonetheless shot at by the Germans in the daily ritual of sniping and shelling. Matej and his buddy, Sergeant Louis Cukela, a Croatian Catholic who, like Matej, had emigrated to the States from the Austro-Hungarian Empire and signed up with the marines, broke the tension with their own skirmish of pranks. One night they accidentally triggered a gas alarm by firing off what they thought was a flare but turned out to be the green six-star signal that a gas attack was in progress. Witnessing the pandemonium that ensued, Matej hatched the idea of making a sneak attack on a YMCA worker he particularly despised. Rushing into the YMCA hut hollering "Gas!," Matej proceeded to help himself to cigarettes and chocolate while the hated clerk was fussing with his gas mask.

The French who had seen their comrades die by the thousands at the hands of the Verdun Boche looked on and shook their heads. When would the Sammies start to fight?

The truth was, the Sammies were doing plenty of fighting that winter, but mostly with each other at boot camps in the States. Tony Pierro was not the only one being called a wop in the crowded, freezing, badly supplied camps. Hunky, dago, kike, yid, guinea, chink, jap: "regular" American guys came to the army with a ready supply of names for the aliens, and a set of stereotypes to match. Italians were volatile, lazy, thieving, armed with knives, willing to fight only when you played on their emotions; Irish were daring and pugnacious, but they would never fight on the side of the British; Jews were hangdog, cowardly, intractable, clannish. In the camps, ethnic groups tended to band together, mutter about the other groups, and blame each other for petty crimes. And, inevitably, when rage overwhelmed them, they'd fight. An Italian recruit with 91st "Wild West" Division training at Camp Lewis got so angry when someone called him a wop that he crept up to the guy while he was asleep and kicked in him in the throat. In the fight that ensued, the Italian was on the verge of being choked to death when another soldier broke it up. Ashad G. Hawie, a Syrian Christian who had emigrated as a young man to Alabama and enlisted in the Alabama National Guard when the war broke out, was hazed mercilessly by "a hulk of a man" named Matt Riker, "a mountaineer, belligerent, harsh." One day on the chow line, Riker got pissed off when he spotted the wiry, dark-complexioned, foreign-accented Hawie standing ahead of him. "Move back and let a real man in where he belongs," Riker muttered as he grabbed the immigrant and shoved past him. Hawie, amazed that the other recruits just stood there watching, sputtered, "I enlisted to fight the Germans—not Americans, you fool." He appealed to his captain to get Riker off his back, but the officer wanted no part of the conflict. In the end, the only way Hawie could hold his head up was to challenge Riker to a boxing match. After the fight, as Riker lay unconscious on the ground, Hawie overheard one of the other men say in admiration, "Spunky little devil. Didn't think he had it in him." He had crossed the line from alien to comrade. Jews had similar experiences. Lieutenant Jacob Rader Marcus, a rabbinical student serving in the army, insisted that the only effective way to counter anti-Semitism—which he said "exists at all times and under all circumstances"—was with your fists.

Mike Valente, training with the New York National Guard at Camp Wadsworth outside Spartanburg, South Carolina, was not the type to get into fights on the chow line. But like everyone else at Camp Wadsworth, he was hopping mad over the new War Department policy of disbanding and splicing together long-established National Guard regiments to form large mongrel divisions made up of men from different regions and backgrounds. Particularly galling was the merger of the 1st Infantry Regiment—the regiment of upstate New York farmers, immigrants, and factory workers that Mike had signed on with—and the venerable 7th "Silk Stocking" Regiment, with its heavy contingent of society boys and millionaire scions from New York's Upper East Side. The two disparate National Guard units were fused into the 107th Infantry of the 27th Division. It's worth noting that the men of the 107th Infantry, Park Avenue bankers' sons and immigrant hospital attendants alike, pulled together when a company of black recruits from New York City—the 15th New York Infantry, soon to be dubbed the Harlem Hellfighters—arrived for training at Camp Wadsworth and stirred up Spartanburg's white racists. The Hellfighters' white commander, New York attorney Colonel William Hayward, gathered all the men at the camp, white and black, officer and conscript, to warn them of the threats made against the black soldiers and to make them promise to stay out of fights in town. "If wrong by disorder is to occur," Hayward boomed to the assembled troops from the roof of a camp washhouse, "make sure and doubly sure that none of the wrong is on our side." Shortly afterward, when a couple of bullies picked a fight with a black enlisted man in downtown Spartanburg, two silk stocking soldiers came to his aid. In the scuffle that ensued, the New Yorkers sent the South Carolina bullies sprawling into the same gutter that the black soldier had just pulled himself out of. "Perhaps the difference between soldier and civilian was beginning to seem more important than differences among soldiers wearing the same uniform," one historian commented on the incident.

In time, Mike Valente and the rest of the men of the 107th settled down and submitted to the seemingly interminable routines of infantry training. Pershing wanted the American army to abandon the stasis of trench warfare in favor of a swift, aggressive, fast-moving fight in the open, and as a result drills at Camp Wadsworth focused on the rifle and the bayonet thrust. Valente and his army buddy Joseph Mastine, a small, quiet, seemingly meek Italian American cigar maker from Ogdensburg, sweated through long days of target practice and slashing dummies with their fixed bayonets. "When they fired they not only missed the targets," York remarked incredulously of the immigrant soldiers he trained with, "they missed the backgrounds on which the targets were fixed. They missed everything but the sky." The men made forced marches through the Carolina woods with 70-pound packs; they tossed stones that were supposed to be grenades and practiced pulling on gas masks; eventually they mastered the art of automatically saluting officers and unhesitatingly obeying their commands. What they hated most were the endless, endless drills in tight formation—three hours in the morning, three more in the afternoon. A break in the monotony came in the middle of February 1918, when Valente and Mastine's unit was ordered to the camp's training trench for seventy-two hours of simulated trench warfare. One of their comrades wrote in amazement of his first experience with grenades—"the real ones": "They kill everything above ground within a hundred-yard radius. We stand in a trench, throw them as far as possible and duck out of range. If you put your head above the trench where it went off, you would look like a sieve." On the third day in the trenches it poured rain and the men had a taste of Flanders mud. Two months later, on April 10, came an even bigger thrill: a simulated infantry attack staged with a real artillery barrage. The men were marched with full packs thirteen miles from camp to Glassy Rock in the foothills of the Blue Ridge Mountains and made to wait in formation while the big guns to their rear let fly a "creeping barrage" of live artillery shells. "It was the most marvelous, miraculous and impressive thing I ever witnessed," Lieutenant Kenneth Gow wrote his parents. "The deadly accuracy of the artillery was wonderful. The guns were four miles away from us, and had to fire over two mountain ranges at a target they could not see. A slight mistake would have meant death to dozens of us." Mastine, who had been nicknamed "Cuckoo," carried a chunk of exploded shell casing off the field as a souvenir. He told the other guys that he had grabbed it in midair as it whizzed by his head.

Farther south at Camp Gordon, Tony Pierro was going through the same drills with a few local variations tossed in by inventive officers of the All-Americans. "Two features of the early training ... will always be remembered by the troops," wrote one officer, "the emphasis on road marching and organization singing." Hundreds of foreign-born soldiers learned to speak English at Camp Gordon's language school that winter; scores of enemy aliens were discharged. When two batteries of 3-inch American-made guns arrived in camp, Tony finally had a chance to get his hands on the artillery he was being trained to service and fire. His unit assembled at the artillery range in Marietta and let off several thousand practice rounds. One day, all the men, machine gunners, infantry, and artillery alike, were made to enter a "gas house" filled with a weak concentration of chlorine gas: officers ordered them to remove their gas masks briefly and inhale so they would know what the stuff smelled like. Then they did the same thing with tear gas. One whiff, army brass reasoned, and the men would remember the smell forever.

Sam Dreben, late of Pershing's Punitive Expedition in Mexico, was sitting out the Great War. Nearing forty, Sam had decided to retire from his career as a soldier of fortune and settle down. So settled that he even got married. Sam proposed to a girl named Helen Spence, a Texas honey who at nineteen was just shy of half his age, and Miss Spence consented. The newlyweds were happily nested at 2416 Montana Street in El Paso, Texas. Mrs. Dreben was expecting a baby. The Fighting Jew had hung up his sword and shield.

But on February 12, 1918, less than a month after his daughter was born, something made Sam change his mind. Patriotism? Frustration at being cooped up at home while American boys were training for war? Unwillingness to miss out on the action? Pressure from army buddies? Sam never said, though Captain Richard F. Burges, the El Paso attorney who commanded the unit Sam signed on with, offered his own explanation: "Undaunted courage, unimpeachable honesty and unlimited loyalty were the outstanding qualities of his character. Patriotism to Sam Dreben was not an idle phrase but a deep and abiding devotion to the country of his adoption." Comradely gush, perhaps—but Burges was not the only one to laud Sam's love of country. "Sam's two most cherished possessions were his Jewish ancestry and his American citizenship," wrote a newspaperman long acquainted with his character. Sam was a born fighter, an American by choice, a soldier by calling. This was the war to end all wars. How could he *not* fight?

Despite his long experience in the military, Sam insisted on enlisting as a buck private. Reporting for duty with the Texas National Guard at Camp Bowie northwest of Dallas, he was assigned to Company A, 141st Infantry, 36th Division. Captain Burges, delighted to have a professional soldier in his company, rapidly promoted Sam to corporal and then sergeant. The step up to first sergeant—known as the top kick, the noncommissioned officer who was really in charge of whipping the conscripts into shape—came two weeks later when another first sergeant was called out to officer training camp. According to Burges, the inevitable grumbling at his fast-tracking of Dreben "melted away as the men realized his qualification for the job."

The short, stout, middle-aged immigrant with the Yiddish accent had his work cut out for him: not a single other recruit in Company A had a lick of military experience.

I t is quite apparent that the physical condition of the men as they file past, stripped, is poor," one medical officer wrote after he inspected the new recruits. "Their pale skins and flabby tissues bespeak lack of tone, and indicate the absence of any kind of exercise." Immigrants were among the worst of the lot. Having grown up poor and often malnourished in Europe, they made stunted, scrawny soldiers. Those who worked indoors as tailors, barbers, store clerks, laundry or restaurant employees were narrow-chested, slump-shouldered, stoop-backed. Factory and mine workers, though they had better muscle tone, bore the scars and deformities of a lifetime of crushing physical labor. Officers, despairing of ever making warriors of these guys, were amazed at how quickly the transformation occurred.

The change was more than physical. The routines of boot camp left immigrant soldiers a lot of time to reflect on the anomaly of their situations—their divided loyalties, their reasons for fighting, their uncertain place in the military and in the nation. Those inclined to introspection brooded over the perplexities of the hyphen: Which side counted more? How could they erase the hyphen and become "100 percent American" when the guys in camp constantly reminded them they were kikes or wops or micks or bohunks? Going from civilian life to boot camp in wartime was a wrenching process for anyone, but for a lot of the foreign-born guys, it was like emigrating all over again. The army was their first exposure to the real America—the customs, language, food, holidays, and attitudes of the Protestant majority. The fact that this forced integration happened under military command made the impressions that much deeper. Six months of basic training was more than enough to launch a guy on that strange reordering of brain cells and body chemistry that happens only in the armed forces—either that or break him. By the time the immigrant soldier was ready to ship out with his company, he had either cracked or cohered. Deep down, he was still whoever he had been when he entered the service—in France he would wear his Star of David or his holy medal blessed by the parish priest next to his dog tag—but day to day, hour by hour, he had learned to act like, and maybe to feel like, a comrade, a buddy, a cog in the military machine, a warrior on fire to stomp the kaiser's throat. In short, one of the guys.

"The foreign element is taking hold like real Americans," wrote Irving Crump in his diary after a few months in training—the same Crump who had wondered glumly on the first day whether all these jabbering aliens would ever be soldiers:

> It is interesting to get their slant on the whole affair. Many of them didn't want to come. They had their own ideas of army life, suggested, doubtless, by tales they have heard of service in the European armies of former days. But when they were called they came; and behold, when they arrived and lived through the first days, they were surprised to find that they were still treated like human beings, had certain indisputable rights, were fed well and cared for properly and worked under officers who took a genuine interest in their welfare. This was something most unexpected. Right off they decided that they were going to get all they could out of this new life and give in return faithful and honest service.
>
> "It's fine, I like it," assured a little Italian friend of mine in the infantry. "I like it because it help make me spick good English, make-a me strong, make-a me beeg an' best-a what is, make-a me good American, jus like-a de boss Lieuten."

Boleslaw Gutowski, a Polish immigrant training with the 10th infantry at Fort Benjamin Harrison in Indianapolis, explained in his broken but vivid English why he had enlisted: "It is a question why joine army. For Here its I joine army not for fun guess only for one raison for raison why—becuase I don lieke King our empral—only Repabic I do love. An my grand father fight for it."

Jack Herschkowitz, a Rumanian-born Jew training at Camp Upton, put it this way: "I didn't feel like going—who wants to get killed?—so I first tried to get out of it—but vonce I was in there, I did it right."

By the spring of 1918, those pale, slack, slouched bodies were hardened for war. Their accents, their suspicions, their longing for families or girlfriends, their fears and dreams, their foreign souls and infinitely varied personalities remained intact—but each of them carried something new inside. Before shipping out, every Doughboy was issued a blue laundry bag into which he was ordered to put all his personal belongings and extra equipment—everything except the 70 pounds of army issue clothing and supplies that he would be carrying on his back into combat. The army would keep their stuff safe, they were told, and they'd get the bag and its contents back when they returned at the end of the war. *If they returned,* each man added silently. Maybe that "if" was the new thing they carried inside—that and their surrender to its terms and consequences. Whatever they felt about their ultimate loyalty, each foreign-born man knew that from now on he was an American soldier. In the eyes of the enemy who would soon be shooting at him, that was all that counted, even if the enemy's blood flowed in his veins.

CHAPTER 7

I GO WHERE
YOU SEND ME

On the first day of spring 1918, the Germans launched an offensive that they desperately hoped would win the war. With the neutralizing of the Eastern Front after the Bolshevik revolution, fifty German divisions had been freed for combat in the west—and General Erich Ludendorff was determined to take advantage of his army's numerical advantage quickly, "before America can throw strong forces into the scale." The opening attack came before dawn on March 21. German forces massed behind the Hindenburg Line between Cambrai and St. Quentin laid down a hellish bombardment of poison gas and artillery shells; when the shelling ceased, the German infantry moved. By this stage in the war, barrage followed by push was a familiar maneuver—really just about the only ground maneuver—and usually when the smoke cleared and the bodies were counted the attacker had little or nothing to show for it. But this time, miraculously, it worked. In a matter of hours the Germans tore a ragged nineteen-mile-long gash in the British line. Twenty-one thousand British soldiers were captured in the course of the day; seven thousand British infantrymen died. The British Expeditionary Force "had suffered its first true defeat since trench warfare had begun three and a half years earlier," in the words of historian John Keegan. The German offensive continued throughout the spring along a north-south front that stretched from Ypres in Flanders down to Rheims in the Champagne region, but no gains were as dramatic as the opening days. As March bled into April, it became increasingly apparent that the armies of both sides were near the breaking point. Demoralized, running out of options, both sides had their eye on the Americans.

The Americans' turn finally came on April 20, a month into the spring offensive. The Germans decided that the time had come to "teach the Americans a lesson," and in a dawn attack their forces moved against the green recruits of the 26th "Yankee" Division of the New England National Guard stationed at the village of Seicheprey to the south of the St. Mihiel salient. It was a quick, painful punch in the nose. In the course of an hour, the Germans advanced nearly a mile into the American position; then they dug in just north of the ruined village and congratulated themselves. But in truth the Germans had had an easy target. Lacking even a semblance of field communication, accidentally shelling its own lines, hopelessly screwing up the counterattack, the Yankee Division acquitted itself dismally. Eighty-one American soldiers died in the attack and 187 more were captured or missing in action. "Seicheprey was taken by storm and was found full of American dead," German commanders crowed in a telegram that the *New York Times* printed on April 23. "Bitter hand-to-hand fighting ensued around the dugouts, vantage points, and cellars, whose occupants were killed almost to the last man. . . . Our losses were slight while those of the untrained Americans were most severe." "Germans Exulting in 'Lesson' to U.S.," ran the *Times* front-page headline the following day. "Gloats over our losses."

In was in this atmosphere of bloody-nosed humiliation that the first substantial waves of American soldiers began to arrive overseas. From 318,000 at the end of March 1918, the number of Doughboys stationed in France rose to 430,000 by the end of April and 650,000 by the end of May. Max Cieminski, Tony Pierro, and Michael Valente were all part of this wave of new arrivals, shipping out within weeks of each other.

By March, Tony and his polyglot comrades in the All-American Division at Camp Gordon had been inspected and deemed combat-ready by War Department brass. The guys from all forty-eight states, foreign- and native-born alike, had shaped up, figured out how to march and sing at the same time, and mastered enough English to follow commands, get along with each other, and snap out responses when officers asked questions. The War Department declared the All-Americans good to go—the second National Army division cleared for departure for France. The first trainloads left Georgia on April 10 en route to Camp Upton on Long Island, and two weeks later the Division Headquarters embarked for Liverpool. Long after the war, Tony reminisced about the excitement of landing in England and getting whisked by train to London. With wildly cheering crowds lining the streets of the capital, the All-Americans marched in review before King George V. "We saluted the old King, up there in the balcony," recalled Tony. "We marched for old King George. While we were marching in front of the King, all of a sudden somebody brought me a bottle of wine, but the sergeant said, 'Give back the bottle.' So she took it back." It's hard to know what to make of this memory. One unit of the All-Americans, the 325th Infantry, was indeed sent to London on May 11 to march in review before the king. "In brilliant sunshine between serried ranks of cheering citizens, these sturdy sons of the New World tramped to the throbbing call of the drums," reported the London *Times*. "Very workmanlike they looked carrying their full kit. . . . Tall they were, clean-shaven almost to a man. . . . Every State in the Union had its representative . . . they were the vanguard of the New Army, that almost numberless force which America is raising to crush for ever the evil spirit of Prussian militarism." The problem with Tony's claim that he was among these "serried ranks" is that he was serving with the 320th Field Artillery, not the 325th

Infantry—and the divisional records show that his unit arrived in England at a later date and was entrained directly to Southampton for departure to France. No doubt the All-Americans who marched through London that day boasted about the roaring crowds and the "old King up there in the balcony," and it's possible that Tony got wind of the boasting and, in the course of time, remembered it as something that had happened to him. But what about the bit with the woman and bottle of wine—a detail too odd to borrow? Maybe Tony Pierro really *did* march for the king through the sunny streets of London—clean-shaven, crisp, workmanlike, ready for war. Military records have been known to be wrong.

It was Max Cieminski's fate to be a rolling stone in the U.S. Army. Two weeks before he shipped out on April 19, he was transferred from the 87th Division to a replacement unit—recruits earmarked to fill out regiments that had been depleted by casualties. Max would be transferred yet again in France before getting his permanent assignment on May 13 with the Yankee Division's 102nd Infantry (which had had its baptism by fire the month before at Seicheprey). The same thing was happening to thousands of other guys that spring, and all this moving around was bad for morale. Soldiers always resented the raw recruits rotated in to take the places of fallen buddies, and the replacements were acutely aware of the resentment. Everything was that much worse if you barely spoke English. Though nobody's English was so bad that he didn't know when he was being laughed at, sneered at, and told to go to hell.

At the beginning of May, the 27th Division was ordered to pack up, vacate Camp Wadsworth, and ship out to France. After six months of training and five years in the United States, Mike Valente would be returning to Europe the same way he had come—in the steerage of a luxury liner. On May 9, Valente boarded the *Susquehanna* at Newport News, Virginia, along with the rest of Company D, 107th Infantry, and set out across the Atlantic. Two weeks later they arrived at Brest harbor on the coast of Brittany. So many American troop transports crowded the rock-bound harbor in the last week of May that one soldier in the regiment remarked that Brest looked like it was hosting a nautical gala instead of an army. Valente, like everyone else, was itching to go ashore and get his feet on solid ground. Back on European soil.

Maybe war was hell, but now that they were "over there," they might as well fight.

As soon as the 27th Division cleared out of Camp Wadsworth, the 6th Division moved in to take over the block of vacant tents—and Joe Chmielewski was among the fresh batch of tenants. Joe had had a rather dreary time of it since enlisting back in June 1917, nearly a year ago now. Assigned to the 16th Machine Gun Battalion of the 6th Division (Regular Army), he ended up training at Camp Forrest, Georgia, along with a bunch of guys from Minnesota and Wisconsin. Only there wasn't much machine-gun training going on at Camp Forrest since there were precious few guns available to train with. So Joe spent the winter of 1917–1918 drilling in the slush and mud of northern Georgia and sharing around a single practice weapon. The tedium was crushing; the weather was cold and wet. In March, the 16th Machine Gun Battalion was reorganized from four companies to two, and Joe was assigned to Company A under the command of Lieutenant Kinloch.

The transfer to Camp Wadsworth on May 9 was a welcome break— at least the men got a train ride through Georgia and South Carolina in the spring. On May 26, two weeks after the move, Joe's unit was ordered to hike twenty-five miles to the machine-gun training center near Landrum, South Carolina—and finally they began training in earnest. A year in the army and Joe was finally doing something other than march, drill, and scrounge for clothing and equipment. Carl J. Lukens, a private in Joe's company from Kalamazoo, Michigan, said that eventually the guys learned to dismantle and reassemble the guns while blindfolded. On June 19, the Landrum interlude came to an end, and Joe and the other gunners of the 16th Machine Gun Battalion broke camp, hoisted their packs, and set out on the two-day march back to Camp Wadsworth. "None will ever forget the downpour of rain and the mud of the second day's march." Back at boot camp, officers culled the men they deemed unfit for service overseas—either because they were too weak or sickly or because their English wasn't good enough or their allegiance was suspect—and sent them to the camp's Development Battalion. Joe made the cut and remained with his company. Now he just had to survive the steamy South Carolina summer until the orders came to ship out.

Famished for ever more men to feed into the "almost numberless force" that the London *Times* raved about, the War Department instructed draft boards to launch a massive new round of recruiting that spring—but this time, the process went forward with breakneck speed. Local Board 430 in Brooklyn summoned Epifanio Affatato to appear on April Fool's Day—but it was no joke. The examiner duly noted his particulars: height, 5 foot 2 inches; age, 23 years, 2 months; black hair, brown eyes, dark complexion; occupation, machinist. Before he had a chance to digest his change of status, Epifanio was ordered to report to Camp Upton, where he was assigned to a provisional depot brigade—essentially a holding tank to process new inductees in the National Army. By April 26, Private Affatato had been given his permanent assignment—not with the National Army after all, but in Company C, 107th Infantry, 27th Division, the same regiment that Mike Valente had been training with all winter in South Carolina. Of the 221 men in Company C, 94 were from New York City. Harry

Stratton, Albert C. Smith, and Edwin S. Munson were the first lieutenants in charge of the company; but the enlisted men had names like
Gaines Gwathmey, James J. Kelly, Torquino Zucco, Michele Chiorini,
Walter Hanrahan, Bruno Koenig, Louis Mirsky, Wasyl Kolonoczyk,
and Epifanio Affatato. Two weeks after his assignment, Epifanio left for
France on board the same ship as Valente. The army was so desperate
for men that it had dispensed with its lengthy stateside training period:
the thinking now was that the new recruits could learn on the job in
France. Where Valente had gotten six months of drill and indoctrina-
tion, Affatato got six weeks. Six weeks between punching the clock as
a machinist in Brooklyn and tossing his pack on an iron bunk belowdecks on a converted German liner bound for Brittany. Even though
he'd been living with his father in Brooklyn for the past seven years,
Epifanio instructed the clerk to put down his mother as the person
to be contacted in case of emergency: Mrs. Rosa Fiore, Scala Coeli,
Italy—spelled right for once. As long as he was in France, he'd be a
hell of a lot closer to his mother in Italy than his dad in Brooklyn. If
anything happened, she might as well be the first to know.

Tommaso Ottaviano up in North Providence, Rhode Island, was
inducted and processed at the same time and just as swiftly. As the sole
support of his widowed mother, Tommaso was entitled to an exemption, and Antonia Ottaviano fervently hoped he would exercise his
rights. More than hoped. Tommaso, after all, was her firstborn and
the de facto head of the family. *How dare the government take him away?*
In fact, the government would not have dared had Tommaso not
insisted on waiving his exemption and serving his adoptive country.
His cousin Carlo was in the army, the guys from the blanket factory
were getting drafted right and left, and so, when the call came, Tommaso decided to go too. Whether he was motivated by patriotism, a
sense of adventure, a desire to bring honor on himself and his family,
a feeling of duty, a need for the army pay and benefits—Tommaso
didn't say. His mother, Antonia, had three other able-bodied sons
under her roof—Giacomo and Domenico were old enough to work

and Ascanio would be soon—so Tommaso tried to ease her worry on that account. It's also possible that the draft board pressured him, since the bulk of the exemptions were going to married men and workers in essential industries like coal, steel, and agriculture. Why encourage a healthy unmarried immigrant who made supersoft Jacquard baby blankets with adorable bunny logos to stay home with his mother? Tommaso did have one other option: since he wasn't a U.S. citizen, he could go fight in the Italian army. That's the route some of his Italian buddies in North Providence chose to take when the induction notice came. But for some reason Antonia could never understand, her sweet, handsome, warmhearted oldest son decided to fight for *l'America*. On April 27, Tommaso left the Esmond Mills and started his new life as a private in the U.S. Army.

Like Epifanio Affatato, Tommaso was shipped out just weeks after induction and with the barest minimum of training. On May 1, he sent a postcard (in Italian) from Camp Dix in New Jersey to his brother Domenico: "Here it's very beautiful because all the trees are in bloom. All of us from North Providence are together. But then, there are many many Italians here, also from other states—and all of us get along well. Don't worry about anything in my regard." After two weeks in a depot brigade, he was deemed sufficiently fluent in English to be given a permanent assignment with Company I of the 310th Infantry, a heavily Italian unit of the National Army's 78th Division. Tommaso left for France with the regiment on May 20.

The army would pay dearly for this haste. As one military historian notes, the use of green depot brigade recruits as quick "fillers" for units going overseas was "one of the reasons that National Army divisions were unready for combat for many months." As the fighting intensified, a report came back from the army's inspector general of untrained recruits with rifles in hand asking to be shown "how to work this thing so that they could go up and get a boche." Sending these men into battle was "little short of murder," in the words of a division commander. Immigrant soldiers had it especially tough on account of language and cultural differences.

But the army kept on drafting them and shipping them out. Meyer Epstein got his call-up papers from New York City's Local Board 93 and reported to Camp Upton on April 27, the same day Tommaso Ottaviano was drafted. Having filed his declaration of intention to become a citizen back in 1915, Meyer was eminently eligible for military service—single, strong, twenty-six years old, long accustomed to roughing it. After a short spell at Camp Upton, he was sent to Camp Devens in Massachusetts and assigned to Company F, 301st Infantry, 76th Depot Division, an entire division of "fillers" who were briefly trained and then moved out into combat units. Meyer would be transferred twice in the coming months, but he was with the 301st Infantry long enough to make a friend. Someone snapped their photo in June at Camp Devens—two short, glum recruits standing side by side at attention and looking clean and uncomfortable in their boxy khaki tunics, khaki trousers ballooning out at the thigh, white canvas leggings (which would be replaced once they were in France by the hated puttees—leg wraps that had to be wound around the calves like bandages), and wide-brimmed Smokey the Bear caps (replaced in France by steel helmets). The nameless buddy looks bland and bored, but Meyer stares at the camera with an expression of solemn, almost wistful resignation—not thrilled to be standing ramrod straight by a barracks in the hot, late spring sun, but determined to do his duty when called on. As indeed he proved to be.

S ince Sam Goldberg, at eighteen, was too young to be drafted, he decided to enlist. "I liked the idea of being in the army," is how he remembered it. "I was a tough kid working at an Atlanta automobile factory—I'd been on my own since sixteen after my son-of-a-bitch father kicked me out. I figured the army would be a good experience." He showed up at the army recruiting office in Atlanta on May 6, explained in his squeaky Irish-Yiddish accent that he wanted to fight for Uncle Sam, braved the inevitable sneers over being named Goldberg and looking like a weedy schoolboy (five feet, three inches; 104 pounds; silver blond hair; piercing blue eyes). Sam had intended to go into the Signal Corps—see some action, maybe learn something useful about telephone and telegraph equipment—but the recruiting officer talked him out of it. "Why not join the cavalry instead?" the guy asked. *I'm afraid of horses,* Sam was thinking. "It's a hell of a lot safer than the Signal Corps," the officer continued. That settled it.

Sam took the train from Atlanta up to Chattanooga, Tennessee, and then hopped a trolley out to Fort Oglethorpe. *Private Goldberg reporting for duty in the 12th Cavalry Regiment,* which triggered the inevitable outburst of derision. "I remember roll call my first day. The sergeant was a Polish guy, a handsome bastard—you know a handsome Pole is handsome—and he does a double take when he gets to Goldberg. 'Jesus Christ, a Jew in the cavalry,' he says. 'I'm surprised they let a little guy like you in. You know, when I was serving in the Philippines, they'd issue guys like you a baby blanket.' Well, I was a goddamn street-smart Jewish kid and I wasn't going to take that. 'Where I come from,' I tell the sarge, 'a remark like that would make a guy feel that you were a sissy.' The sergeant, he walks over to me, snarling, 'What did you say, Goldberg?' 'I heard what you said and I don't like it,' I tell him. 'And you don't like what I said.' That pretty much ended it." Sam heard the sergeant muttering, "From now on don't fool with this Jewish boy—he thinks he's got all the answers." But he didn't have any more trouble.

At first Sam was so small and weak he had trouble picking up a cavalry saddle. The other guys at Fort Oglethorpe started calling him the Flying Ghost because of his pale complexion and silvery hair. He never did get comfortable around horses. But he learned to shoot a rifle and he excelled at saber drill—running the cavalry saber blade through the head of a dummy. Once he heard another soldier say,

"That Jewish kid is the best saber dummy guy in the army." Nobody was laughing at Goldberg anymore.

The cavalry turned out to be a wise choice, at least as far as staying safe. While hundreds of thousands of guys in the infantry, artillery, engineers, and Signal Corps were shipping out to France that spring and summer, Company M Troop, 12th Cavalry Regiment, was posted to Hachita, New Mexico, to patrol the border in case of a German attack launched from Mexico. There was a lot of talk in the unit about Pershing's Punitive Expedition against Pancho Villa, a lot of speculation about what would happen if Germans tried to infiltrate, a lot of words but not much action. It was, as they said in the army, a quiet sector.

On May 9, three days after Sam Goldberg enlisted in the U.S. Cavalry, Congress passed an amendment to the nation's naturalization laws that essentially allowed alien soldiers serving in the war to become citizens at once. By the terms of the newly amended law, alien soldiers in military or naval service "in the present war" could file for citizenship "without making the preliminary declaration of intention and without proof of the required five years' residence within the Unites States." It was a nod to the growing number of nonnaturalized soldiers who were about to face enemy fire for the first time. The gift of American citizenship in exchange for the possible sacrifice of their lives.

In the late spring of 1918, the U.S. Marine Corps was ready to fight in every sense—pretty much the only branch of the American Expeditionary Forces in France that could make that claim. Marine standards, as the officers never tired of pointing out, were a notch higher than those of the regular army. For starters, every marine was in the service because he chose to be: while the bulk of U.S. forces in

France (72 percent) was conscripted, the marines were 100 percent volunteers. To qualify for the Marine Corps, you had to be conversant and literate in English, stand at least five-feet-four, weigh at least 124 pounds, and possess at least twenty teeth. Then as now, marines were not renowned for their modesty, but in the course of June 1918, five miles northwest of the French town of Château-Thierry, they would earn some bragging rights.

"Wherever he goes, whatever he does, the most pronounced characteristic of the marine is a snappy self-confidence, justified by performance," wrote Theodore Roosevelt Jr. (the president's son and a decorated AEF regiment commander) in a book about World War I heroes. This was Matej Kocak to a tee. After the Battle of Belleau Wood, Kocak was finally in a position to justify that snappy self-confidence.

By May 27, the third phase of Ludendorff's spring offensive brought German forces to within fifty miles of Paris. While the French army retreated west in disarray, preceded by long lines of refugees fleeing smoldering villages, Clemenceau's government frantically prepared to abandon the capital. General Ferdinand Foch, who had recently taken over as supreme commander of the Allied armies, signaled Pershing that now if ever was the time for the Americans to enter the fight in earnest—and in response Pershing ordered the 2nd and 3rd Divisions, the sole divisions he considered ready for combat, to head for the Marne River. The 4th Marine Brigade, now part of the 2nd Division, was mobilized to meet the German attack, and Corporal Kocak was mobilized along with them. The journey to the front was a nightmare. Confusion and fear mounted as Kocak's unit approached Château-Thierry by truck and finally on foot. Near the Marne River, the Americans passed more and more French soldiers going the other way—haggard, bandaged, exhausted, mute. The Americans reached their positions on the last day of May. Early on the morning of June 1, the men of the 5th Marines were told to prepare for battle. For Kocak the rush of adrenaline was tempered by a burst of personal pride: that was the day he was promoted to sergeant.

The marines' objective at Belleau Wood sounded simple enough when the orders came down: advance through the patch of woods surrounded by wheat fields; capture the village of Bouresches at the eastern edge of the wood and secure the heights to the north and west; halt the German advance on Paris and keep the enemy from gaining control of the vitally important road leading east to the city of Rheims. In fact, carrying out those orders consumed most of the month of June and more than half of the marines in the field—some 5,600 officers and enlisted men killed, wounded in battle, or missing. Belleau Wood was also the occasion of an exchange that would enter marine mythology. Three days into the assault, as a troop of battered French soldiers straggled to the rear through the American lines, one of the French officers advised Marine Captain Lloyd Williams to follow suit. "Retreat?" the captain bellowed. "Hell, we just got here." Captain Williams would die in the fighting to come.

Kocak's unit suffered as much as any. The 66th Company took part in the ferocious counterattack that the marines mounted on June 6, which went down as one of the deadliest days in Marine Corps history. Marine casualties (killed and wounded) totaled 1,087 on that day alone. Half of Kocak's company was killed or wounded in the dense pitted wood that, to the Americans' surprise, proved to be thick with German machine-gun nests and barbed wire. "Men dropping, men dropping, men dropping," was how 2nd Battalion commander Lieutenant Colonel Frederick May Wise described the agonizing advance through Belleau Wood. German machine-gun fire inside the woods was so dense that men were literally decapitated by bullets.

One snapshot of Kocak at Belleau Wood has surfaced. Amid the myriad World War I documents slowly crumbling to dust in the National Archives there is this paragraph written by Captain (later Major) G. K. Shuler: "Add the following recommendations to those already sent in: Cpl Kokac [sic] 66th Company took charge of a party of 23 men and crawled with them through an open field exposed to snipers and machine gun fire—to the assistance of an isolated group of Marines who were in danger of being surrounded." Captain Shuler got the rank and spelling of his name wrong (Kocak had just been made sergeant)—but he nailed Kocak's coolness under fire and his instinct for "taking charge" when it counted. Nothing came of this recommendation—no medals, no citations for valor. But it wasn't the last time Kocak would distinguish himself for extraordinary bravery.

Belleau Wood was finally cleared at terrible cost in human life and sanity. "It was enough to break your heart," said Lieutenant Colonel Wise, whose men fought as hard and suffered as much as any. In nearly a month of fighting, from June 1 to June 26, this scant square mile of French forest became a wasteland of rotting corpses, exploded shells, garbage, blood, excrement, and disease. The cost in human life and suffering was appalling, but the Americans prevailed. Even the Germans admitted that the American "attacks were carried out with dash and recklessness." For the Americans, for the marines, Belleau Wood was as much a psychological as a tactical victory. A U.S. military intelligence report concluded that after Belleau Wood the French public was "unanimous in believing that when *our* armies are large enough, the war will be brought to a victorious conclusion."

Local draft boards, boot camps, and divisional headquarters were working as fast as they could to make that happen.

Andrew Christofferson joined the army on June 25, the day marines cleared the last remaining German machine-gun nests out of Belleau Wood. Twenty-eight years old, still single, still hardworking, still God-fearing and straight as a die, Andrew walked the twenty-two miles from his Montana homestead to the local draft board in Chinook. A long lonely walk in the endless light of new summer. Andrew had read President Wilson's vow that this was "the war to end all wars," and he believed it. Pious as he was, he never doubted for a moment that his draft notice was the will of God. By temperament and faith, Andrew was a man of peace—but if God wanted him to fight in the American army, so be it. Andrew locked the door of his claim shack, walked to Chinook through the short grass of northern Montana, and never looked back. Weighed, measured, examined, approved for service, he was put on a train bound for South Carolina. Uncle Sam had decided that Andrew Christofferson of Haugesund, Norway, and Chinook, Montana, was going to serve with the Carolina and Florida boys of the 81st "Wildcat" Division, training at Camp Sevier, near Greenville, South Carolina.

Though Andrew had been in the United States for seven years now, he still spoke with a pronounced Norwegian accent—*dese und dose und dee udder tings.* At training camp in South Carolina, when Andrew heard his name called by the drill sergeant, he was in the habit of responding with a hearty "Yah!" as only a Norwegian can say it. One morning, after the Norwegian farmer-turned-soldier belted out his inevitable "Yah!" yet again, the sergeant narrowed his eyes and growled "Christofferson, can't you say anything besides 'Yah'?" To which Andrew replied without a moment's hesitation, "Yah!" The sergeant walked away shaking his head. It was a story Private Christofferson never tired of telling to anyone who would listen.

After Belleau Wood, it was no longer possible to cling to the belief that American boys were somehow going to get off easy in a war that had already decimated Europe. Doughboys shipping out, trained or not, had no illusions about what they were heading into.

Samuel Levin, a Jewish tailor in Philadelphia whose father died in a Russian pogrom, was drafted on May 31, the day American marines were taking their position at the edge of Belleau Wood. One of the reasons Sam had emigrated was to avoid being drafted into the Russian army—but the draft had followed him halfway around the world. Sam was ordered to Fort Bliss near El Paso, Texas, and assigned to the 18th Field Artillery, 3rd Division, one of the two divisions engaged in the slaughter at Belleau Wood. He trained that June while scores of American boys were being cut down each day by German machine guns. At some point, Sam's captain called him out of the ranks. "You're an alien," the captain told the twenty-two-year-old Jewish private, whose uniform fit him like a glove thanks to his abilities with needle and thread. "When we ship out, you don't have to go overseas." "It's my country," was Sam's response. "I go where you send me."

CHAPTER 8

JULY 4, 1918

May 25, 1918: On the eve of the battle of Belleau Wood, four young conscripts stood huddled together on the rail platform at Parkston, South Dakota, and waited for the westbound train. Three of the men were brothers in their early twenties, Joseph, Michael, and David Hofer; the fourth was their brother-in-law Jacob Wipf. All four had received induction notices from the Hanson County draft board with orders to proceed to Camp Lewis outside Seattle to begin training. Even though they were just standing there quietly minding their own business, these four young men were getting pelted with suspicious looks and wisecracks from the other recruits waiting on the platform. Their beards and homemade clothes marked them as Hutterian Brethren—members of the Rockport Hutterite colony, one of seventeen South Dakota communal farms run by German-speaking immigrants from Russia. Hutterites, Hutterish, Hutterian Brethren—call them what you would, every red-blooded South Dakotan in 1918 knew that they were little better than Huns—in fact, they *were* Huns, only they wouldn't admit it. They kept to themselves on vast tracts of choice farmland that they bought up on the cheap from the government or squeezed from poor

American farmers. They were rich as Croesus when everyone else was struggling. They spoke and prayed in German, grew beards, and wore funny clothes. They were "poor citizens" who refused to buy their share of government war bonds. And, worst sin of all, they refused to fight. While other American boys were shipping out by the tens of thousands to the battlefields of France, these Huns claimed that their religion forbade them to serve in the military, even invoking the Bible to justify their cowardice. "Any other country is welcome to them," railed one South Dakota newspaper, these "so-called men who will let others shed blood to make this a land of freedom."

When the train pulled in, the four bearded men got on together and were placed in an empty compartment along with another young Hutterite conscript. They decided to barricade themselves in. They knew it would be a long ride and they didn't want trouble. But trouble came knocking in the shape of a fellow recruit named William Danforth. Danforth, a thirty-year-old local lawyer from the nearby town of Alexandria, stood at the barricaded door and begged them to open up. He said he just wanted to talk to them. At first the four made no response, but as the lawyer kept knocking, they finally decided to comply. After all, he was a lawyer. No sooner had they opened the door a crack than a gang of recruits stormed in and set upon them. Taunting the Hutterites, they grabbed Jacob Wipf and Michael Hofer and cut off hunks of their hair and beards. They beat them. They dared them to speak another word of German. It was as if these four quiet bearded men were the enemy—or prisoners of war.

All four of them were married with children: Jacob Wipf, thirty, was the father of three children, ages three to seven; David Hofer had five children ranging in age from six years old to four months; his brother Joseph, twenty-four, had a young son and daughter and his wife back at the Rockport colony was pregnant; Michael's wife had given birth to a daughter two months before his induction notice came. They were all simple, humble farmers, brought up in a strict faith that had taught them to serve God and live peaceably with all people. Their religion

expressly forbade them to hold public office, to serve in the military of
the country they lived in, or to wear the uniform of the state. All their
lives they had lived with their own kind in a sheltered colony, working
the fields, speaking German, upholding the faith and the customs that
their families had brought over from Europe. When the draft notices
arrived from the Hanson County board, the four had talked with their
minister about what they should do. They knew that other Hutterite
brethren had fled to Canada, which accorded Hutterites and members
of other so-called peace churches the status of conscientious objectors—but they feared if they followed that course they might never see
their wives and children again. The minister helped the Hofer brothers
and Jacob Wipf work out a compromise position: they would submit
to the draft notice and report to Camp Lewis as ordered—but that was
all. They would enter the army but not serve. It was the only way they
saw to obey the law and remain true to their faith.

On the train, the four Hutterite men covered their faces and wept
as the other recruits went after them with scissors and fists. As they
rattled west for two endless days, they trembled to think what would
happen to them when they arrived at Camp Lewis and fell into the
hands of the U.S. Army.

German, German-speaking, German by descent or heritage,
name or association—May 1918 was not a good time to be
German *anything* in America. For going on four years now, ever since
the rape of Belgium in August 1914, hostility to all things German
had been mounting in the United States. But when American soldiers began dying of German bullets, shells, bayonet blades, and gas in
April 1918, anti-German hostility boiled over into rage. Some of the
rage took ridiculous forms. Sauerkraut was renamed liberty cabbage;
local patriots mounted machine guns in front of Milwaukee's Pabst
Theatre to prevent a troupe from performing Friedrich Schiller's play
William Tell; people caught speaking German or people with German

names who refused to buy liberty bonds had their houses or churches painted yellow. German books were torn from library shelves and burned; German-language classes were banned in secondary schools and universities; German and Austrian music—even Beethoven—was barred from the nation's music halls. But sometimes the rage turned from ridiculous to violent, even deadly. A South Dakota rural mail carrier with a German name was caught by a mob of his neighbors and horsewhipped for his lack of zeal for the American war effort. On April 4, 1918, a thirty-one-year-old baker named Robert Paul Prager who had emigrated from Dresden and settled in East Saint Louis was seized by a mob, stripped of his clothing, paraded through the streets barefoot, bound with an American flag, and made to kiss an American flag. The fact that Prager had attempted to enlist in the U.S. Navy but was barred on medical grounds made no difference to the mob. Fearing for the man's life, the police took Prager into custody and hid him in the local jail—but the mob, now five hundred strong, found him, dragged him out, and lynched him after midnight at the outskirts of town. "Pray for me, my dear parents," Prager managed to scrawl on a piece of paper before he died. At the ensuing trial, the lawyers defending the mob ringleaders claimed it was a "patriotic murder" and the jury voted to acquit.

Almost as shocking as the lynching itself was the reaction of the federal government. U.S. attorney general Thomas W. Gregory turned the Prager murder into an argument for cracking down harder on dissidents: "Until the federal government is given the power to punish persons making disloyal utterances," Gregory declared, "Department of Justice officials fear more lynchings." In fact, the federal government had already given itself extraordinary powers over "persons making disloyal utterances" and would shortly give itself more. The Espionage Act, passed at the urging of President Wilson in June 1917, criminalized the spread of information intended to obstruct the American military in any way. Wilson's goal was to outlaw protest against his war policy, but the Espionage Act proved to be a handy way of legally spying on aliens, particularly Germans, silencing political opponents, and removing from circulation publications that the government found objectionable. Socialist presidential candidate Eugene V. Debs was arrested and sentenced to a decade in jail under the Espionage Act in 1918 when the government claimed that he was trying to obstruct military recruiting. The law acquired a sharp new set of teeth in May 1918 with the passage of the Sedition Act: now it was a crime to speak out against the United States government, flag, or military with "any disloyal, profane, scurrilous, or abusive language." Historian David Kennedy notes that "commentators ever since have rightly viewed [the Sedition Act] as a landmark of repression in American history." In essence, the Espionage and Sedition Acts put a legal muzzle on freedom of speech for the duration of the war. But the pernicious effects went beyond an evisceration of the First Amendment. Under cover of these laws, Wilson's government launched a massive surveillance and propaganda campaign directed against aliens, socialists, pacifists, labor leaders, Wobblies, really anyone perceived or accused of being less than 100 percent American. Loyalty Leagues were set up in states to enforce patriotism in the ethnic communities; citizens were actively encouraged to spy on their neighbors, root out foreign spies, turn in slackers and doubters. "Not a pin dropped in the home of anyone with a foreign name but that it rang like thunder on the inner ear of some listening sleuth," boasted George Creel, the head of a wartime federal propaganda agency known as the Committee on Public Information. By the middle of 1918, the Department of Justice was receiving fifteen hundred letters *a day* related to loyalty charges. "Never in its history has this country been so thoroughly policed," proudly insisted Attorney General Gregory.

Germans bore the brunt of this thorough policing. German Americans were arrested and in some cases imprisoned for "crimes" like calling the war "all foolishness" or suggesting that the United States might lose. German Americans who had not yet become U.S. citizens were forced to register as enemy aliens and subjected to travel restrictions. A German pastor was imprisoned in South Dakota for urging his congregation not to buy war bonds. It was increasingly dangerous, and in some states illegal, to speak German, either over the phone or in public. The ban on preaching in German was especially hard on older congregants who had never learned English and for whom, as one put it, "German had become synonymous with religion itself." One Montana preacher wrote to the state's governor in his imperfect English, "I cannot believe that it is the intention of the government to leave the people in these sad times of war without any consolation from the word of God, while their sons are offering their lifes for the country."

The lash fell especially heavily on the Hutterite colonists of the upper Midwest because they had two sins to answer for: not only were they of German origin, they were also conscientious objectors. The Hutterites of South Dakota traced their ancestry to a group of Swiss, German, and Tyrolean farmers who split off from the followers of Martin Luther at the dawn of the Protestant Reformation. They were called Anabaptists, meaning "to baptize again," because one of their fundamental beliefs was that the rite of baptism should not be conferred automatically at birth but rather chosen by aware believers as a confession of faith. Even more insidious in the eyes of their fellow Christians was the Anabaptists' doctrine of nonresistance: citing Matthew 5, Romans 12, and Acts 5, among other biblical texts, Anabaptists refused to bear arms, swear allegiance to the state or wear its uniforms, pay war taxes, or hold public office. For these beliefs, and particularly the doctrine of nonresistance, they had been burned at the stake, sold as galley slaves, branded, flogged, and hounded from country to country through Europe. Hutterites, who separated themselves from other Anabaptist sects (Mennonites, Amish, Swiss Brethren) in the 1520s under the leadership of Jakob Hutter, were persecuted with particular savagery because they added a third offensive tenet: the community of goods. The story goes that in 1528, the founding group of Hutterites had become so impoverished that they agreed in desperation to place

their few possessions on a blanket and then divide their pooled resources among themselves. It was an early form of communism, and it
was feared and hated as much during the Reformation as the godless
Marxist version would be in the twentieth century. In the long years
of their wanderings through central Europe, the Hutterites did enjoy
periods of peace and prosperity. They lived freely in Moravia for a
long stretch before being forced to relocate to Slovakia, Transylvania,
and Romania. Finally, in the late eighteenth century, they found what
they hoped would be a permanent haven in the Ukrainian region of
the Russian Pale. Granted the right to speak German, follow their
faith, and remain exempt from military service, Hutterites prospered
in the southern Ukraine for three or four generations.

The Russian idyll ended abruptly in 1870 when Czar Alexander
II withdrew their rights and privileges as part of a policy of Slaviciz-
ing the Germans in Russia. Under the new regime, teaching their
children in German was no longer allowed and military service was
henceforth required. It was the prospect of seeing their sons drafted
that convinced the families of Michael, Joseph, and David Hofer and
Jacob Wipf to emigrate to the United States from their villages north
of the Black Sea in 1877. Three Dakota colonies were established initially—at Bon Homme, Wolf Creek, and Elm Spring near Ethan—and
these prospered so well that over the next half century fourteen satellite colonies were spun off in South Dakota and two in Montana. It
was never easy to make a living off the American prairie, but the Hutterites proved that it was easier to survive communally than on isolated
homesteads. The strain of wheat that they brought with them from
Russia—a tough hard-kerneled winter wheat variety known as turkey
red—proved to be perfectly adapted for cultivation on the Midwest
prairie. Centuries of persecution prepared them for the vicissitudes
of a strange, unpredictable climate. The Hutterites clung to their old
customs and traditions, speaking German exclusively at home and in
church; making their own clothes; celebrating holidays and festivals as
their ancestors had in Europe. Men grew beards after they married;

families tended to be large and tight-knit. Even though by 1918 many families, including the Hofers and the Wipfs, were starting on their third generation in the United States, in a sense these people were perpetual immigrants since they had no desire to mingle or assimilate with the mainstream. Unlike the Amish, they did not eschew modern inventions—they were happy to use telephones and cars and tractors so long as they could use them in their own way and for their own ends. But they insisted on living apart—and by living apart they prospered. Life was good in South Dakota until the United States declared war on Germany and instituted the draft.

When Jacob Wipf and the Hofer brothers received their induction notices in the spring of 1918, the U.S. Army had yet to implement a clear or consistent policy toward conscientious objectors. Officially, Secretary of War Newton Baker had declared that conscientious objectors should be allowed to request noncombatant service, be permitted to live apart from combat soldiers, and be treated with "tact and consideration." But secretly, in a confidential memo, Baker urged training camp commanders to pressure COs to serve—the types of pressure being left to the discretion of the officers in charge. In March 1918, President Wilson attempted to resolve the problem by defining "noncombatant service" as work in the medical or quartermaster corps or the corps of engineers. On paper, this sounded like a workable approach. But many COs—Hutterites among them—took the position that Wilson's noncombatant options contributed to the war effort and were therefore a form of military service. In any case, once they got to training camps, they were at the mercy of army officers who had been given free rein by Baker. "To hell with your conscience," one lieutenant told a Hutterite draftee who refused to work at Camp Funston in Kansas—this was typical. Verbal abuse, coercion, and beatings were common; many were tried and sentenced to lengthy prison terms; some were found guilty without a trial. The pressure to serve was relentless and cruel. One conscript who wouldn't sign for his uniform reported being dragged to a bathhouse, stripped naked, shoved under the shower, and scoured with a broom. "Part of the time they had me on my back with face under a faucet and held my mouth open," he said of the ordeal. "They got a little flag ordering me to kiss it and kneel down to it." Hutterites reported being hung upside down in a tank of water until they were on the verge of drowning and being stood on their head in a shower with water running into their noses. Those who refused to obey orders were sometimes told they would be shot: they had their heads covered with sacks and were forced to listen to an officer give the command to fire, whereupon a soldier would clap two boards together.

This was the pit that the Hofer brothers and Jacob Wipf fell into when they arrived at Camp Lewis on May 28, 1918. Shortly after they got off the train, they were assigned to the 29th Company, 8th Battalion, 166th Depot Brigade and ordered by the company's second lieutenant, Robert S. Shertzer, to fall in with their squads. Shertzer and his sergeant, Reynolds B. Hilt, explained that the order to fall in had "nothing to do with fighting," but still the Hutterites wouldn't budge. They stood apart and conferred with each other in German, and then Jacob Wipf stepped forward as their spokesman. "We can't do it," he told the officers. "Our conscience won't allow us."

They were immediately imprisoned in the Camp Lewis guard-house. Five days later, on June 2, the four were ushered into the camp's muster office and ordered to fill out enlistment and assignment cards— a standard form on which all enlistees were required to state their place of birth, occupation, and marital status and then sign their names next to the words "Declaration of Soldier." The four Hutterites refused to fill out or sign the form, and they were returned to the guardhouse. A week later, at one thirty in the afternoon of June 10, their court-martial began.

The trial was brisk and routine, though the outcome was any-thing but. Officers testified that the four Hutterite conscripts refused to obey orders as soon as they got off the train and repeatedly refused to sign any military papers or fall in with their squads. "Their attitude

[was] 'This is as far as we go,'" said one officer. "They intended not to obey anything. But they were meek about it."

The four Hutterites were put on the stand, one after the other, and asked about their background, the history of their religion, their family situations, the nature of the Rockport Hutterite colony, why they spoke and wrote in German, whether they were in fact German partisans. The men, who spoke English haltingly and crudely, had trouble understanding some of the questions.

From the testimony of Jacob Wipf:

> Q. Are the members of your church permitted by your church
>   principles to engage in war?
> A. They are strictly against war. That is why we left Russia.
>   . . .
>
> Q. Are you willing to take part in any noncombatant branch of the service
>   of the army?
> A. No; we can't.
> Q. What are your reasons?
> A. Well, it is all for war. The only thing we can do is work on a farm for
>   the poor and needy ones of the United States.
>   . . .
>
> Q. Does your religion believe in fighting of any kind?
> A. No.
> Q. You would not fight with your fists?
> A. Well, we ain't no angels. Little boys will scrap sometimes, and we are
>   punished; but our religion don't allow it. . . . We can't kill. That is
>   strictly against our religion.

From the testimony of David Hofer:

> Q. Have you any sympathies—I mean, do you favor the cause of Ger-
>   many as against the United States in any way?

A. *Well, I ain't got nothing for Germany. I wouldn't talk one word for Germany, because they done our old folks.*

Q. *They persecuted you?*

A. *Yes; killed just thousands of our old folks four hundred years ago.*

. . .

Q. *Do you pay taxes?*

A. *Yes, sir.*

Q. *Do you vote?*

A. *No; never voted. None of our members never voted. We don't take any part in this world; not a bit. If we are persecuted out of this land, we go to another one. If we are persecuted there, we go on to another one, just like Christ said. If the government wants us to go out, we leave our property right there and go out.*

From the testimony of Joseph Hofer:

Q. *Do you personally believe that you could not take part in war in any form?*

A. *I do believe that we can't take any part in the war.*

Q. *Why do you believe that?*

A. *Why do I believe that? Well, Christ says that you shall not kill. We confessed that this is right, and my conscience tells me that.*

The court was adjourned at 4:20 P.M. by Lieutenant Colonel Harold D. Coburn, 363rd Infantry, 91st Division.

Five days later the division's commander, Major General H.A. Greene, reported the outcome of the trial in a memo to the judge advocate general of the army:

The findings of the court in the case of each of the accused was guilty of all specifications and charges. Each of the accused was sentenced by the court to be dishonorably discharged the service, to forfeit all pay and

*allowances due and to become due, and to be confined at hard labor at
such place as the reviewing authority may direct for twenty (20) years.*

The "reviewing authority" directed that the three Hofer brothers
and Jacob Wipf be confined in the U.S. Disciplinary Barracks at Alcatraz, California.

The government, in a stroke of public relations genius, dedicated
the Fourth of July that year to immigrants. Pageants and parades in native costume, long windy speeches and newspaper articles,
banners unfurled on Main Streets across the country, public vows of
unswerving loyalty—from sea to shining sea, the nation mounted a
daylong celebration of the wartime contribution of Americans of foreign birth or descent. This brilliant bit of political theater had been
brought to America by the Committee on Public Information—"the
world's greatest adventure in advertising," as the committee's head
George Creel called it. For over a year now, Creel and his public relations army had been "engineering consent" among the immigrant
communities with patriotic stories planted in the foreign-language
press, prowar pamphlets (often horrendously translated) in all the
major immigrant languages, short flag-waving speeches delivered at
public gatherings nationwide by a cadre of "four-minute men." Yiddish-speaking orators were dispatched to the Lower East Side, Poles to
Buffalo and Chicago, Spanish speakers to the Texas-Mexican border
towns. Simultaneously, the CPI mounted a campaign against all things
German—broadsides depicting Germans as bloodthirsty monsters and
baby killers; CPI-sponsored hate films like *The Kaiser, the Beast of Berlin,*
with footage of the rape of Belgium and the sinking of the *Lusitania;*
a whisper war of paranoia and suspicion and neighborly spying. But
for one day, July 4, 1918, all were welcome in the CPI's great patriotic
tent—even Germans—so long as they supported America's push to
win the Great War. "Turning over" the day to immigrants was, boasted

Creel, "one of the great ideas of the war." In fact, the July Fourth "festival of loyalty of the foreign-born" was a war rally disguised as a celebration of diversity.

The day's showcase event was a "pilgrimage" from Washington, D.C., to George Washington's Mount Vernon plantation, and Creel and his team choreographed it for maximum PR punch. President and Mrs. Wilson agreed to host a pageant with representatives of thirty-three nationalities—Albanians, Armenians, Assyrians, Belgians, Bulgarians, Chinese, Czechoslovaks, Costa Ricans, Danes, Dutch, Ecuadorians, Filipinos, Finns, French, French Canadians, Germans, Greeks, Hungarians, Italians, Japanese, Lithuanians, Mexicans, Norwegians, Poles, Rumanians, Russians, Spaniards, Swedes, Swiss, Syrians, Ukrainians, Venezuelans, and Yugoslavs. On the morning of the Fourth, the president, his wife, and their daughter Margaret boarded the riverboat *Mayflower* in Washington and received the crowd of stocky, serious foreigners with long names and florid signatures, many of them dressed for the occasion in silk hats and frock coats. As the boat made its stately way down the Potomac, President Wilson "moved from group to group," in Creel's reporting, "laughingly suggesting that they put their high silk hats to one side . . . and giving every man and woman the feeling of being a sovereign citizen in a free country."

The floating Babel docked at Mount Vernon and the passengers disembarked. "The scene," wrote Creel, "was one that etched itself in memory. The shining stretches of river, the walk up the winding path through the summer woods, the hillsides packed with people, the beat of their hands like the soft roar of a forest wind, the simple brick tomb of the Father of Our Country overhung with wisteria in all the glory of its purple bloom." Irish tenor John McCormack, once he caught his breath, crooned "The Battle Hymn of the Republic" "while each of the thirty-three immigrant representatives walked into the tomb, one by one, laid a wreath upon the grave, and offered a prayer to the 'august shade of the departed.'" Then Belgian-born Felix J. Stryckmans, chairman of the Committee of the Foreign

Born, stepped forward to address the assembled: "In my own city [Chicago], 800,000 foreign-born men and women are at this moment lifting their hands and renewing their vows of loyalty. . . . When, to-morrow, the casualty list brings heaviness to some homes and a firm sense of resolution to all, we shall read upon the roll of honor Slavic names, Teutonic names, Latin names, Oriental names, to show that we have sealed our faith with the blood of our best youth." Finally, Wilson made a speech insisting on the "destruction of every arbitrary power" that threatens "the peace of the world" and lauding the right of every people to freely settle questions of "territory, of sovereignty, of economic arrangements, or of political relationships" in the manner they chose. "John McCormack then sang 'The Star-spangled Banner' as it was never sung before," according to Creel. The *New York Times* reported on the front page that as McCormack "held the last ringing note" of the phrase "for conquer we must," "the President extended his hand. A thousand soldiers of the nations allied against Germany who were in the crowd stood rigidly at attention." Creel declared that "From that day, a new unity was manifest in the United States."

That month, Jacob Wipf and David, Michael, and Joseph Hofer, with hands and feet shackled, were sent under guard from Camp Lewis to San Francisco to begin serving their twenty-year terms at Alcatraz. David Hofer later gave an account of what happened to them to a man named J. Georg Ewert, and Ewert wrote this narrative of the ordeal:

> When they arrived at Alcatraz, their own clothes were forcefully removed and they were told to put on military uniforms. They refused just as before. After that they were brought into the dungeon, into dark, dirty, stinking cells for solitary confinement. The uniforms were thrown beside them with the warning, "If you do not conform you will stay here until you die, just as the other four we dragged out of here yesterday." So they were locked in there in their light underwear.

The first four and a half days they did not get anything to eat, only half a glass of water every twenty-four hours. During the night they had to sleep on the wet and cold cement floor without blankets. . . . The last one and a half days they had to stand with their hands over their heads crosswise chained so high to the iron bars that they could barely touch the floor with their feet. . . . They could not talk to each other during this time since they were separated from each other. Just once he [David] heard Jakob Wipf cry out, "Oh, Almighty God!"

After five days the four men were brought out of the lower part of the prison into the prison yard where more prisoners were standing. Some of those were overcome with compassion when they saw the Hutterians. One commented with tears in his eyes, "Is it not a shame to treat human beings in such a way." The men were covered with boils, bitten by insects and their arms were swollen so badly that they were unable to slip the sleeves of their jackets over them. They had been beaten with sticks. Michael Hofer was beaten so badly that he, on one occasion, became unconscious, and fell over.

The "dungeon" referred to here is a group of cells on the lower level of Alcatraz known as "the citadel"—off-limits in today's tours of the prison. Originally, these cells were open niches in the prison wall probably used for storage: at some point bars were added to convert them into cramped, damp, suffocating cells. Since the citadel lacks a natural source of light, when the electricity is switched off the darkness is complete.

Wipf and the Hofer brothers had in fact gotten off with a comparatively light sentence. Of the approximately 500 conscientious objectors who were court-martialed during the war, 7 were sentenced to death (though none was actually executed) and 142 to life in prison.

Peter Thompson spent the night of July Fourth marching with a full pack through dense humid woods. He and the other men of his company had been rousted out at 11:00 P.M. with shouts of "Everyone out," given coffee and sandwiches, and then ordered to get moving. "The packs became unbearable," wrote a sergeant marching with the unit, "the feet hurt miserably—sweat ran off the face in streams—the rifles were heavy and cumbersome—the overcoats and slickers slung on the arm were constantly slipping from the almost nerveless arms— it became unbearable." All this terrible misery and pain and Peter and his comrades in the 362nd Infantry, 91st "Wild West" Division were still in New Jersey. The men were humping their "unbearable" packs not to the front lines in France but to Alpine Landing on the New Jersey side of the Hudson River, where riverboats were waiting to take them to a troop transport docked in New York Harbor.

Peter had spent the previous eight months training with the Wild West Division at Camp Lewis, and his last days in boot camp had over-lapped with the imprisonment and trial of the Hutterites. In fact, Cap-tain Joseph W. Sutphen, an officer with the 362nd Infantry (though not Peter's battalion commander), sat on the court-martial that convicted Wipf and the Hofer brothers of willful disobedience. Had the Hutter-ite men chosen to obey orders instead of their consciences, they would have been marching alongside Peter that July Fourth night through the steamy Jersey woods. Peter had no strong feelings about the war one way or the other. He had enlisted not out of any burning desire to slit the kaiser's throat but as a fast track to U.S. citizenship. And lucky for him, that track proved even faster than he had anticipated thanks to the recent amendment to the naturalization act. On June 1, 1918, Private First Class Peter Thompson, all five feet, three inches and 115 pounds of him, raised his right arm and swore allegiance to the United States of America. Irish no more.

It had taken Peter's regiment five days to move itself from Camp Lewis to the embarkation center at Camp Merritt in Bergen County, New Jersey. Trains packed with soldiers rattled east through Idaho, Montana, North Dakota, and Minnesota, then southeast through Wis-consin until they reached Chicago, and finally made a straight shot to the coast. "Guards were stationed at each end of every coach to prevent the men from leaving the train," wrote Zenas A. Olson, a sergeant with the division's 361st Infantry. "The opportunity of seeing the American girls along the route was one that could not be missed and the side of the train that faced the station at each stop was crowded and heads filled the window. Handshaking was one of the big orders of the day." Another recruit described the train trip as "one mad panorama of noise and excitement." The madness reached a fever pitch at Camp Merritt—Olson described it as a "nightmare"—as officers worked frantically to get the men equipped, inspected, showered, and issued new boots and clothing prior to departure.

That miserable Fourth of July night march down the Palisades with full packs ended at Alpine Landing on the Hudson, where the men boarded boats bound for New York harbor. The last glimpse of America was stirring—the wall of the Palisades rising to the west, the glittering pinnacles and canyons of Manhattan to the east, the river echoing with the shouts and songs of excited men. When the men caught sight of the Statue of Liberty, they let rip with a chorus of "Over There"—"And we won't come back till it's over, over there!"—the immigrants among them no doubt remembering how they had gazed in silence at the "goddess" the last time they had sailed these waters. Peter and the men of the 362nd Infantry were herded onto a liner called the *Empress of Russia*.

In the peak immigration years before the war, arriving ships had to wait for hours or even days to unload their human cargo on Ellis Island. Now, in the frenzy to get American soldiers overseas, the same thing was happening in reverse, often with the same ships. More than 1.6 million American soldiers shipped out between April and October, 1918; in July, when the mass military exodus reached its crescendo, some 306,350 soldiers crossed to Europe. July 5, 1918, the day Peter Thompson's infantry battalion boarded the *Empress of Russia*, was also the day Meyer Epstein shipped out with the 76th Depot Division.

July 5, wrote Olson, was a full-blast New York summer day: "below deck was stifling, above it was crowded beyond comfort." The *Empress of Russia* sat motionless in its berth through interminable sweltering hours. The steerage quarters began to reek as they filled with sweating men and their equipment—or rather to reek worse. Weak electric bulbs barely penetrated the brown air. The percussion of army boots on the decks and packs hitting the bunks echoed through the chamber. Late in the afternoon, as the gangplanks were finally raised, a current of excitement surged through the ship. "The boat drifted slowly out and away, swung into the current and dropped down the river to anchor in the lower harbor," wrote Olson. "We had made the start; we had broken the last ties that bound us to the old life; we had shoved off for France. The light of the early morning of July 6th, revealed through the fog a flotilla of ships, resting at anchor, all camouflaged in weird colors and designs, all in readiness for the start."

One of the twenty-one ships in that weird flotilla was the British troop transport *Belgic* carrying the 6th Division's machine-gun battalions, Joe Chmielewski among them. The anonymous officer who wrote the history of the 16th Machine Gun Battalion (Joe's unit) described the explosion of joy that morning from men who couldn't wait to go to war: "At 9:05 [A.M.], a mighty whistle blew two long blasts, and like a snail we began to move out into the Hudson. Who will be the first to forget the joyous yell that went up! All New York city must have heard it! Soon we were in mid-stream, the faithful tugs released their grip, and the mighty ship steamed ahead under its own power. It was a good time to sing 'Goodby Broadway, Hello France.' The Band played it and a few more than five thousand on board sang it. Soon we passed 'The Lady' and joined in the convoy of 21 other ships loaded with 'Kahki.' [*sic*] Out, out, out, we went, till America was lost to view."

The battalion historian claimed that that afternoon a detail from Company A, Joe's company, was assigned to man the ship's 6-inch guns, though surely he was mistaken, because American machine gunners would never have been enlisted to operate British artillery. Still, given the constant threat of attack from German submarines even this close to home, ships' guns had to be at the ready.

Leonardo Costantino, an affable Italian immigrant transplanted to San Diego, kept a journal of his experiences with the Wild West Division's 364th Infantry—just a couple of lines every few days describing what he did, what he saw, how he felt. In his entry about shipping out, Leonardo complained of the lousy conditions on board the troop transport: "This Vessel here is very large but soldiers do not like the way they run it. Both eats and sleep were very unsanitary. The officers and nurses had staterooms and had a real time on the way. Many Boys got sea sick but I was feeling good all the way." Leonardo, who had emigrated at the age of sixteen from the southern Italian town of Canetto near Bari, did not mention how the "eats and sleep" on the troopship compared with what he'd gotten in 1909 on board the immigrant ship *Re d'Italia*. The lower decks of the *Empress of Russia* on which Peter Thompson shipped out reeked of hard-boiled eggs. But the stench didn't stop him and his buddies from playing hands of panguingue (punished for betting, since gambling was against army rules, they had to resort to placing bets on the number of kittens a pregnant stowaway cat would bear). The nights were long and dull and devoid of even the comforts of smoking and cards. "As soon as darkness comes every light goes out," one Doughboy wrote. "Not even a match may be scratched. The transports then resemble great black lizards, prehistoric monsters, crawling over a grey desert."

Leonardo and Peter wrote nothing in their letters or journal about being afraid or lonely or proud or bitter about being shipped back to Europe to fight. If Joe Chmielewski wrote letters back to Fifficktown, his family has lost them. But in his novel *Three Soldiers,* John Dos Passos, who served in the war with the French ambulance corps and then as a soldier in the American army, captured something of the snappish anxiety that permeated the "dark pit" of steerage quarters on these departing troopships:

Fuselli [an Italian American from San Francisco] sat on his bunk looking at the terrifying confusion all about, feeling bewildered and humiliated. For how many days would they be in the dark pit? . . .

"An' if we're torpedoed a fat chance we'll have down here," he said aloud.

"They got sentries posted to keep us from goin' up on deck," said someone.

"God damn them. They treat you like you was a steer being taken over for meat."

"Well, you're not a damn sight more. Meat for the guns."

A little man lying in one of the upper bunks had spoken suddenly, contracting his sallow face into a curious spasm, as if the words had burst from him in spite of an effort to keep them in.

Everybody looked at him angrily.

"That goddam kike Eisenstein," muttered someone.

"Say, tie that bull outside," shouted Bill Grey good-naturedly.

"Fools," muttered Eisenstein, turning over and burying his face in his hands.

Later in the novel, the guys speculate about whether the "goddam kike" Eisenstein is in fact a spy:

"He's foreign born, ain't he? Born in Poland or some goddam place."

"He always did talk queer."

"I always thought," said Fuselli, "he'd get into trouble talking the way he did."

"How'd he talk?" asked Daniels.

"Oh, he said that war was wrong and all that goddamned pro-German stuff."

"D'ye know what they did out at the front?" asked Daniels. "In the second division they made two fellers dig their own graves and then shot 'em for sayin' the war was wrong."

"Hell, they did?"

"You're goddam right, they did. I tell you, fellers, it don't do to monkey with the buzz-saw in this army."

Such were the rumors circulating belowdecks as the troopships zigzagged across the Atlantic in patterns intended to elude German submarines. It took the *Belgic* and the *Empress of Russia* eleven days to cross the ocean to Liverpool harbor. There was no cheering from either ship when the first green smudge of the English coast came into view.

CHAPTER 9

THESE FOUGHT
IN ANY CASE

N one of them were ready for war. No matter how long they had trained, no matter what newspaper they had read or rumor they had heard, no matter how eager they were to fight, none of the men on those boats were really prepared for what awaited him in France. "We can now see that things are getting worse and worse every day," wrote Leonardo Costantino in his log as the 364th Infantry approached the front a few days after landing in France. And then, on July 23: "Are still in this horrible . . ." That was the entire entry. There was nothing else to say then—and later, after the war, nothing he wanted to say. In their journals (forbidden by the army but kept secretly by those with an urge to write) and heavily censored letters home from "somewhere in France," the men wrote of "eats and sleep," rats and lice, battlefield souvenirs, weather, the suffering of the horses that dragged the big guns from battle to battle, mud, wet clothes, sore feet, the crushing weight of fatigue. They did not write about what they could not have imagined beforehand and did not want to remember afterward. *This horrible*. They clung to details and data, no matter how trivial—how many miles they marched, how many hours they went without taking their boots off, how many days in a row it rained—but "this horrible" was beyond them. "I do not recall nothing," Corporal Charles Bellizzi with the 27th Division wrote in a questionnaire about his experiences in the war. "Only that duty called and had to leave behind the whole family, and go." To the question "what do you recall you were thinking and experiencing at the time [of combat]," Bellizzi answered: "to come out alive and do my duty."

They said that you died if a shell had your name on it—you lived if it didn't. And it helped if you kept your head down and your gas mask on. That was as deep into it as most of them wanted to go. Like the God of the Jews, the war itself could not be faced or even spoken of by name.

Ezra Pound wrote of the delusions and self-delusions of men who "walked eye-deep in hell" in the first section of his poem "Hugh Selwyn Mauberley":

> These fought in any case,
> and some believing,
> pro domo, in any case . . .
>
> Some quick to arm,
> some for adventure,
> some from fear of weakness,
> some from fear of censure,
> some for love of slaughter, in imagination,
> learning later . . .
> some in fear, learning love of slaughter . . .

Whatever they had brought with them of love or fear or belief, they learned in "this horrible" to discard or hide or see in a strange new light.

In the heat of battle, when it came, they fought not to make the world safe for democracy or because the Hun had raped Belgium or to end all wars: they fought because the lieutenant gave them orders that he got from the major. They fought to survive and to save the life of the guy next to them—and of course they fought to kill. Not a cause, not a regime, but to kill the invisible guy at the top of the ridge firing a machine gun at their guys. In battle, all that mattered was the moment, even if the moment stretched to eternity. Memory, identity, background fell away: body and soul, there was nothing else.

For the foreign-born guys who had been ridiculed, exploited, or oppressed because of their identity, the nakedness of combat proved to be something of a boon. The "Izzies, Witzers, Johnnies, Mikes, and Tonies" who were laughed at when they shuffled off the train at Camp Upton fought as bravely, as selflessly, as selfishly as anyone else. They still got called kikes and wops and polaks despite the army's campaign to stamp out ethnic slurs—but after the first time in battle together, no one was laughing anymore. This was one reason why the war became a source of pride for so many of the immigrant soldiers.

In the theater of death—death on a scale the world had never seen before—men turned religious or superstitious. Those who had never prayed in their lives went to services in YMCA huts or asked the blessing of priests or rabbis. They clung to signs and portents, crosses, Jewish stars, dog tags that had been blessed back home. Faith was often the most precious thing the immigrants had brought with them from the old country—and the one thing they brought back when they returned to fight. The War Department, to its credit, made some effort to help these foreign-born soldiers feel closer to their God in the hell of war. For the first time in an American war, the army appointed Jewish chaplains—eventually six rabbis ministered to the hundred thousand Jewish soldiers serving in France—and, starting in February 1918, the graves of slain Jewish soldiers were marked with the Star of David. Jewish, Catholic, and Protestant chaplains learned each other's prayers and last rites, and dying soldiers thanked God for blessings bestowed by clerics they would have cursed before the war. As Rabbi Lee J. Levinger, a chaplain with the 27th Division, wrote, "I, for one, have read psalms at the bedside of dying Protestant soldiers. I have held the cross before a dying Catholic. I have recited the traditional confession with the dying Jew. We were all one in a very real sense."

Rabbi Levinger could not explain how the war changed the men he had seen fight in it—no one could—but he was convinced that these "world-shaking events have become a part of their very being. . . . War gave the world a new angle of vision on life and death, on good and bad. The deepest impress of this new viewpoint is on those men who were themselves at the front, who underwent the most extreme phase of it in their own persons." For the foreign-born, the change was perhaps most profound. These men from the ghettoes and mines and factories went to war, willingly or not, with so little to lose—often nothing, really, aside from their lives. If they were lucky and their lives were spared, they took home something they could never have gotten in peace. All of those who made it were grateful to be going home—and proud, maybe for the first time, of the home they were going to.

Matej Kocak was among the men who, in Rabbi Levinger's words, "underwent the most extreme phase of [the war] in their own persons." As Americans took their place at the front in the bloody months of June and July 1918, Kocak revealed his genius for modern warfare—braving the fiercest fire, acting alone or taking charge of comrades as the need arose, making the right choices under deadly pressure, coming out alive. Kocak, in short, became a hero in that crucial summer of war, though he was never commemorated in carved marble or celebrated beyond a passing mention in the pages of books. All he did was to help win battles, inspire the admiration of those he fought beside, and raise a flicker of pride among his fellow Slovaks before vanishing with barely a trace in government documents and family stories.

After the month of punishment at Belleau Wood, the men of the

5th Marine Regiment who had come through alive and intact believed they were due some time off. At the very least, Kocak and the other survivors desperately needed hot meals, showers, new uniforms, delousing. A rumor made the rounds that the entire 2nd Division was going to be sent to the south of France for two months while new recruits were rotated in to bring the depleted units back to strength. Wishful thinking, as it proved. On July 15, General Ludendorff let loose the fifth and, as it turned out, final phase of his spring offensive with a thrust at the ancient cathedral city of Rheims, and the Allied command snapped back with a response that put the marines on the front line once more. The French generals had identified a point of extreme vulnerability in the German line—a semicircular bulge, or salient, that stretched from Noyon in the northwest, down to Château-Thierry in the south and just shy of Rheims in the northeast. If the Allies could cut through the salient at Soissons, they would disrupt German transport and communication links at a strategic point and eliminate the threat of a German march on Paris. But just as critical was the symbolic value of this action: for the first time since the spring, the Allies were going on the offensive. July 18, the jump-off set by French commander Marshal Foch, would be the day the pendulum reached its apogee and began to swing back. Soissons was to mark the great reversal—the battle in which the ever-growing American presence finally made a difference—and because of this, Allied command deployed the best forces at their disposal. The French Moroccan Division (with which the French Foreign Legion fought) was to be flanked to the north and south by the American 1st and 2nd Divisions (which included the marines). Far from getting a vacation, Kocak was heading back into battle with the 5th Marines.

Heading back with the utmost haste and stealth. Total surprise was of the essence of Foch's strategy at Soissons—and total surprise in this war meant superhuman speed. The plan called for loading all combat divisions into trucks the night before the attack and transporting them in one immense sweep to hidden drop-off sites within marching distance of the salient. Then, before daybreak, they were to move out. Without German trenches or barbed wire to break their momentum, the Allies believed they could roll right through the stunned German line and keep going.

Kocak's platoon arrived at the drop-off site twelve miles from the German line on the afternoon of July 17. "A sort of sixth sense seemed to predict a hard battle," a marine officer recalled later, "and I was surprised to see the men and officers exchanging valuables, letters, and addresses of loved ones. . . . It occurred to me that the experience in Belleau Woods had really driven home the knowledge that war was not a respecter of persons." A steady drizzle set in around four o'clock, and by the time Kocak led his men into the Retz Forest that night, the drizzle had become a soaking downpour. In the forest it was so dark and wet that men had to grab onto the packs or coattails of the guy in front in order to stay together—and whole companies ended up wandering off course. All night long the muddy road filled with a pandemonium of motorized transport, tanks, wagons, gun carriages, ammunition trains, staff cars; the infantry, staggering under sodden packs heavy as lead, were forced to march in a streaming ditch at the side of the road. The combination of foul weather, the exhausting night-long march, lack of food, and the impossible orders made the 5th Marines suffer "extreme agony getting to their appointed position," according to one military historian. The marines had orders to move out at 4:35 A.M., but as of 4:30 A.M. July 18, Kocak's platoon was still tangled up in the quagmire on the single road feeding men to the front. First Marine battalion commander Major Julius S. Turrill gave the order that the stragglers should proceed at double time. Kocak, at the head of his sodden platoon, broke into a trot. As one of the men wrote, the 5th Marines literally ran into battle at Soissons, "covering the last few hundred yards in double time, [reaching] their positions just as the first streaks of dawn strike the eastern fringes of the forest behind them and as the artillery barrage comes down with a crash upon the enemy's lines in front of them."

Without pausing for breath or for orders, marines swarmed through the exploding forest. Every two minutes the rolling artillery barrage jumped another hundred meters forward, and the infantry followed in the wake of the shells. In the predawn twilight it appeared that Foch's strategy had paid off: the surprised Germans offered little resistance and Kocak's 66th Company was able to move quickly past machine-gun nests that had yet to be manned or put into action. Scores of Germans surrendered, some running into the arms of the captors. The marines with the 1st Battalion (of which Kocak's company was a part) were now far ahead of the rest of the regiment and hoping they could clear the forest unimpeded.

That hope died in a burst of German machine-gun fire. The forward line of Kocak's platoon crumpled, and half of the men went down in an instant. The rest pressed on until another round of bullets came from somewhere between the trees and halted them. Kocak, as a newly appointed sergeant, had been briefed on how critical it was to keep his men moving forward in sync with the rolling barrage—one hundred meters every two minutes. If the artillery barrage got ahead of them, they stood to lose twice as many men to German snipers and machine guns. Kocak realized that the only way to advance was to take out the machine-gun nest—and he decided on the spot to go in silently and alone and get it done. He signaled to the other men in his platoon to hit the ground and keep still; then, flat on his stomach, he used his elbows and knees to wriggle snakelike to the German position. The ground was soft and wet from the night's rain, so there was no crunching of leaves to give him away. Luck or shrewdness was with him. He managed to approach the nest from the rear, so the Germans were all looking the other way, the gunner firing sporadically into the underbrush. Kocak got close, so close that when a branch snapped under his weight, the Germans heard it and swiveled in his direction. The machine gun fell silent. All the soldiers in the tight little clearing froze. For a heartbeat the German soldiers stared transfixed at the Slovak marine staring back at them from the forest floor. Then the heartbeat ended and the marine went insane. "I saw him jump up and rush forward with his bayonet like a crazy man," Leonard J. Wilson, a marine private who was present that morning, testified later. "The Boche stopped firing then and started to run but he struck some of them before they could get away. Why he didn't get killed I don't know because the Germans kept up a terrible fire nearly all the time and we couldn't help him with our own fire because the machine guns were too well hidden." Kocak made swift precise use of his bayonet. The first German he came upon he caught in the throat with the tip of the blade and pushed until he heard the choking gurgle of blood. He used the butt of his rifle to crush the skull of the next German to come within range. Then, with his finger on the trigger, he faced down those who remained. The surviving Germans chose to drop their weapons, put up their hands, and surrender.

Kocak had just become a new member of an unofficial but highly regarded marine organization called the Solo Club—membership open exclusively to those who had taken out a German machine-gun position unassisted.

Kocak confirmed his membership a few minutes later when he knocked out another nest nearby. This time the operation proved a little trickier. Kocak lost his helmet in a spray of German bullets and was reduced to using his rifle as a bludgeon when a crowd of Germans closed in on him. He managed to fell three of the enemy, but he was still outnumbered six to one. Kocak owed his life that morning to the timely arrival of a group of guys from his platoon—and to the ineffable power of fear. Word had spread through the German lines that American marines were "inhuman" and "headlong in battle," and this crew had no stomach for any more displays of American insanity. Those who could still walk after Kocak's assault surrendered and were escorted to the rear.

Kocak's buddy and comrade Sergeant Louis Cukela, a native of the Croatian city of Split who had emigrated to Minneapolis at the age of twenty-five, was also inducted into the Solo Club that morning. Cukela had joined the U.S. Army soon after emigrating to the United States in 1913, and he was stationed in the Philippines when war broke out in Europe. Eager to get into action, he managed to obtain an honorable discharge from the army and then enlist in the marines, and he ended up in the same company as Kocak. Stubborn, aggressive, hotheaded, blind to danger, Cukela turned out to be a born marine—and he came into his own that morning in the sodden Retz Forest. When his troops ran into persistent German machine-gun fire, Cukela decided to take matters into his own hands just as Kocak had done. Tapping two men in his platoon to go with him and ignoring the warnings of the others who stayed behind, Cukela crawled toward the offending machine nest. His two comrades were killed in seconds, so Cukela continued on alone, becoming aware as he drew nearer that there were actually three enemy machine guns hidden by brush. When he came in range, he tossed a grenade to one side of the first nest, anticipating that the gunner would swivel in the direction of the explosion. That's exactly what happened, and Cukela used that moment of distraction to rush in and go to work on the gunner with his bayonet and rifle butt. When the other German soldiers fled, Cukela grabbed the hand grenades they had left behind and used them to take out the other two machine-gun nests. The Germans who survived Cukela's assault surrendered gladly.

Cukela, though technically an enemy alien from the Austro-Hungarian Empire, had entered the U.S. military expressly to topple the regime that was oppressing his people. On his enlistment in the marines, Cukela told the recruiting sergeant that his father was in prison in Croatia for his anti-German, anti-Austrian views and that other family members were fugitives fighting a guerrilla war against the Huns. "I want to go and fight those Huns," Cukela declared, figuring correctly that the marines would give him his best shot at action. As it turned out, the marines got more soldier than they bargained for. Though Cukela claimed to be able to speak six European languages, he never quite mastered English and his butchery of the language became something of a marine legend. "Next time I send damn fool I go myself," Cukela once shouted at a hapless subordinate who reported back after screwing up a simple mission. The phrase became a byword in the AEF, repeated endlessly by exasperated squad leaders and inspiring a series of cartoons in the *Stars and Stripes* army newspaper. Supposedly even Pershing was heard to let fly with these words when pushed to the wall by damn fools. Cukela's heroism on July 18 won him a promotion to second lieutenant and two Congressional Medals of Honor—one from the army, the other from the navy.

The marines managed to push through the Retz Forest in the course of that morning, but they ran into trouble as they emerged onto the rolling wheat fields that opened out east of the woods. Orders called for the men to turn south through the fields, but without landmarks or adequate field communications, they became disoriented. American marines and French Moroccan troops wandered into each other's assigned zones and became hopelessly ensnarled. By midday it was hot as blazes and the men were weak from hunger and thirst. German resistance intensified as the day wore on and Allied morale took a hit when a squadron of German aircraft—the recently killed Baron von Richthofen's notorious Flying Circus—arrived and began strafing the infantry below. "The enemy planes shot at or dropped a bomb on everything that moved," according to one marine officer. "Dimples" appeared "in the smooth surfaces of the wheat," wrote another American officer. "In each dimple lay a dead soldier, his uniform purple-stained." American machine-gun units failed to show up in time to do the infantry any good. At the height of this melee, Kocak became separated from his men. He must have strayed into the French zone, for he came upon a troop of twenty-five Senegalese soldiers whose officers had all been killed in action. Though they had no language in common, Kocak took command of the unattached Senegalese and gestured for the men to fall in behind him. The stocky, bronzed Slovak American and the lanky black West Africans formed what was surely the most cosmopolitan platoon in the great war, but it worked. Advancing together and without words, the Africans under Kocak took out two more German machine-gun nests.

The 5th Marines chalked up July 18 as a victory—by nightfall they had pushed the Germans out of what was left of the town of Vierzy, sent long lines of weary German prisoners and seventy-five pieces of German artillery to the rear, captured hundreds of German machine guns, and positioned the division to continue the drive south and east. But as a fighting force they were finished. In the platoon Kocak led, nearly half the men had been killed or wounded. The commander of the 2nd Division, Major General James Harbord, surveying the wreckage of a farm that was being used as a field hospital that night, wrote of "the hundreds of wounded and dead men, infantry, marines, artillery, Moroccans, Germans, and Americans all lying on the ground in the common democracy of suffering and death."

With the 5th Marines decimated by losses, other units of the 1st and 2nd Divisions carried on the attack for the next three days, and at the end of July 21, the Allies had succeeded in their mission of slicing through the Marne Salient at Soissons. But, as at Belleau Wood, the price of victory was nearly ruinous. The 2nd Division as a whole lost some 5,000 men, and in some battalions the casualties were 70 percent; in the course of just a few hours, the 6th Marines suffered some 1,300 casualties—more than half of the 2,450 men who had gone into battle. The 1st Division fared even worse: 1,252 men killed in action, 8,365 total casualties, 50 percent of the infantry felled, 60 percent of the infantry officers lost.

There is no record of newly minted heroes Matej Kocak or Louis Cukela complaining about the strategy that led to such staggering numbers of casualties. Tactics were not the concern of marine sergeants: their job was to carry out and pass along orders. If they went above and beyond those orders in the line of fire, so much the better. For their "extraordinary acts of valor" they'd get a medal, a slap on the back from the captain, a respectful glance from their fellow soldiers, a line or two in a regimental history. Kocak and Cukela, the heroes of

Soissons, were cogs in the great machine of war, just as their fellow Slovak and Croat immigrants were cogs in the machine of industry back in America. They were honored by their officers for their success and daring in killing—and above all for their willingness to risk everything to push the marines a few yards forward through the woods and bleeding fields of central France. The officers who awarded medals—and those who wrote the histories in which Kocak and Cukela figure briefly—recognized those qualities and were lavish in their praise. But there was just a touch of condescension mixed in. Like the Senegalese fighters with their "splendid white teeth" and their "shrill barbaric yapping," these stubborn Slavic warriors fought like animals—something commendable in war so long as the animal is on your side and under your control.

As for what Kocak and Cukela made of their heroism at Soissons, one can only surmise that they were glad to be alive, proud of their nerve, bent on doing the same next time they were called to fight. Since they were buddies, it's likely they congratulated each other on being inducted into the Solo Club. At the very least Kocak and Cukela could now lay claim to the term *battle-hardened*. They had fought as marines and they knew they would fight again as long as the war lasted. Meanwhile, they were grateful to get their first hot meal in three days.

The action south of Soissons that consumed so many thousands of men in the 1st and 2nd Divisions between July 18 and July 22 was but one wedge of a massive French-American advance into the Marne salient. Six other American divisions were deployed in the last weeks of July and the first days of August in what became known as the Aisne-Marne operation—and among them was the 26th "Yankee" Division. When the 26th Division went into battle on the morning of July 18, Private Max Cieminski, Company C, 102nd Infantry, went with them. This gentle, soft-spoken, Kaszubian youngest son got his

first taste of war fighting uphill into heavily fortified terrain under the
blazing sun of mid-July.

It was a bitter taste in every way. Max had been with the 26th Division for just two months at the start of the Marne offensive. Shortly
after arriving in France at the end of April, he was placed in a depot
division, and then, a week later, rotated into the 102nd Infantry to fill
a spot left empty by a soldier who had been shot or shelled at Seicheprey.
They didn't even issue him a new uniform. Max still wore his old
tunic with the collar discs of the 345th infantry (the regiment he had
trained with back at Camp Pike in Arkansas), something the guys of
the 102nd were quick to notice and remark on with a sneer. As Max
found out the hard way, two months in a quiet sector was not enough
to turn a replacement into a comrade. It didn't matter that the 102nd
had a large contingent of immigrants, including a core of seventy-five
Italian machine gunners who had been recruited from New Haven's
colonia and sent to France with gold crucifixes blessed by the city's Italian priests. The men of the 102nd had suffered together through the
humiliation of Seicheprey—and any Johnny-come-lately greenhorn
who hadn't endured what one officer described as the "awful shelling and grenading" or heard the "screams and screeches" of the "half-
drunken Germans" or seen their buddies lying dead "in windrows,"
wasn't part of the club. "Goddamn replacements" one marine private
wrote, were always "outside the pale of genuine approval" until they'd
been through at least one major battle. Goddamn replacements only
reminded survivors of how much they missed their fallen buddies—
and how fresh and young they had been themselves just a few months
back. The fact that Max Cieminski was a crack shot from his years of
hunting in the fields and woods of central Wisconsin was no doubt
lost on the veterans of Company C. On July 18, as the 102nd prepared
to go under fire again, the battle-hardened survivors of Seicheprey
closed ranks against Max and his fellow replacements. Privates and
noncoms wrote off the new guys as yellow sons of bitches. Officers
like Company C's commander, Captain James A. Haggerty, shook their

heads in despair over how "criminally unprepared" the replacements were. In Captain Haggerty's eyes, Private Cieminski was just another raw recruit shuffled in from a depot division. He barely spoke English; the only thing that made him stand out was the missing trigger finger. How the hell did he get into the army anyway? *Criminally unprepared.* Max was certainly not the perpetrator of this crime—but he was about to be punished for it.

Max went into battle for the first time within a few hours of when Kocak and Cukela won their membership in the Solo Club. The 5th Marines and the 102nd Infantry were both elements in the multifaceted push of the Aisne-Marne operation, though the two units were stationed about twenty-five miles apart. To visualize this vast, complex battlefield, imagine the salient that the Germans took during their spring offensives as a tattered pair of boxer shorts hanging off a clothesline: Soissons near where the marines fought was at the far west edge of the waistband; Rheims, the cathedral city the Germans were so keen on taking, was where the clothespin would have clipped the east edge of the waistband to the line; Bouresches, the village where Max and the 102nd Infantry were stationed, was at the bottom of the west leg inseam. On July 18, the day that Kocak and Cukela and the 5th Marines drove east across the waistband near Soissons, Max Cieminski and the men of the 102nd began to move northeast up the west leg from Bouresches at the bottom toward the village of Epieds near the crotch.

Tidy wheat fields filled most of the rolling country around Bouresches, with patches of woods standing like dark islands in the swells of ripening grain. In *Three Soldiers*, Dos Passos makes this countryside of midsummer war look like a painting by Cézanne:

On both sides of the road were fields of ripe wheat, golden under the sun. In the distance were low green hills fading to blue, pale yellow in patches with ripe grain. Here and there a thick clump of trees or a screen of poplars broke the flatness of the long smooth hills. In the hedgerows were blue cornflowers and poppies in all colors from carmine to orange that danced in the wind on their wiry stalks. At the turn in the road they lost the noise of the division and could hear the bees droning in the big dull purple clover-heads and in the gold hearts of the daisies.

It might have reminded Max of the Wisconsin farm country where he had grown up, only the villages were tighter, older, and built of stone and brick, not wood. Unfortunately for his regiment, the golden fields sloped up to the north and east—nothing precipitous, but enough to give whoever commanded the high ground a distinct advantage. And of course the Germans had staked out every ridge and hilltop and tucked machine-gun nests into the most advantageous spots. Beautiful countryside to look at, but hideously exposed if you were moving uphill against those guns. This was the theater where Private Max Cieminski fought his first battle.

The jump-off on the morning of Thursday, July 18, went fast. There would have been a religious service before the men set out—probably the evening before since jump-off generally happened hurriedly and before dawn. Max, a devout Catholic all his life, would have taken communion; if there had been time, he would have confessed his sins to the regiment's Catholic chaplain. It was a comfort to go into battle knowing that you were "ready to meet God," as one Catholic soldier put it. The 102nd Infantry moved north from Bouresches; before the sun was very high, the villages of Torcy, Belleau, and Givry fell to them with little opposition. Not much happened the rest of that day, and the following day Max's company was in reserve while other units moved forward. On Saturday, July 20, Company C advanced to the front with supplies of food and ammunition; there was some German machine-gun fire late in the day, but casualties were light. Company C was ordered to push north the following day, Sunday, July 21. The plan was to cross the open wheat fields and continue into the heavily shelled thickets of Borne Agron and Rochet. Captain Haggerty had the men form two long lines and spread out, so each man was standing a few yards away from the soldier beside him. As they swept through the fields, they passed "a few dead Germans," wrote regimental historian Captain Daniel W. Strickland, "evidently killed a day or two before ... black and bloated like balloons, a great cloud of greenbottle flys [*sic*] buzzing around them as the sun came up." Now that they were actually in combat, the veterans of Company C had softened a little toward the damned replacements. The "green draft-men" were "standing up in good shape," according to Strickland, "steadied and directed by the initiated veterans." The morning was going well. The few Germans they came upon hiding in the woods were eager to surrender—not fight. "They had had enough of the war and were simply quitting." The men of the 102nd took heart from the sheer numbers of Americans advancing through the fields.

Around eleven o'clock, as they approached the main north-south road linking Soissons and Château-Thierry, an enemy machine gun opened fire from the woods east of the road. Company C's Second Lieutenant Orlando G. Hills Jr. ordered his men to form a skirmish line and advance to the edge of the woods. When they reached the machine-gun nest, they found ammunition and equipment in abundance but no Germans—evidently the gunners had been ordered to retreat. The men cheered when they crossed the Soissons–Château-Thierry highway, their assigned objective for the day. Fighting had been "negligible," reported Strickland, "not even enough to satisfy the curiosity of the new men."

The cheers and self-congratulation turned out to be premature. In a pattern that would become horribly familiar, the Germans had fallen back only to regroup at a more tactical position. Epieds and Trugny were a pair of insignificant Picardy farm villages, two dots about eight miles apart just east of the Soissons–Château-Thierry Road. But in a few days, hundreds of American families, including the Cieminskis back in Michigan and Wisconsin, would come to recognize and hate these two French names, because it was here that the Germans took their stand against the 102nd Infantry.

So far this battle had been easy—too easy. By Sunday night, when the men of Company C bivouacked along a farm road and the adjacent Bois de Breteuil, everybody was saying that the fun was over and hell was waiting for them the next day. Word came down that the Germans had a machine gun on every ridgetop. The flash and boom of artillery continued sporadically through the long midsummer twilight. The men dug themselves shallow pits—they called them graves—to afford a little protection from shell fire while they slept.

The guys in those "graves" repeated what they heard from the guy next to them, added a bit to make it sound even worse, and secretly hoped that tomorrow wouldn't be all that bad. But their officers knew exactly how bad it was going to be. "It was an ideal position to defend," wrote Captain Strickland of the terrain ahead of them, "as machine-guns could cover every inch of ground on both slopes and with the numbers employed could have converted the ridge into a veritable fortress." In all there were fifty-two German machine guns along the Trugny-Epieds front; as the Americans were beginning to learn, German machine gunners tended to keep firing until they killed every advancing infantryman or were killed themselves: retreat or capture were not options. Company commanders warned headquarters that there would be a "terrific expense in man-power" if they moved their men into this trap, and they pleaded for a revised battle plan. But according to Strickland, "No argument on the part of the company commanders could effect a change."

The attack began at dawn on July 22 with no advance artillery barrage. The orders from on high, said Strickland, were "to set out in columns of squads and simply march through any position of the enemy at either Trugny or Epieds." The four companies (A, B, C, and D) of the 1st Battalion were to spearhead the advance: Company B to the right, D to the left; Max's company, C, was in the second wave with Company A. As the waves of infantry moved out into the rising mist, German machine guns opened fire immediately and "the ripe heads of wheat spouted like fountains in the air." The advance broke down at once. Frightened by "the withering fire," the replacement troops panicked and dove for cover. Lieutenants and sergeants screamed at the men to get up and move forward—and when that didn't work, the squad leaders threatened to shoot the men who wouldn't fight. "Greenhorns in the rear tried to fire through the ranks ahead and increased the casualties," wrote Strickland. "It developed that morning that the last batch of replacements sent up could not even load a rifle, much less fire it."

Eventually, "by threats and by encouragement and by the assistance of the few veterans," companies A, B, and D of the 1st Battalion took Epieds and forced the Germans back to the high ground behind the village. But as the captains had feared, the "expense in man-power" was ruinous: in less than half an hour, the first wave of infantry had been slashed from five hundred men to fifty. The attack that Company C mounted on Trugny to the south was even more costly. Second Lieutenant Orlando Hills, the leader of Company C's fourth platoon, said that the only way his men could advance was "by crawling hidden in the wheat." The men were carrying their gear in packs and whenever the hump of a pack appeared above the wheat, the Germans let loose a spray of machine-gun fire. When they took off their packs, they were riddled with bullet holes.

By midmorning, Lieutenant Hills' platoon was pinned down by German machine guns outside of Trugny. As the men sweated it out in the wheat and wondered how they were going to penetrate this wall of bullets, the field around them suddenly began to erupt in volcanoes of earth and smoke and fire. The Germans were laying down a massive artillery barrage. Among the soldiers crouching in terror and praying for deliverance from those exploding shells was Max Cieminski.

An officer with Max's regiment carefully noted in his account of the battle that at midmorning "the enemy started throwing shrapnel and high explosives over the wheat." Though the distinction between shrapnel and high-explosive artillery has become blurred with time, it would have been sharp and clear to anyone out in that field. Soldiers in combat quickly learned to tell the difference between the two kinds of shells by how they sounded, how they exploded, and, when they hit you, how they made you suffer and die.

Shrapnel, named for Major-General Henry Shrapnel (1761–1842), the English artillery officer who perfected this type of explosive, technically refers to the small lead musket balls that are packed into the shaft of an artillery shell. Shrapnel shells explode in the air on the downward arc of their trajectory: a timed fuse in the shell's nose activates on firing and ignites a load of gunpowder packed in the shell's base. When the powder blows, hundreds of lead marbles—half an inch to an inch in diameter depending on the size of the shell—are forced out through the shaft of the shell. For those on the ground it's as if the sky is raining bullets from a cloud of acrid smoke. Since shrapnel balls come at you from above rather than sideways like a machine-gun round, their signature wounds are to the head, face, neck, and shoulders.

In the Great War, shrapnel was ideal for killing men out in the open or on the move without cover—like Company C on July 22—but it was useless for soldiers dug into trenches. Hence the introduction of high-explosive shells. Bigger and cruder than shrapnel, the HE shell was essentially a case of steel packed with an explosive (lyddite, TNT, and Amatol were all popular during World War I) that detonated when the shell hit the ground (though HE shells could be time-fused as well). The signal difference between HE and shrapnel was what happened to the shell: unlike a shrapnel casing, which remained intact while spewing out musket balls, the HE shell was designed to shatter into thousands of ragged shards. What killed the most men when HE detonated was not the force of the explosion but the shards of the shell itself. "The earth booms. Heavy fire is falling on us. We crouch into corners. We distinguish shells of every caliber. Each man lays hold of his things and looks again every minute to reassure himself that they are still there. . . . We look at each other in the momentary flashes of light, and with pale faces and pressed lips shake our heads. . . . When a shell lands in the trench we note how the hollow, furious blast is like a blow from the paw of raging beast of prey. Already by morning a few of the recruits are green and vomiting." So Erich Maria Remarque described a barrage of high explosives in his classic *All Quiet on the Western Front*. "Nobody could stand more than three hours of sustained shelling," a British private wrote. "You're hammered after three hours and you're there for the picking when he comes over." Shrapnel and shell-fragment wounds tend to be more ragged, more prone to infection, and slower to close and heal than gunshot wounds; in the filth of the Great War, with antibiotics not yet available, the odds of surviving a hit were low. Between them, HE and shrapnel accounted for nearly 70 percent of the casualties in the war—much higher than the number of wounds inflicted by machine guns.

HE shells—American soldiers nicknamed them "ash cans," "trolley cars," and "Jack Johnsons" after the celebrated black boxer—were used to blow up buildings, collapse trenches, and cut through barbed wire. When a Jack Johnson landed, trench walls caved in, black fountains spewed into the air, the earth cratered. Shells exploding at night set the horizon on fire. But what haunted soldiers' nightmares was that split second after an HE shell disintegrated and a thousand hot splinters of steel flew at them. It was the image of those mangled ax blades hurtling through the air at five thousand feet a second that terrified men hunkered down in trenches—or crouched in fields of summer wheat.

Two of those fiery steel splinters tore into Max Cieminski's body during the advance on Trugny on July 22. The exact circumstances of Max's death will never be known. The only record of his involvement in the battle outside Trugny that morning is the "burial case file" at the National Archives detailing the condition and disposal of his remains, and even this brief file has some contradictory pieces of data. But from the battle reports and the evidence inscribed on what was left of his flesh, a likely scenario of Max's final hours can be reconstructed. It's possible that Private Cieminski died at a different time or at a different spot on the battlefield: all that is certain is that the twenty-six-year-old son and brother and uncle fought with Company C near Trugny and perished in the battle. This is the most plausible version of how it happened.

At nine o'clock on the morning of July 22, Max was one of the men in Second Lieutenant Orlando Hills's platoon pinned down by German machine-gun fire in the wheat field outside of Trugny. Hills's men were spread out in the field and crouched low to avoid being targeted by the German machine gunners, which meant they couldn't see each other when the artillery barrage opened up. The only way they knew if one of their guys was hit was if they heard him scream. Max was lying on the ground with his teeth clenched and his heart pounding when an HE shell exploded a few yards away. The impact of the explosion lifted him into the air and blew him onto his side. He felt something hot tear into his stomach and the left half of his body and, a split second later, his ears roared from the blast. Max had taken two shell fragments: one chunk shattered his left shoulder blade and chopped away his upper left arm; a smaller piece lodged in his stomach. For however long he remained conscious, the pain from torn flesh and severed nerves would have been excruciating—and the thirst unbearable—but it's unlikely Max remained conscious for more than a few moments. With his brain concussed and his lungs damaged by the shock waves from the blast, Max went into severe shock—the color drained from his skin, his pupils dilated, his breathing became ragged, his blood pressure dropped, and his pulse slowed. In time, Max would have contracted an infection from the contaminated material—the cloth of his uniform, dirt, bits of animal manure, the shell fragments themselves—that entered his arm and shoulder wounds. In time, this infection would likely have killed him. But the volume of blood pour-ing out of the axillary artery in his left shoulder and the puncture wound to the stomach cut the time left to Max to minutes or even seconds. His temperature plunged, his pulse diminished to the vanishing point, his face took on the waxy pallor of death.

Max lay unconscious in the wheat while his blood and his life drained away.

Did a buddy see Max crumple, hear him cry out in agony, and come to his aid? Or was he too weak to call for help? Did a medic crawl through the wheat to try to bandage his wounds? Did one of the regiment's chaplains appear by his side to administer the last rites before Max's spirit left his flesh? Did anyone even realize that Private Cieminski had been hit when Lieutenant Hills gave the order to retreat? Impossible to answer any of these questions and futile to speculate.

When there was a break in the shelling, the survivors of Company C began crawling single-file through the wheat and took shelter in the ravine near the village of Trugny. The relentless machine-gun fire made maneuvering next to impossible. The wounded were carried or dragged off the field; but the dead, or those who appeared dead, were left behind—not worth risking life to move them or bury them. Max was one of the men of Company C left behind in the wheat field.

For the remainder of July 22 and all that night, Max's corpse lay in the wheat, uncovered and untended and swarmed by flies like the German corpses he had marched past the day before. "We could not get back to help our wounded," wrote an infantryman named Gordon A. Needham who fought in that battle in the same division as Max. "Their cries ceased after about 48 hours." At some point in the course of that long, hot, pitiless summer day, a German bullet entered the base of Max's skull—probably from a random spray of machine-gun fire, though it's possible a German gunner took pity on the severely wounded man, or decided to dispatch him rather than take him prisoner.

The next day, Tuesday, July 23, the fields around Trugny were again swept by fighting as elements of the 101st Infantry tried to advance through the position where the 102nd had been stopped the previous day. The 101st made little progress, but a French division positioned to the right of the Americans broke through and finally cleared the Germans out of the woods behind Trugny. By afternoon the wheat had grown quiet enough for the Americans to start collecting their dead.

Company C of the 102nd Infantry, depleted by the severe losses they suffered the day before, had done no fighting that day. The survivors had regrouped in the comparative safety of the Bois de Breteuil, though Germans continued to shell them sporadically. But late in the day, when word came that the Germans had retreated, a chaplain led a detail of men out of the woods and back into the field where Max and the other dead from his platoon lay. The chaplain and a burial party went from body to body. The men examined each corpse carefully, reading the dog tags (each soldier wore two), cutting open pockets to search for scraps of letters or a Bible, unbuttoning the tunics to look for a chain bearing a crucifix or a Jewish star—anything to indicate what sort of prayer was appropriate. The bodies of the dead, after surrendering their last shreds of identity, were wrapped in army blankets or shelter halves—or nothing—and carried to the spot that the chaplain designated as the field cemetery. Was the chaplain able to ascertain that Max was Catholic? He almost certainly wore a crucifix blessed by the priest in Bessemer; perhaps he had a rosary or a holy medal as well. The name on the dog tag, though Kaszubian, would have been taken for Polish. So it's likely that the chaplain said the Catholic prayer for the dead before the burial party placed Max's body into a four-foot-deep hole and covered him with earth.

In manus tuas, Domine, commendo spiritum meum. Domine Iesu Christe, suscipe spiritum meum. Sancta Maria, ora pro me.

Into your hands, O Lord, I commend my spirit. Lord Jesus Christ, receive my spirit. Holy Mary, pray for me.

Inclina, comine, aurem tuam ad preces nostras, quibus misericordiam tuam supplices deprecamur: ut animam famuli tui, Max Cieminski,

quam de hoc saeculo migrare iussisti, in pacis ac lucis regione constituas,
et Sanctorum tuorum uibeas esse consortem. Per Christum Dominum
nostrum. Amen.

O Lord, hear our prayers, in which we humbly ask of your mercy
that you would give your servant, Max Cieminski, whom you have
commanded to leave this world, a place in the land of light and peace,
and bid that he be made a companion of your saints. Through Christ our
Lord. Amen.

The chaplain may have closed with the prayer for the dead from
the *Catholic Prayer Book for the Army and Navy*:

Out of the depths have I cried unto Thee, O Lord: Lord, hear
my voice.

They marked the grave with a simple wooden cross and fastened
one of Max's dog tags to the wood; the other tag remained with the
body. There was the inevitable army form—a "grave location blank"—
which was to be filled out at the grave site and then forwarded to the
chief of Graves Registration Service. This was the army's procedure
for keeping track of the bodies of men buried on battlefields. Judging
by the spelling and the handwriting, Max's grave location blank was
written out by a soldier without much education.

> *Place of death: Trugny*
> *Cause of death: Schrapnel in stomack*
> *Date of burial: July 23, 1918*
> *Grave number: Not known*
> *Nearest relative: Unknown*

The soldier who filled in this form was mistaken when he wrote
that the nearest relative was unknown. In due time, those relatives

would be found and notified, and an eight-year-old girl in Besse-mer, Michigan, would learn that she was never to see her favorite uncle again.

There was no mention of Private Cieminski in any of the regimental histories or unpublished typescript accounts of the fighting around Trugny. His name does not appear in the company rosters published in Captain Daniel W. Strickland's 404-page volume *Connecticut Fights*, probably because he was a replacement soldier. At three o'clock on the morning of July 25, two days after the explosion of the HE shell that ripped off Max's left arm and shattered his shoulder, another HE shell landed on the troops of the 102nd Infantry while they were camped at the edge of a woods. Major George R. Rau, the thin, handsome, melancholy-looking leader of the regiment's 1st Battalion, the battal-ion to which Company C belonged, died instantly in the blast. Strick-land's moving account of the death of Major Rau will have to stand in for the official recognition that Max Cieminski never received:

> *Major Rau, possessed by that intuition which time and again warned of death, had spoken about it to several officers, among them Captain Bissell, then in command of the 2nd Battalion. He had stated that a peculiar sense of loneliness and despair had settled down upon him that morning and that he had been unable to shake it off. He said he supposed that his time had come and that he must simply meet it. It was just after the terrific bombardment of gas and shrapnel of a few moments later, that the group of officers on checking up after hurriedly taking what shelter was to be had, found that Rau was missing. A shell had made a direct hit and blown him to pieces. The fragments of his remains were gathered together and wrapped in his blanket by the same officers that had served with him for years past and left for the burial detail to care for. Thus ended the life of a genuinely humane and sym-pathetic yet capable and efficient leader, a man whose heart sorrowed as much over a dead or wounded enemy as it did over one of his own command.*

On July 25, just hours after the shell exploded on Major Rau, the Yankee Division was removed from the line and the 42nd Division—the so-called Rainbow Division composed of National Guard units from all over the country—took its place. By American standards, the men of the Rainbow Division were battle-hardened veterans, having been among the first American troops to arrive in France back in November 1917; but they had never been in action as fierce as the fighting they encountered at the Ourcq River in the last days of July. The Rainbow Division's Fighting Irish 69th—an infantry regiment with a core of Irish immigrant kids from the toughest neighborhoods in New York City—lost nearly half its men in the push to cross the Ourcq River (the River O'Rourke, the Irish boys called it). Among the Fighting Irish who died in that battle was the well-known poet Joyce Kilmer, shot through the head while scouting out German machine guns.

"They marched in wearied silence until they came to the slopes around Mercy Farm," wrote the regiment's beloved Catholic chaplain, Father Francis Duffy, of the survivors' withdrawal. "Then from end to end of the line came the sound of dry, suppressed sobs. They were marching among the bodies of their unburied dead. In the stress of battle there had been but little time to think of them—all minds had been turned on victory. But the men who lay there were dearer to them than kindred, dearer than life."

The Aisne-Marne offensive raged on for another week after Kilmer died and finally ended on August 6, when the Germans completed their withdrawal from the Marne salient and regrouped at a new defensive line along the Aisne and Vesle rivers. On paper, it was billed as an Allied success—and it was much touted as the first bold display of U.S. military prowess. But in the three weeks of fighting the American Expeditionary Forces had suffered some thirty thousand casualties. General Pershing's insistence on open warfare was proving to be just as costly in human life as the agonizing trench warfare that had come before.

So many who fought that summer in the farm fields of France remarked on the birdsong. For minutes or hours or days on end, the air reverberated with the boom and roar of artillery or the swarming rush of machine-gun fire, a "queer *zeep-zeep*, like insects fleeing to the rear," as one soldier described it. But when the guns fell silent at last, the song of birds resumed—perhaps it had never stopped?—as if nothing had happened, as if just another summer day was coming to an end, as if everything under the sun was as lovely and sweet as it had ever been.

CHAPTER 10

THE JEWS AND THE WOPS AND THE DIRTY IRISH COPS

The immigrant soldiers learned to fight like Americans that summer—but when they weren't fighting they learned to fool around like Americans. Guys were thrown together who wouldn't have given each other the time of day in civilian life, and once they got over the shock, a lot of them became buddies—first because they had to in order to survive and then because they had survived. They were hungry together, tired together, scared and lonely together; together they blew through their money in French estaminets and fell in love with French girls. War is hell, but war is also marching, smoking, foraging for food, scrounging for a place to sleep, getting drunk whenever you could, washing yourself and your clothes, keeping boredom and fear at bay during the stretches between battles. War is a dialect of dirty slang and catcalls and foul curses and muttered complaints about the rules and regulations and the damned officers who enforce them. For the immigrant soldiers, this sudden total immersion in American military vernacular was as strange as anything else. Back home, *they* were the foreigners, but when a bunch of them sat down together in a noisy smoky bar full of Aussies and Brits and French and Belgians, the dagos and yids and hunkies were all Yanks no matter how thick their accents.

They ate and swore and drank and bitched together, but most of all they sang together. It was a war in which men were forever raising their voices in song. American popular music was sweeping the country in those days, and all the catchiest jazz melodies and Broadway show tunes and toe-tapping little ditties inevitably swept the army. "Good Morning Mr. Zip-Zip-Zip," "Goodbye Broadway, Hello France," "How Ya Gonna Keep 'Em Down on the Farm (After They've Seen Paree)," "K-K-K-Katy," "Mademoiselle from Armentières," "There's a Long, Long Trail A-Winding," and "Over There" resounded loudly wherever American soldiers went. Songs were often the first thing the guys from a company had in common, the first thing they did in unison, the rhythm by which they learned to fall in step together. The foreign-born soldiers learned to speak English by learning to sing it. One private from a heavily Italian company of the 107th Infantry bragged that his comrades "sang the loudest, shouted the lustiest, and played the most." New York City's 77th "Melting Pot" Division marched and drilled to a song written by one Corporal John Mullin with the fitting refrain:

Oh the army, the army, the democratic army,
All the Jews and Wops, the Dutch and Irish cops
They're all in the army now.

Jewish recruits, who had been policed on the Lower East Side, liked to change the refrain to "the Jews and the wops and the dirty Irish cops."

One of the great movie moments of the war occurred that summer in France when the 77th Division was called in to relieve the Rainbow Division, and the Irish lads of the Fighting 69th and the East Side Jews and wops passed each other marching in opposite directions under a full moon in the Baccarat Sector. "Yesterday was New

York 'Old Home Day' on the roads of Lorraine," Father Duffy wrote in his diary. "There were songs of New York, friendly greetings and badinage, sometimes good humored, sometimes with a sting in it." Ethnic jokes and slurs flew back and forth. Veterans and rookies razzed each other as only New Yorkers can razz. "We're going up to finish the job that you fellows couldn't do." "Look out for the Heinies or you'll be eating sauerkraut in a prison camp before the month is out!" "Well, thank God we didn't have to get drunk to join the army." "Anyone there from Greenwich Village?" As Father Duffy reported: "One young fellow in the 77th kept calling for his brother who was with us. Finally he found him and the two lads ran at each other awkwardly and just punched each other and swore for lack of other words until officers ordered them into ranks, and they parted perhaps not to meet again." Inevitably, the two endless columns began to sing as they tramped past each other:

> East side, West side, all around the town
> The kids sang "Ring-around-rosie," "London Bridge is falling down"
> Boys and girls together, me and Mamie O'Rourke,
> We tripped the light fantastic on the sidewalks of New York.

Major General John O'Ryan, commander of the 27th Division (and son of an Irish immigrant), thought it would do his men good to have some musical entertainment; so during a break in the action in Flanders in mid-July, he ordered them to revive the play *You Know Me, Al* and the musical vaudeville routines that had been such hits in boot camp back in South Carolina. "It is strange to see this organization performing at a point where a 'Jack Johnson' or some similar bird might land at any minute and spread performers, stage and audience all over the landscape," one soldier wrote home. But the guys were lucky and the show went on without being shelled. General O'Ryan remarked that wide-eyed Flemish kids packed the outdoor theater every night even though they didn't understand a word of it: "At first they gazed in almost dumb wonderment at the nonsense of the clowns or the dancing of the 'girls' and listened intently to the divisional jazz band and the popular Broadway songs. ... Before the division had been at Oudezeele a month one could hear the boys of the village whistling such songs as 'Wait Till the Cows Come Home,' 'My Heart Belongs to the USA,' 'Mother Machree,' and other melodies. In two months they were singing the songs in English." Such was the irresistible power, then as now, of American popular entertainment.

Among the men tapping their toes, whistling at the "girls," and picking up Broadway slang in the outdoor theater at Oudezeele were Mike Valente and Epifanio Affatato, both with the 27th Division's 107th Infantry. By mid-July Valente and Affatato had been in France for nearly two months, most of it spent marching from one training area to another. So far, pretty much the only thing they had to show for their time "over there" was sore feet. The division was on the road so much that the guys started calling it O'Ryan's Traveling Circus. "The worst part of the war is getting to it," one soldier remarked. "I feel as if I have walked all over this country," Private First Class Angelo Mustico wrote his Sicilian immigrant parents in Newburgh, New York. "We have passed through a lot of citys [sic] and towns and most of them have been bombed from the air and very few people live in them." The few inhabited villages they did encounter struck the American soldiers as unbelievably primitive. French peasants in 1918 still looked and acted like something out of a Breughel painting, throwing slops in the street, wearing clogs in the muddy lanes, piling manure right outside their houses, cooking their meals in cast-iron pots hanging in open fireplaces, living cheek by jowl with their farm animals. "The farmers are busy haying," Private Matt E. Palo, a second-generation Finnish American farmer, wrote to his sister back in Fruitdale, South Dakota. "It seems kind of funny to see hay hauled in two-wheeled wagons and one horse hitched to it." The stone houses packed together, the daily trek to fetch water, the smell of animals mingling with the smoke of cook fires, the fields going blond under the sun of midsummer: to immigrants from the south of Italy like Epifanio Affatato, it was exactly what he had grown up with in Calabria and never wanted to experience again. Even the language the French peasants spoke had a familiar ring—certainly more familiar than English had been when he landed at Ellis Island seven years earlier.

Epifanio hadn't seen much action yet, but he had been close enough to the front to learn to recognize the smell of poison gas and the sound of an incoming shell. He had been in war long enough to take pity on the German prisoners of war with their dirty unshaven faces and haunted eyes. Always kindhearted, Epifanio shared his cigarettes with the prisoners. One time, when he came upon a bunch of German POWs suffering in the cold, he struck up a barter—his warm hat for an officer's silver belt buckle stamped with the motto GOTT MIT UNS—God with us. It was more humane than scavenging German corpses for souvenirs as a lot of guys did. Accustomed all his life to prizing any useful object, however humble, Epifanio was shocked at how careless American soldiers were with their government-issued equipment. Once he picked up an ax that another soldier had tossed—he didn't really need an ax, but he wore it around his belt anyway because he couldn't bear to see it go to waste. The day came when the thing would save his life. He was squatting down to relieve himself when German artillery shells began landing in the vicinity followed by sprays of machine-gun fire. An enemy bullet caught Epifanio literally with his pants down—but by some miracle the bullet hit the ax blade and ricocheted off. Epifanio got up and ran with his trousers around his ankles—a stunt his buddies never let him forget. But for that bit of luck, he might have ended up like the guy next to him on patrol one time whose comrade fell writhing to the ground, turned purple, and the spot. Maybe there was something about his name Aff it was a story to share with "under a spell" or "bewitched." war and had children. his children, assuming he sur

In the course of that season in France, Epifanio learned to shed the last of his peasant superstitions. Back in Calabria, *le fate*, the spirits or fairies embedded in his last name, were believed to haunt wild waste places and caves and linger around anything that belonged to the dead. Epifanio was convinced that to take the blanket off a dead man would rouse his spirit to come after him and seize his soul. But one night when he was half-dead of cold and wet on the battlefield, he made up his mind to put this fear to rest. Brutal as it sounds, he crawled out to where a comrade lay dead and got the man's blanket and wrapped himself in it. Crazy as it sounds, he waited in terror for the spirit to come. When *la fata* failed to show, Epifanio slept, and woke up the next day a changed man. "That's that," he said to himself. "No spirits." One more door shut on his Calabrian childhood; one more thing he had in common with the guys huddled next to him.

Leonardo Costantino and Peter Thompson, both serving with infantry units of the 91st "Wild West" Division, spent that July drilling and marching through France and wondering if they'd ever see any action. In his log Costantino kept a running account in his imperfect but vivid English of what mattered to him most—how much ground they covered, how little they had to eat, how bored and hot and frustrated they felt as the real war went on without them.

July 25 . . r one long and hard Hike we arrived in Milliers, France. Am very tired . many boys had to fall out and taken easy for it sure was one terrible Had no supper to-night.
July 28 We a . ing on the floor and Ho my back! Eats are very bad, some of us are . nothing at all. This evening we had one spoon of tomatos and cup o . No Bread so only I took it was coffee, no sugar or milk. Went to . this after noon, Water was very cool so no one of us had regular . e first one we had since we left Camp Mirrett [on July 11]

July 29 Last night we heard the big guns play their game from far away. Are now 40 miles from the front. Every body is hungry and here you can't buy a thing expect wine. We have to start training once more and expect to keep it up until we go to the front.

Aug. 9 Granade practice, all day along. Private Aiello [another Italian American recruit and a buddy of Costantino] threw one and nearly Kill Lt. Fitts. He was nerves [nervous?]. Real action.

Aug. 16 Our Company have turn in one suit of everything. Are getting ready to move for the front. All of us are anxious to get a crack at the Huns.

Aug. 17 There is a french girl live next to our Billett and the owner of the place. She often come to me and ask when we will move.

Aug. 20 I do not talk any-more with this French girl she make me sick try to get your last penny out you. Am too wise at these French people.

"The French people are certainly rubbers [robbers]," Leonardo scoffed in one entry. In the course of that summer, he came to appreciate how much more he had in common with his buddies in the Wild West Division than with these French "rubbers" with their lousy wine and shifty women. It was a case of us against them, and Leonardo was now clearly one of us. On a day when the sun was shining and the weather not too hot, he wrote yearningly of "the old dear California" he had left behind when he was drafted. Not dear old Italy.

Meanwhile, Peter Thompson was picking up jokes and slang from the regiment's hard-bitten Montana cowboys. One of the guys remarked that marching under the summer sun with a steel helmet on his head was like "wearing the kitchen stove with the fire drawing well and the coffee boiling over." A homesick cowboy christened Peter's regiment the Powder River Gang after the legendary western watercourse that supposedly ran a mile wide and an inch deep. "Powder River to Berlin" got chalked on the side of one of the hated *40 hommes 8 chevaux* French boxcars. "Wild and woolly, full of fleas. Fright and frolic, as we please. Powder River—let 'er buck! Wade across it you're in luck" became their marching ditty. But as of late August, the Powder River Gang had done more buckin' than fightin'. Peter and his cowboy comrades were starting to wonder if they would ever see any action.

E pifanio Affatato had to work a double shift in assimilating to army life that summer, because his division had been folded in with the British army. So in a sense, he had to learn how to be both an American and a British soldier. In fact, the 27th had been serving with the British army ever since arriving in France back in May, though initially it was only for training purposes. The idea was that men would get up to speed under British tutelage, and then, when the time came to move to the front, they would fight on their own as Americans. Pershing had been adamant about this arrangement all along: American forces, at whatever cost, must preserve their integrity and independence. It was fine for Doughboys to train with the Allies; but the idea of using American troops as cannon fodder in decimated foreign units or putting American divisions at the disposal of foreign generals was utterly repugnant to the commander. But now, as the summer wore on and the Allies contemplated new campaigns against the weakened German army, Pershing came under increasing pressure from General Haig and General Foch to supply American manpower to their depleted British and French armies. Knuckling under, Pershing reluctantly agreed to temporarily lend both the 27th and the 30th divisions to the British. The American units would not be broken up and dispersed into the British forces but would retain their identity as American fighting forces. But as far as the men in the field were concerned, the Brits would be running the show, giving the orders, calling the shots. And so, in a sense, all the guys in the 27th and 30th were in the same boat as Affatato and Valente and the thousands of other immigrants: all of them were foreign soldiers that summer and fall.

"The line isn't so bad," the seasoned Tommies were always telling the green Americans of the 27th Division, "unless you're at Ypres. Don't go near Ypres." But that's exactly where Affatato and Valente ended up with the 107th Infantry. On August 12, Affatato in Company C and Valente in Company D were posted to a section of the front line at Dickebusch Lake a couple of miles south of Ypres. It was a flat fetid landscape of muddy fields, poisoned water, trench lines zigzagging in perplexing mazes. The terrain was so monotonous that one of the intelligence officers advised the men that the best way to avoid getting lost and wandering into a German trench was to memorize the names on the crosses marking the graves of Allied soldiers buried on the battlefield. The dominant feature on this prairie of Flanders mud was the ridgelike knob of Mount Kemmel where the forces of Crown Prince Rupprecht of Bavaria were dug in. "It was a bare, bleak earthen mound," wrote one soldier; "not a tree nor a bush could be observed on top. It had been a target for artillery for years, and was battered and banged as no other in France." Another soldier wrote his family, "You cannot conceive of the suffering, the wreck and ruin for miles and miles. Love seems to have vanished from the face of the earth." Company C got its first real taste of war amid this desolation— a taste that lasted eleven days.

Company D was placed in reserve with the East Kent regiment called the Buffs before being rotated to the front with a Shropshire unit. "The English regiments were mixed with us to teach us the fine points of the game," wrote the Company D historian. The men, however, took a dimmer view of this business of being mixed in. American soldiers grumbled that they were taller, fitter, and better fighters than their English counterparts; they hated the "iron rations" on which British soldiers were expected to subsist during battles (canned pork loaf or corned beef; biscuits, sugar, and tea; a couple of cubes of meat extract; some cheese if they were lucky); they looked down on the Brits for having failed to get the job done. The many Irish immigrants in the 27th were especially vociferous about having to answer to the

English, whom they considered more their enemy than the Germans. They'd be damned if they were going to fight for the king who had his heel on the throat of their people. The same resentments seethed in the Rainbow Division's Fighting Irish 69th. When the regiment was issued British uniforms because there weren't enough American khakis to go around, the Irish soldiers refused to put them on because the buttons were stamped with the insignia of the English crown. "These buttons were a hated symbol of their former oppressors," one member of the regiment wrote. "Some hotheads . . . built a fire in the main street of the village and started to burn the British issue, and there was great excitement as one after another joined in." A full-scale anti-English riot was averted only when Father Duffy appeared on the scene to calm the lads down. "We have our racial feelings," the priest declared, "but these do not affect our loyalty to the United States."

The prevailing attitude among American soldiers, Irish and otherwise, was that if you were going to be slapped together with foreigners, it was better to serve with Aussies, who at least had a bit of pluck and dash. But it was best to stay with your fellow Americans, even if they had funny last names and strange accents. Some of the men of the 107th worried that while they were stuck in Flanders under the thumb of the Cockneys, the real show was going on elsewhere. That attitude would change shortly.

"We were given our line training at Ypres," wrote Private Albert G. Ingalls in the Company C history, "that storm center of the world's greatest passion, where fields of golden grain, mingled with myriad poppies, mask uncounted graves of fallen Australians, Canadians, French, and English." Both companies C and D suffered losses at the "storm center" before being relieved by other units of the 27th on August 24. "I can't describe the awfulness of the war," one soldier wrote home the following day. "As most of the men say, it is not war, it is slaughter." "Don't go near Ypres" proved to be a prophetic warning. The 27th took nearly thirteen hundred casualties before being rotated out on September 3. In the end, the advance that the combined

British-American forces made in the sector between August 18 and September 4 could be measured in yards.

Officially, Private Antonio Pierro was serving in an artillery unit—Battery E, 320th Field Artillery, 82nd Division—but in reality, Tony was spending a lot more time with horses than with high explosives that summer in France. After all those months of training and drilling at Camp Gordon, here he was wandering around the French countryside like a peasant with a horse and wagon. At least Tony knew something about animals. Back in Forenza, the family kept a mule to haul firewood and water and baskets of laundry to and from the stream, so Tony had grown up shoving and shouting at a stubborn beast of burden. One more thing he didn't miss about Italy. Horse, mule—it made no difference to Tony. If that's what the army wanted him to do, that's what he did.

Anyway, he was partly to blame for getting put on equine detail. One day when he was ordered to clean up the cigarette butts littering the encampment, Tony made the mistake of raising an objection. "I don't smoke," he told the sergeant, "and I'd rather clean up after the horses." Whether the sergeant laughed—or more likely cursed—in his face, Tony never said. But the officer took him at his word and gave him a horse to clean up after—and a lieutenant too into the bargain. So now he was a lieutenant's orderly and in charge of the lieutenant's horse. Somehow that led to Tony getting a horse of his own—by far the best thing that had happened to him since being drafted. Whatever pride he felt in his mount, however, was quickly soured by a terrible accident. One day a soldier in his unit walked around behind Tony's horse and slapped its rump as a practical joke. The horse spooked and kicked the guy in the head and killed him. Not the kind of death you wanted to be remembered for.

You'd think that nationality wouldn't make much difference to a horse, but Tony found out otherwise. On the trip across the Atlantic,

German torpedo sank the ship carrying the regiment's horses, so when the 320th Field Artillery got to France, they had to procure new steeds from the French—French steeds, naturally. When Tony landed the job of looking after the two horses, he discovered that they didn't respond to commands in English. Curses didn't work either. Nor did Italian. These were French horses and they obeyed only if you addressed them in their native tongue. So Tony picked up some French. Actually it wasn't all that hard for him. Different accent from Italian but a lot of similar words.

Tony's regiment was on the move a good deal in the first days of August—from a training area at La Courtine well south of the front to a series of villages in the Marne district and then, on August 12, over to Château-Thierry. On August 13, the 320th Field Artillery was ordered east to the Meurthe et Moselle district between Verdun and Strasbourg, and it was here in the dense woods shrouding the Moselle River north of Nancy that Tony got his first real taste of combat. German planes flew over every night to launch bombing raids on the American line. During the day the guys worked on their unit's 75-millimeter guns—the standard smaller field artillery used by the AEF—along with nine 90-millimeter guns on loan from the French. One of Tony's jobs was to carry crates of high-explosive shells to the gun emplacements, and he did this job so often that he wore out the front of his tunic—particularly irksome for a guy who took such pride in his appearance. August 21 was a day worth remembering: at 5:00 P.M., the 320th FA fired its first shot of the war at a German airplane that was harassing an American observation balloon. A week later, word came down that a major offensive was in the offing and the regiment snapped into feverish activity. Tony and his fellow privates were put to work digging gun pits, dugouts, and shelters; in order to conceal the impending attack from the Germans, all the work had to be done at night or behind camouflage nets. Nonetheless, the Germans got wind of the fact that something was up behind the American lines, nd in the first days of September they became "vigorous with their harassing and searching fire," in the words of the regiment's historian. On the night of September 5–6, the 320th was severely gassed—and they responded immediately and lethally. Battery E, Tony's battery, was moved forward and ordered to fire four hundred shells of high explosives and gas into the enemy trenches "at a rapid rate." (By now, after the initial fumbling attempts to deliver poison gas from spray canisters, both sides had figured out how to load gas into artillery shells and fire them from the big guns along with rounds of HE and shrapnel.) "This fire had the desired effect," the anonymous regimental historian noted tersely.

Whatever was afoot behind the American lines in that first week of September, Tony could only guess at from the stepped-up pace of work. "The nearer we get to the trenches the less we think about the war in its larger aspects," one soldier wrote home. "Our own particular job fills our time and thoughts." A common complaint of the enlisted men was that they understood even less about the progress of the war than the folks back home reading newspapers. They were the army's muscle—and they found out about a new push only when their sergeant told them to pack up and head out.

Rosh Hashanah—the Jewish New Year—started at sundown on Friday, September 6, that year. It was a sign of how culturally sensitive the army had become that Jewish soldiers were granted furloughs (whether they wanted them or not) from noon on September 6 to the morning of September 9.

Samuel Goldberg, serving with the 12th Cavalry Regiment in Hachita, New Mexico, was called in by his lieutenant colonel the day before the Jewish New Year and given a three-day pass to go to El Paso, Texas, about 150 miles east. "What for?" Private Goldberg asked. "You're Jewish, aren't you?" the officer shot back. "Goldberg is a very Jewish name. I just got a telegram instructing me to give furloughs to all Jewish soldiers for the weekend. You're going to El Paso. Get y

than hauling junk around the shtetls of the Pale with a horse and cart. Even the replacement stigma didn't set him back for long. Some of the other Jews complained about anti-Semitism, but it was never an issue for Meyer. Whatever they threw at him, he'd know how to handle it.

On August 19, Company H moved into the hamlet of Liffol-le-Grand, just east of the provincial border between Champagne-Ardenne and Lorraine, seventy-one kilometers southwest of Nancy, and here they remained for the next twelve days. Meyer and the other 279 replacements with the second battalion were given a round of intensive training. Another 125 replacements arrived on August 25. The lieutenant colonel let it be known that before moving on he wanted every man in the battalion to have experience firing a Chauchat, the French-made light machine gun that the AEF was acquiring as fast as it could. So in the last days of August, Meyer became adept—or at least competent—at shooting. On September 1, the men of Company H were piled into trucks and driven a hundred kilometers north to Remebercourt, arriving at their new post at three in the morning on September 2. Four days later was Rosh Hashanah, and that night, starting at ten thirty, Meyer's company was ordered to pull out and head north toward Verdun. It was a miserable, fouled-up exodus. The rolling kitchens and water carts weren't ready, so the men departed without provisions. Drivers got lost or separated and had trouble turning around. According to the company historian, at some point in the night, "The truck drivers were . . . lined up and told that the whole trouble was because they were afraid. One replied that who wouldn't be afraid if one was going to the front for the first time." Finally, at six the next morning, they arrived at a woods east of Fort d'Houdainville—practically at the outskirts of Verdun.

Instead of praying and meditating to the blast of the ram's horn that ushers in the Jewish New Year, Meyer had spent Rosh Hashanah huddled in a truck on his way to the front. Private Epstein knew—everyone in Company H from the scared truck drivers to Captain Bertram M. Cosgrove knew—that it was only a matter of days before the big offensive was launched and they'd be ordered into the fire of German machine guns.

On September 9, three days after Rosh Hashanah, General Pershing's headquarters issued an order allowing alien soldiers serving in Europe to file for immediate citizenship without returning to the United States. The May 9, 1918, amendment to the naturalization act had already expedited the citizenship process for aliens serving in the armed forces by waiving the residency requirement and the need to take out first papers—but now General Order 151 made it even easier and quicker. Under GO 151, immigrant soldiers needed to file but a single form (in duplicate) comprising "the Petition for Naturalization, the Affidavit of Witnesses and the Oath of Allegiance," and they were in. Pershing further ordered that "all commanding officers will lend their active aid to aliens under their command, and will see that the naturalization papers are supplied, understood, properly executed and promptly forwarded." The accompanying Bulletin No. 68 written by Pershing's chief of staff more or less rolled out the red, white, and blue carpet: "After any alien has made out his 'Petition' and taken the oath of allegiance before the proper army officer, he may for all purposes be regarded as an American citizen. There is nothing further for him to do so long as he remains in the army."

Pershing's headquarters was immediately flooded with petitio
Fridolin Blanchard, Frank Zappala, Vincenzo di Clemente, Olaf P
Jake Goldberg, Lars J. Rindal, Michael Christian Schouten, C
Lombardo, Aristotile Zoccoli, and Michael Francis Walsh
the names sent in with the first flush. Private Nich
pended this note to his petition: "I don't exactly r
but over five years ago I took out my first na
New York City. I was born in Locovis, Maced
then Turkey, now Greece, by Greek pare
April 17, 1912 as a Turkish citizen and

I had to denounce the Sultan of Turkey." Private Carras was General Pershing's butler.

The timing of GO 151 was revealing. With the AEF about to mount the first all-American offensive, Pershing had flung open the doors to citizenship as wide as they would go. A symbolic, perhaps even a cynical gesture—but immigrant soldiers nonetheless streamed through by the hundreds.

The big offensive—the drive to clear out the St. Mihiel salient south of Verdun that the Germans had held since 1914—was the first action in the Great War entirely planned and led by American forces, and it had all the signs of a young, untried power flexing its new muscles. Pershing laid on ten American and three French divisions, with four more American divisions in reserve—more than a quarter of a million men in all—against a force of perhaps twenty-three thousand war-weary Germans. The logistics were mind-boggling, especially since the American command was frantic to keep the attack a secret. In this they were at least partly successful. Though the Germans realized the Americans were up to something in the sector, they were taken by surprise when American artillery opened fire at 1:00 A.M. on the morning of September 12.

The All-American Division was on tap for the initial assault, and Tony ·rro was there on the front line feeding his battery's 75-millimeter ...vith unbelievable quantities of shells. While the opening barrage ...ɔgress, the 320th Field Artillery was under orders to fire at the ...and two-thirds rounds per gun per minute: every thirty- ...rtillery piece would rear back and vomit forth a high- ...led with either gunpowder or poison gas. Tony's tunic ...·readbare patches that night.

...Thompson, Leonardo Costantino, and Tom- ...·antrymen at St. Mihiel, but this first brush ...·mething of a letdown. Meyer's unit had been assigned to support the 59th Infantry—and "the duties consisted of watchful waiting and constant preparedness to act," in the words of the company historian. The 91st Division was held in reserve, so Peter Thompson and Leonardo Costantino sat out the battle perched on a hill in the rain watching smoke rise from the villages below. Leonardo jotted down in his log it how it looked and felt to him:

> *Sept. 11 Are now in reserve for the Saint Michel Drive. Went through many towns but cannot see much. Pitched a tent at day light. Spent a very cool night. SOME life into woods.*
>
> *Sept. 12 Again we moved last night. Can't stay in one spot very long for fear of the enemy. Are only 9 miles from the line. Rain all the time out here.*
>
> *Sept. 13 The Americans took 20,000 prisoners in two days. Eats are O.K. Sleep pretty warm in my tent with my friend Emil C. Berhard.*

That pretty much said it all as far as the Wild West Division was concerned.

The first great American battle of the war was over in a matter of hours. "One of the darkest nights ever seen," wrote the Catholic chaplain serving with Meyer Epstein's regiment on Friday, September 13, "and yet biggest trick pulled off in war to date." It was not, however, entirely clear who had been tricked. The Germans had been intending to abandon the salient anyway, and the American push merely served to hasten the inevitable. As Pershing's mighty army advanced, the Germans staged a swift, devastating scorched-earth retreat, burning villages, fouling water supplies, and exploding bridges as they went. In the two-day campaign the AEF managed to seize some 13,521 German prisoners (Leonardo's figure of 20,000 was a touch inflated), 466 guns, and a large smoking triangle of mud that the Germans had occupied for four terrible years. Allied casualties totaled around 7,000. There was mu[...] weeping and embracing from liberated French villagers. Germa[...] diers were also falling into the arms of the advancing Do[...]

Many Americans noticed the grins on the faces of the tattered, hungry German prisoners as they surrendered. "Can I go back now to Sharon, Pennsylvania?" pleaded one Austrian prisoner who had been working in the States when the war broke out. The American command, desperate to show the AEF in the best possible light, boasted of the lightning-fast decisive victory—one general crowed over the "vim, dash and courage" of the army's "splendid young manhood"—but our partners were less impressed. The French sniffed that the battle was no more than a German retreat disguised as an American advance. French military observers faulted the American army for its lack of foresight, poor map-reading skills, wretched traffic management, and mediocre liaison. And some officers wondered why, with the Germans on the run, Pershing had failed to press his advantage and continue east to the strategic city of Metz on the Moselle River.

With both the Marne and the St. Mihiel salients pinched off, the war had in a sense come full circle: after the German retreat was completed on September 13, the Western Front ran through more or less in the same territory as in the autumn of 1914 when the conflict had first become stuck in trench warfare. The signal difference was that the German army had steadily and massively beefed up its defensive position in the interval. In the winter of 1916–1917, the Germans commandeered gangs of prisoners of war and French civilians to fortify a hundred-mile stretch of the front from east of Arras to Soissons with an elaborate network of triple trenches, barbed wire, concrete bunkers, observation posts, and machine-gun emplacements. In their pride, the German command named sections of the line after the Norse gods and heroes of Wagner's *Ring* operas—from north to south, Wotan Stellung, Siegfried Stellung, Alberich Stellung, Brunhilde Stellung, and Kriemhilde Stellung, with the Siegfried section reputed to be the strongest. The Allies dubbed the entire vicious ─ein of wire and concrete the Hindenburg Line and spoke of it with ─udging awe one accords an enemy's most lethal ingenuity.

─ermans insisted that the Hindenburg Line was an impregnable barrier between the Fatherland and the Allied armies, and so far the Allies had had no reason to write this off as an idle boast. But the word "impregnable" did not figure in the battle plan that British, French, and American generals had come up with to end—and win—the war. Even as American divisions were streaming north and east through the St. Mihiel salient, orders were being sent out to move hundreds of thousands of men into new positions and, in a matter of days, to throw them at the wire, the concrete, and the bullets of the Hindenburg Line.

Epifanio Affatato and Mike Valente had the bad luck to be among those thrown first and hardest at the most impregnable stretch of the line—but no combat soldier, even those who had yet to see a lick of action, got off easily in the autumn of 1918.

CHAPTER 11

THE ARC OF FIRE

Vogliamo sperare che si finesse presto questa guerra," Tommaso Ottaviano wrote to his mother on September 3: We want to hope this war will be over soon. On that day, Tommaso's regiment, the 310th Infantry of the 78th Division, was on its way to the front at St. Mihiel. So far, in three months in France, the 310th had been spared serious fighting—and Tommaso was praying that the war would end before their luck turned. He'd seen enough suffering and devastation in his weeks of marching through the French countryside to know what war was. He didn't need to be a hero. For him, it was enough to be an American soldier and to come home to his mother and brothers and sister in one piece.

The St. Mihiel offensive came and went, and the 310th emerged with barely a scratch—wet, stunned, tired, but intact. Since the division had been held in reserve, Tommaso and his comrades were more or less spectators at the big show. But this first all-American push marked a fundamental change in the division's status, and the guys on ground knew it. The time of training and marching had ended; forth the 78th would be at the front, constantly within range of shells, bullets, and bombs. "We have entered the war zone,"

Tommaso wrote his mother on September 15. "We are in territory won back the day before yesterday by our good Americans who have put themselves near the enemy and we aren't leaving them. Perhaps you all know this as well as I from reading it in the newspapers."

Two streams of pride run together in this letter. Tommaso in the *zona di guerra* has not forgotten his Italian heritage—his letters home, all of them written in Italian, were full of concerns and messages for relatives, both those who remained in the home village of Ciorlano (some of them now at the front with the Italian army) and those in North Providence. But much as he loved his family, Tommaso's immediate loyalty was with *i nostri bravi americani*—literally, "our good Americans" though *bravi*, the plural of *bravo*, also means brave, honest, decent, skilled. These *bravi americani* were the guys he was serving with. *Our guys.* And Tommaso was now one of them. "We're more or less always happy," he boasted in another letter, "and without dark thoughts because we're surrounded by so many buddies—and with one *fellow* who says one thing and another something else, the time *seems like* nothing." On September 29, on the eve of being sent into *combat*, real combat this time, Tommaso wrote again to assure his mother that all was well: "War Zone, 29 September, 1918. Dearest Mother, Here I am with my news, which by the grace of God is still as good as ever because I'm healthy, and I don't need anything else, though I'd sure like to hear from all of you. . . . I have some good buddies and we like each other a lot—my best pal also comes from Providence—he's been my blanket-mate for the last months—we're always together—his name is Antonio Giosi—he actually comes from Glinville (?) RI and the two of us pass the time talking about Providence etc. . . . Be happy about your son because in a short time it will all be finished and we will all return to our homes."

What Tommaso did not write about from the *zona di guerra* was the anxiety that preyed on his mind, that preyed on all of their minds on the eve of battle. How would he act under fire? When the moment came and he was ordered to stride out into that patch woods or through that cloud of poison gas or into the hail of bullets spitting from that concrete bunker, would he be brave and good? Or would he run for his life? Would he be able to look a German soldier in the eye and then run him through with his bayonet or lob his grenade into a trench full of cowering enemy soldiers? If he killed one German, would he crave killing more?

By this stage in the war they'd all heard the horror stories that were going around—stories about soldiers shot by their officers for fleeing the battlefield; stories about guys who gleefully competed to rack up enemy kills, as if war was a blood sport and German carcasses were trophies; stories of Germans who hid in their trenches and shot you in the back as you ran past. They grimly passed on the rumors about the latest kind of gas that smelled like garlic and made burning blisters erupt on your skin. For untried recruits like Tommaso, the sheer animal terror of going into this for the first time overwhelmed all else. Facing it side by side, with any luck surviving it—this is what bound together *i nostri bravi americani* on the eve of battle, and that bond was far more powerful than whatever differences of language or background divided them. "You love your comrade so in war," a French soldier had written in the Middle Ages. "A great sweet feeling of loyalty and pity fills your heart on seeing your friend so valiantly exposing his body to execute and accomplish the command of our Creator. And then you prepare to go and die or live with him." It was no different in 1918.

"*Preghiamo per la fine di questa guerra,*" Tommaso wrote his mother: We pray for the end of this war. Not "I pray," we pray, in English, Italian, Yiddish, German, Polish. We pray as soldiers, as common men being sucked together into a maelstrom. We pray that the war will end before it drowns us all.

No one was more surprised than the men themselves how the differences between them mattered anymore.

had gone into the army expecting Jews to be cowards, Italians to be thieves, Germans to be spies, Poles to be lazy, Irish to be disloyal—but even in the thick of combat they stopped to acknowledge how wrong they had been. General John F. O'Ryan, commander of the 27th Division, wrote of how admirably the German Americans in his division fought, despite the suspicions of "otherwise fair-minded Americans" that they would be traitors or saboteurs: "The contribution to the American military record made by those of German blood should be most gratifying to them. As a matter of fact, if all those with German blood had been released from the American army, the record of that army would have been very different from what it actually was. In battle none were more intelligent and dependable than this class, and in our own division the roster of the dead who gave up their lives at the extreme front fighting the German menace tells the story most convincingly." German soldiers and officers captured in battle revealed how stunned they were at the ferocity with which German Americans fought against them. These men spoke German; they had friends and family in Germany; some of them had relatives in the German army—and yet they were fighting and dying for America. How could they have transferred their loyalty so quickly and so completely?

Charles Minder, a native-born New York soldier of German heritage, spoke for many when he wrote his mother from the front that he viewed immigrants in an entirely new light after living and fighting with men of so many different backgrounds. Minder began by describing how fond he had become of a Jewish fellow in his platoon named Selig, whose hair he had just cut:

> [Selig is] a nice chap and one of the smaller men of the company, who have stood up so well enduring all the hardships of hiking and carrying the pack as well as the bigger fellows. We have about every nationality you can think of in my company. There sure is some mixture, and I think it is about the finest thing in the world for anyone, who like myself, l always suffered with race-prejudice, to be mixed up in an outfit li

The last six months of my life in the army, living and suffering with these fellows, has done more for me to get rid of race-prejudice than anything else could have done. I am beginning to get a better realization of the different things I have read and heard on the "Brotherhood of Men" and "Love your neighbor as you love yourself." I am beginning to see more and more how we are all one common herd, ruled by another class that has more power than we have. We are told to go and fight and kill and we must go, even though it is against our highest sense of right to kill another. They seem to even mock God, the Father of us all, when they make His children slaughter one another.

The slaughter would continue. In fact, for Americans of every nationality fighting in France at the end of September 1918, the slaughter was about to begin on a scale none of them had ever seen or could imagine.

General Pershing and his staff had been working under the assumption that the ever-growing American army would position the Allies to deliver the coup de grâce in the spring of 1919; but by summer's end, the Allied command had begun to press for victory—by which they meant unconditional German surrender—before the end of 1918. The German spring offensives had come to nothing. The enemy army, after three months of setbacks, was clearly hurting—the men in the trenches were weak, haggard, malnourished, either too young or too old for effective soldiers. Word trickled back from the front of the miserable condition of the German prisoners of war. "They are exhausted and emaciated," Tommaso Ottaviano wrote his mother of the prisoners his unit took. "I *have* seen the bread they ɛt and it made a big impression because it consists of absolutely noth-ʃut ground-up grass." Scenting blood, Generals Foch and Haig ʃo move in for the kill that autumn.

As the plan took shape to unleash a massive Allied offensive along the entire Western Front at the end of September, one major issue remained unresolved: how best to use the Americans. Foch and Haig had been pressing Pershing for months to lend them American divisions to use as they saw fit, but Pershing kept adamantly refusing. He had already given the 27th and 30th divisions to the British, and the African American 93rd Division to the French. The remainder of the American forces would fight under American command. Furious cables and memos flew back and forth between Paris, London, and Washington; the generals went toe-to-toe at a series of heated meetings. Finally Foch acceded to Pershing's demand for a separate American force, but at a price: the Americans could be as independent as they pleased so long as they undertook to fight two battles, practically back to back, in September. A scant three weeks after St. Mihiel, Pershing was to shift his army sixty miles to the north and launch a second campaign through the heavily fortified woods and hills of the Meuse-Argonne sector west of Verdun. If the Americans wanted to fight on their own, that was how they must do it.

Pershing was well aware that the two-battle plan was a logistical nightmare, perhaps a suicidal nightmare given how green and undertrained most of his army remained. But as one staffer at headquarters put it, "It was now or never. . . . The time had now come when America was to show its teeth and . . . the great drive must go through." And so, in the second half of September, even as the St. Mihiel operation was still in progress, preparations for the great drive—the final drive, as the Allied generals devoutly hoped—went forward. The Meuse-Argonne operation was but one facet of a Grand Offensive that Foch envisioned as an arc of fire converging on the Belgian city of Liège: at the far western edge of the quadrant, a combined Belgian and British force would push east from Ypres in Flanders; at the center, the American 27th and 30th divisions (under the command of the Australian Corps) would spearhead an attack against the Hindenburg Line the French Fourth Army (with the help of units from the African American 93rd Division) mounted a drive in the Champagne region; and simultaneously the main American force at the arc's southernmost point would slash north through the forests of the Argonne until they reached the strategic transport hub of Sedan. As the Allied vectors converged, the Germans would fall back until they reached the point of collapse and surrender. That was Foch's master plan. The Grand Offensive was the biggest operation of the war—and the Meuse-Argonne sector proved to be the most monumental battle that the American army ever mounted. As historian Edward G. Lengel writes in his recent history, "No single battle in American military history, before or since, even approaches the Meuse-Argonne in size and cost, and it was without question the country's most critical military contribution to the Allied cause in the First World War." By the end, some 1.2 million American soldiers fought in the Meuse-Argonne—22 American infantry divisions supported by 840 planes, 324 tanks, and 2,400 pieces of artillery laying down more than 4 million shells.

On the maps, Foch's Grand Offensive looked elegant and decisive, but American soldiers on the ground had little or no idea of where they were going or what their objectives were. Montfaucon, Blanc Mont, the St. Quentin Canal—sites that would soon absorb unbelievable volumes of American blood—were as yet meaningless placeholders. On the eve of battle, guys were writing home that they understood less about the progress of the war than their families did from reading American newspapers. All they knew for sure was that the big American show was going to happen someplace around Verdun, where there had been a hell of a fight earlier in the war, and that if they weren't near Verdun, they were out of it. Not out of the fighting—God knows there was enough fighting to go around all along the front—but out of the American part of the show. When the time came, every Ameri-
no matter where he'd been born, wanted to be fighting in the
an sector.

On Thursday, September 26, Meyer Epstein, the Russian-Jewish plumber from the Lower East Side, finally got the chance to throw his strong back into combat. The Ivy Division, with which Meyer had been serving for six weeks now, was one of nine American divisions on the Meuse-Argonne jump-off line. Meyer was there when the American artillery laid down the opening barrage in the hours before dawn: for three hours, he hunched his shoulders against the cold and gritted his teeth against the noise while the sky erupted in deafening flame. "There was nothing to do but sit and listen to those shells," wrote one soldier, "and wish and wish and wish, with the strongest wishing in you, that every one of the shells, big and little, meant the less Germans on the advance."

Meyer's unit, Company H, 58th Infantry, was at the eastern third of the American line in a terrain of steep rolling hills, bombed villages, and patches of bare dead woods. The landscape's dominant feature was the 1,122-foot rise of Montfaucon, the site of an ancient ruined abbey that the Germans had held and fortified since 1914. It was from a bunker on this hilltop that Crown Prince Wilhelm of Germany observed the carnage at Verdun in 1916. As the American guns pounded away on the morning of September 26, the men of Company H stood around little campfires they had lit in the trenches. Shivering in the damp air they ate the sandwiches that had been distributed earlier. They had been instructed to leave their packs behind so they could move swiftly and nimbly through the shell craters and tangles of German wire. No bedding, no overcoats, no food aside from the reserve rations they carried on their person. At 5:30 A.M. the first line of men fanned out into the smoke and fog.

Meyer and the couple of hundred men of Company H grabbed their rifles, crouched down with their fingers on the triggers, and out in lines of half-platoons. The land rose in front of them, land of old trenches, shell holes, and destroyed farms, without a single living tree or structure to obstruct their view of what lay ahead. To their right was the ominously named Dead Man's Hill, scene of unspeakable carnage earlier in the war. Far to the left was the German fortress of Montfaucon. And straight ahead the hamlets of Bethincourt, Malancourt, and Cuisy. Meyer's first minutes of combat were eerily quiet. There was little German resistance. The infantry proceeded through open ground that had been fought over and repeatedly shelled in the first months of the war. From the regimental history: "The soil had been churned and rechurned by the bursting shells. Not an inch is undisturbed. The battered trenches tell of the minutely organized system of defense. Here and there a skull or a shoe containing the bones of a human foot, bore mute evidence of the strife. It is said that over one million men lost their lives here."

A shell-torn meadow rolled down to the ruins of Bethincourt. Houses and shops had long since been reduced to stumps, charred timber, piles of rock and debris. The streets had dissolved into weedy paths between the ruins. Vines and wildflowers flourished in the crevices of the rubble. There was no other sign of life. The road out of Bethincourt climbed steeply to the northwest. It was almost possible to imagine how beautiful this countryside once was—the ridges rising and falling like big slow swells of green, dark woods shadowing the watercourses, the delicate Gothic arches and spires of the Montfaucon abbey capturing the day's first and last rays of sun. The men had been warned that whenever they cleared the top of a ridge they would be targets of German machine gunners and artillery. But so far, they had encountered almost nothing. The road cleared the spine of a long ridge, and below them to the west, nestled in the folds of the hills, they saw the village of Cuisy, or what was left of it. Here, in a network of [t]renches that the Germans had abandoned that morning, Company [...]g in for the night. The regimental historian described the scene: [...]ating a meal of our reserve rations, we crawled in the holes [...], to get what rest was possible. During the night, a cold penetrating rain began. We couldn't build any fires. We had no over-coats, and had left our blanket rolls in the Bois de Sivry. Some found overcoats and blankets left by the Boche, and rolled up in those. The army slicker is as good as nothing, as far as heat goes, and as to turning water—well, we who wore them in the Argonne, knew what they were worth. The moisture from one's body collects on the inside of the coat, and as soon as the wind strikes you, you are cold for the rest of the day." So Meyer Epstein passed his first night of active combat.

The great bombardiment start at 11:30 P.M.," Leonardo Costantino wrote in his log on September 26. "At 2:30 A.M. we arrived at Hill 612. The Artillery ceased at 5:15 A.M. The machine gun Battalion put up a Barrage. It last 15 minutes. The 2nd Battalion of all the 91st went 'over the top' at 5:30. Saw the first American wonded at 6:10 A.M."

The 91st "Wild West" Division was stationed southwest of Montfaucon, three divisions west down the line from the Ivy Division. The landscape here was more heavily wooded than around Bethincourt and Cuisy, though most of the trees had been bombed and shelled to smithereens. Otherwise the setup was the same: First the predawn barrage (the 11:30 P.M. "bombardiment" Leonardo mentioned was the preliminary firing of the long-range artillery directed at the enemy's rear), then the infantry attack. Leonardo Costantino (364th Infantry) and Peter Thompson (362nd Infantry) went over the top with the rest of the Wild West boys to shouts of "Powder River, Let 'er Buck!" The Irish miner from Montana and the Italian poolroom proprietor from San Diego were finally seeing some action. Resistance, at first, was light. The Germans had left behind some snipers, but the fog was so thick in the smoldering woods that the men had no idea where the sniper fire was coming from. Every now and then they'd stop shoot blindly into the air. Occasionally a German body wou' from the remains of a tree.

The division hit heavier fire from both snipers and machine guns later in the morning, after the fog had burned off, but they still managed to make good progress. Too good, in the case of Peter's regiment. By evening, the 362nd's first battalion had broken through the German line and closed in on the village of Epinonville about three miles due east of Montfaucon. Peter with the 2nd Battalion moved up to the village before dawn. The 362nd was now so far out in front that its flanks were exposed and it was at risk of being bombarded by American artillery to the rear. No ambulances, medical supplies, or food was getting through to them. Men with leg wounds—or legs blown off—crawled to the rear in the hope of finding a field hospital.

Speed was of the essence in Pershing's plan in the Meuse-Argonne. Everything hung on overrunning the sector before the Germans had time to bring in reinforcements. A swift, decisive breakthrough would be the ultimate justification of Pershing's strategy of open warfare—but time and space were both extremely tight, and the commander knew it. "There was no elbow room, we had to drive straight through," chief of staff Brigadier General Hugh A. Drum wrote after the war. Gains were scheduled by the hour: by the afternoon of day one, less than ten hours after jump-off, the Americans were to have swallowed Montfaucon and formed a great line radiating east and west of the highpoint; by the following morning they were slated to be on the other side of the bristling Kriemhilde Stellung, ten miles inside enemy-held territory. From there it was a sprint to victory. That was the plan. But as General Philippe Pétain (Chief of state of France's Nazi-aligned Vichy government during the next war) helpfully pointed out, if anything broke down, the Americans could spend the winter stuck at the base of Montfaucon.

"This was the most ideal defensive terrain I have ever seen or read wrote Drum of the Meuse-Argonne—and there was no section as ideally defended as Montfaucon. In a countryside of rolling hills and ridges, Montfaucon rose high and clear like an enormous solitary haystack. The Germans had honeycombed the heights with bunkers, pillboxes, carefully placed machine guns and artillery pieces; they had wrapped the slopes in barbed wire and stuffed their hideouts with ammunition and supplies. It was the stronghold from hell, and for inexplicable reasons the task of storming it fell to the totally inexperienced 79th, a division cobbled together from immigrant Pennsylvania coal miners, draftees from Maryland and Washington, D.C., and replacement troops from Ohio and West Virginia. "It was predestined slaughter to put a raw division in such a place," wrote one of the officers.

At nine o'clock on the morning of September 26, two companies from the 79th's "Baltimore's Own" 313th Regiment hit a wall of German machine-gun fire south of Montfaucon and halted the advance. While the Ivy and Wild West divisions were surging ahead on either side, the 79th remained pinned down for five critical hours. By the time its infantry broke free, daylight was fading and the companies were severely depleted. The wounded suffered at overcrowded forward aid stations with no hope of being evacuated since ambulances were stuck in an epic traffic jam at the rear. The dead lay where they fell or were hastily buried on the battlefield. Pershing was unmoved. Fearing that the entire Argonne operation was in jeopardy, he ordered the 79th to continue fighting until they had taken the heights that night. The men did their best, but as of 9:00 P.M., the summit was still over a mile away and ablaze with German fire. The Americans fell back, dragged their wounded to cover, and dug in for a long, miserable night.

At the end of day one of the great American drive, the most critical point on the line was still in German hands. Pershing had failed to break through and the Germans, now fully apprised of the Americans' intentions, were madly regrouping and reinforcing their tions. American soldiers had fought in the open, as their commander insisted they must, and they died in the open, just as British and French and German soldiers had died in the previous four years whenever they left their trenches.

The fighting would go on.

CHAPTER 12

BREAKING THE LINE

Usually they just gave you orders—pack your gear, line up, get moving—without saying anything about what or why. So Epifanio knew that something big was in the works when his company commander, First Lieutenant Ralph Polk Buell, called out the 151 men of Company C on September 26 and gave it to them straight. Buell, a New York lawyer, a graduate of Princeton, and a descendant of President James K. Polk, had a long horsey face with amused-looking eyes and an extra long chin with a cleft at the bottom. Despite his classy background, the lieutenant was a wiry, hearty, down-to-earth fellow, a veteran of the Spanish-American War, the kind of no-nonsense foulmouthed commander soldiers liked and followed willingly—not stuck-up like some of the blue-blood officers in the 107th. As Buell stood lecturing the company with his chin jutted out and his eyes narrowed, Epifanio could see that he was all fired up about something. Maybe the men had heard rumors about the big offensive rolling out that day in th[e] Meuse-Argonne; maybe they were sorry to be stuck here in Pic[ardy] a hundred miles from the show; maybe they were worried th[at those] guys would return to the states covered in glory, while the[y] obscure and forgotten—borrowed soldiers fighting for the Brits and the Aussies. Well, the lieutenant had some news for them. The moment of glory for the 27th Division was at hand. Tomorrow at dawn, their comrades in the 105th and 106th regiments would move out in the first phase of the battle to break the Hindenburg Line. There was a hush as the words "Hindenburg Line" sank in. Those two other regiments had been assigned the job of breaching the German defenses and establishing a nice straight jump-off line. That would happen tomorrow. The following day, September 28, the 105th and 106th would consolidate their position. The day after that, September 29, was their turn. Two companies of the 107th infantry had been chosen to lead the main assault on the Hindenburg Line itself—Company B was one, and their company was the other. It was the hardest task imposed upon any unit in this great attack—those were the exact words of General O'Ryan—but Buell assured his boys they were more than up to the challenge. They had three days to steel themselves. Then they'd have a chance to prove what they were worth.

The lieutenant threw in a lot of names of the landmarks they would be storming—Sart Farm, Guillemont Farm, the Knoll, the St. Quentin Canal—but to Epifanio it was just a bunch of words. His regiment had been stationed in this sector of Picardy for a couple of days now, and whatever names they used for these villages and fields, it was the most godforsaken place he'd ever seen. Low rolling hills lacerated with trench lines; farmland scorched and shelled to a pitted brown desert; no crops in the fields; no trees or shrubs surviving in the hedgerows. The only thing that sprouted on the bare hills were the crosses that marked the graves of the Australians who had tried and failed to take the outworks of the Hindenburg Line in the days preceding. And in the fields surrounding their bivouac, a bounteous harvest of German corpses was strewn all over the place. Epifanio had ... rror of rifling dead bodies for souvenirs, though it was popular ... h a lot of guys—which was why the officers had issued stern ... p their hands off. Word had come down that Germans were mining their own dead in order to blow up souvenir hunters. So the bodies around them rotted undisturbed while the men of Company C counted down to zero hour.

Epifanio and his buddies were still asleep on the morning of the 27th, when the barrage opened to the north at the stroke of 5:30. Nine brigades of British artillery and ninety-six American-manned machine guns hammered away simultaneously for all they were worth. Even a couple of miles to the rear, the roar was loud enough to jolt the men instantly awake. For a moment they lay blinking in their blankets, each locked in his own gray thoughts. Then one of them began to bellow his support for the fellows of the 105th and 106th going over the top, and soon they all joined in, throwing their heads back and shouting at the top of their lungs.

By noon, while the battle raged to their north, Company C had broken camp and was on the march toward the front. Their packs were nearly empty since they had received strict orders the day before to leave behind everything but their emergency rations, a few toilet articles, and light slickers. When their time came, the officers wanted them to go fast and light. But even unburdened, it was a nightmare march. As they moved forward, they passed the remains of the 105th and 106th moving by them in the opposite direction—"an apparently interminable train of ambulances moving slowly to the rear," wrote regimental historian Gerald R. Jacobson, "with men who only a few hours before had been as these men now advancing toward the front—strong, virile, hopeful."

The men of Company C prided themselves on how loudly they sang, but nobody was singing now. "Doughboys when they're going up in the lines they look straight in front of them and they swaller every third step and they don't say nothing," wrote novelist James M. Cain, who was in the Argonne fighting as a private first class with th... beat-up 79th Division. No need to talk, since they were all th... the same thing and they knew it: in forty-eight hours th... haggard parade will be ... or worse. Hadn't General O...

the first battalion was going to carry out "the hardest task imposed upon any unit in this great attack"?

Company C marched through the evening, pausing at dusk for a hasty roadside dinner followed by a prayer service led by Father Peter Hoey. Numbers were always large on the eve of battle. It was some comfort to Epifanio to have a Catholic priest in the regiment, though of course it would have been better if he had been Italian.

When the service was over, the officers got the men in line again and they set off into the night. The shell fire was continous from both sides of no-man's-land. Just east of the quarries and the rubble that had once been the hamlet of St. Emile the march ground to a halt. No one knew what the hell was going on, but they were grateful for a break. Lieutenant Buell told the men to fall out while he went forward to join the rest of the company commanders for a confab at the head of the column. Epifanio threw down his pack, propped up against it, and grabbed what shut-eye he could get. An hour later he stirred awake as word of what was going on rippled through the ranks. Evidently the 106th infantry had not completed their mission: not only had they failed to establish a jump-off line for the 107th, but pockets of their men were stranded or lost on the hills around the villages of Vendhuile and Bony. This balls-up would play out with dire consequences the following day. The assault on the Hindenburg Line, like all American attacks, was to be accompanied by a rolling artillery barrage—but with the 106th scattered all over the landscape, that barrage plan was snarled in an impossible dilemma. Allied commanders had the choice of either bombing the stragglers of the 106th in the first wave of the barrage or sending out units of the 106th with no artillery cover. To bomb their own exhausted, defenseless men in order to safeguard the advance of fresh men or spare the exhausted and leave the fresh ones vulnerable to enemy fire: those were their options. It was one of those terrible dilemmas that haunt commanders for the rest of their lives. O'Ryan, agonizing on the eve of battle, concluded that "no matter how logical it might be in a tactical sense," exposing the 106th to friendly fire "would be repulsive to the mass of officers and men of the division, and destructive of morale." Morale over logic—no American commander would have chosen otherwise. But in the words of British Captain G. H. F. Nichols, it "sealed the fate of the 27th American Division."

The 107th's march continued through the night. Around dawn on September 28, Company C halted and took shelter in a trench. Before they collapsed, the men were keen on getting a glimpse of what kind of terrain they'd be up against—so, despite the nearly cease-less German artillery fire, they cautiously peeked over the top of the trench. What they saw was a distinct disappointment: aside from the unusually thick tangles of German barbed wire, rising eight feet high in spots, the dread Hindenburg Line looked like everything else at the front—wasted, cratered, strewn with the remains of whatever had been alive.

That night final word came down that Company C would be going over the top the next morning. That was when the men learned that they would be starting the assault without artillery cover. There would be a gap of twelve hundred yards between their trench and the line of the rolling barrage, and the two forward companies had been allotted four minutes to cross the gap and catch up with the barrage. "After four minutes that curtain of exploding shells would start for-ward, jumping a hundred yards every four minutes," said one of the sergeants, "and we all knew it was safer for the doughboy to be as close as possible to the barrage clearing the way for him." The men were tense with the preparations for combat. Epifanio collected his extra rounds of ammunition, water, and enough rations to carry him through the following day. He was also issued a spade to carry forward and use to dig in at their next forward position. The officers told the the spades were for digging, but Germans used them instead of nets in hand-to-hand combat. They sharpened the edges in the skulls of the enemy with them. That was why so many French, English, and American helmets turned up with parallel gash marks on the top.

"You can't imagine what emotions I then experienced," French Jesuit priest and stretcher bearer Pierre Teilhard de Chardin wrote from a trench on the eve of combat, "nor what one feels conveyed in the clasp of a man who shakes one's hand while the shells are going across, almost like a solid vault overhead. . . . There's no doubt about it: the only man who experiences right in the innermost depths of his being the weight and grandeur of war, is the man who goes over the top with bayonet and grenade. In that moment, of course, the infantryman leaving his trench for the attack is a man apart, a man who has lived a minute of his life of which other men have simply no conception at all."

"They were quiet now for the most part," the historian of the 107th wrote of the final hours. "They were gathered in little groups in the trenches, saying little, thinking much, and smoking cigarettes." At four o'clock in the morning of September 29, the men of Company C, each one burdened with 220 rounds of ammunition and a pair of Mills grenades, were assembled along the tape line that intelligence officers had laid on the battlefield the night before to indicate their jump-off point. It was almost like the white stripe marking the bounds of a ball field.

N o one ever charges into battle for God and country," writes war correspondent Chris Hedges. Nor does anyone charge into battle bent on heroism—or if they do, they rarely come out again. Chance of course plays a part in the making of war heroes: a moment presents itself and you seize it. But it's the seizing that is impressive and mysterious. Why does one man spring forth while others keep their heads down? It just happened. I wasn't thinking. Time stood still. Something like I was watching myself from the outside. That is what

decorated soldiers say afterward. They don't pin it on God or country
or patriotism—at least not when the memory is fresh. In the heat of
battle, God is far away and country shrinks to the vanishing point.
Country is the buddy in the shell hole next to you; the guy out in
no-man's-land moaning for water; the lieutenant with his leg blown
off; your own precious skin and the beating heart inside it that you
want to preserve more than anything in the world.

Epifanio Affatato, son of Calabria, Brooklyn machinist, disap-
pointed seeker after the golden pavement of America, had seen only
a few days of combat since arriving in France that spring. Line training with the British near Ypres, skirmishes in Flanders, dodging shells
fired from Mount Kemmel—until the morning of September 29, the
violence of his young life had been like so many rocks smashing glass.
What awaited him now was violence of a different order of magnitude.
The fighting in the next few hours would be among the worst in the
worst war the world had ever known, and Epifanio's unit had been
assigned the most exposed and perilous spot on the field of battle.
"Our company has been honored above all other companies in being
awarded the danger post," a machine gunner with the 107th wrote to
a friend back home on the eve of battle. "Our division has had a rare
honor laid upon it, because we are the best, and we are going in to
sacrifice it all if necessary. A big game can be played but once." General
O'Ryan himself admitted that none of his officers truly "believed at
the time that any single battalion of troops could fight its way" through
this terrain "under the conditions as they existed on the morning of
September 29th." Nonetheless, O'Ryan insisted, "it was essential" that
the attempt be made. And so it was.

Six months earlier, Epifanio Affatato had been an immigrant civilian, living with his father, doing his job, studying English, healthy,
happy to be alive and in America. He was not by nature or by choice
a soldier. And yet, within minutes of jumping off on the morning
September 29, he distinguished himself from the other three
men in the battalion. Epifanio wasn't even an American

that morning, while performing the impossible task of trying to break the Hindenburg Line, he became an American hero.

The assault began in the cold quiet before dawn. It was a Sunday, doubly holy because it was Michaelmas, the feast of St. Michael the Archangel, the heavenly warrior who had rallied the good angels when Satan rebelled against the rule of God. "O God, who dost establish the ministry of angels and men in a wonderful order," the faithful would pray later that day in their churches, "graciously grant that Thy holy angels, who ever serve Thee in heaven, may also protect our lives on earth." The soldiers on the line prayed much the same, only in rougher language. Epifanio stood in the dark, five yards away from the next man, and waited. Some of them trembled, from cold or fear or both. Companies C and B had been stationed at the farthest left verge of a front that stretched for miles paralleling the Hindenburg Line. The soldiers didn't yet have enough battle experience to realize how vulnerable this spot was. But the officers, anticipating trouble, had brought up Company D in support. Among the men shivering there in the cold that Michaelmas morning was Company D's Private Michael Valente, named for the warrior archangel. He too would come face-to-face with a part of himself he had never known in the hours ahead.

A few minutes before zero hour the light came up enough so that a cigarette could be lit without attracting German fire. Word got passed from ear to ear that it was okay to have a last smoke. Mist hung over the ruined land. The men could just make out the ground they'd soon be overrunning—a smooth bare rise that was like it might have been Knoll. But for the shell craters and barbed western New York State. hillside pasture in Ohio, Pennsylvania, or were hidden on the ty knew how many machine-gun nests rise in their orders amid the ruins of Guillemont Farm to its right Canal, he hill and the farm, continue past the St. Quentin

ABOVE: Immigrants disembarking in the United States around 1900. Most of the 23 million people who emigrated between 1880 and the 1920s crossed the ocean in the overcrowded steerage of large ocean liners like this one.

LEFT: An Italian family on board the Immigration Service ferry that transported newly arrived immigrants from New Jersey steamship piers to the processing center on Ellis Island.

A family with eight children—a daughter and seven sons—arriving at Ellis Island during the peak immigration years before the outbreak of war in Europe in 1914.

In 1912, when this photo was taken, three-quarters of New York City's Jews lived on the Lower East Side. "America was noise," one Jewish immigrant wrote of the neighborhood, the city's most densely populated.

LEFT: In 1910, thousands of immigrants slept and ate like this Italian family, in windowless, unventilated tenement rooms on New York's Lower East Side.

BELOW: In 1911, when this photo was taken, children earned 50 cents a day for ten-hour days in the coal mines. These breaker boys, many of them from eastern and southern European immigrant families, picked slate from coal in chutes at the Pennsylvania Coal Company's Ewen Breaker in South Pittston.

A bonanza wheat farm in North Dakota around the turn of the last century. Andrew Christofferson and thousands of Scandinavian and Slavic immigrants like him provided seasonal labor on these vast spreads, some comprising nearly 100 square miles.

American marines at Veracruz, Mexico, during the American occupation that began in April 1914. A contingent of marines, among them Private Matej Kocak, was stationed at Veracruz when war broke out in Europe that summer.

LEFT: Andrew Christofferson worked on a fishing boat off the west coast of Norway before emigrating to the Great Plains in 1911. His devout Christian faith sustained him through the vicissitudes of homesteading in Montana and the hardships of service overseas in the "Wildcat" Division's infantry.

ABOVE: Italian-born Michael Valente enlisted in the New York National Guard in 1916, three years after emigrating. He became a hero at the Hindenburg Line on September 29, 1918, when he stormed two German machine-gun nests.

RIGHT: Sergeant Matej Kocak (at left), a Slovak from the Austro-Hungarian Empire, posing in a French photo studio with two buddies. Kocak enlisted in the marines a few months after arriving in the United States in 1907, and he served with the Marine Corps for the rest of his life. He fought in every major American engagement of the war.

LEFT: Italian-born Tommaso Ottaviano, though exempt from military service because he was the sole support of his widowed mother, nonetheless chose to serve when he was drafted in the spring of 1917. His unit, the heavily Italian 310th Infantry, suffered terrible losses in the nineteen-day battle to take the Bois des Loges.

TOP RIGHT: Meyer Epstein (right) with his younger brother Alexander and their aunt, in the Russian Pale of Settlement around 1900.

BOTTOM RIGHT: After emigrating on board the *Lusitania* in 1913, Meyer Epstein worked as a plumber in New York City. Drafted in the spring of 1918, he saw action in the massive Meuse-Argonne offensive that began on September 26, 1918.

TOP AND BOTTOM RIGHT: Joseph Chmielewski (top) and his older brother Frank and sister-in-law Mary on their wedding day (bottom right). Frank emigrated from Poland in 1907 and went to work in a coal mine in South Fork, Pennsylvania; five years later he brought over Joseph and got him a job in the mine. Joseph enlisted shortly after the United States entered the war in order to become an American citizen.

BOTTOM LEFT: Max Cieminski, the son of immigrants from the Prussian sector of Poland, was exempt from military service because he was missing his trigger finger. He nonetheless served with the "Yankee" Division infantry, fighting in the Aisne-Marne offensive in July 1918.

Epifanio Affatato emigrated from Calabria, Italy, to Brooklyn, New York, with the firm belief that the streets of America were paved with gold. He won the Distinguished Service Cross for his "extraordinary heroism" on September 29, 1918, in the battle to break the Hindenburg Line. Private Affatato's unit, the 107th Infantry, lost more men that day than any other regiment in the war.

Peter Thompson left Ireland in 1914—five months before war broke out in Europe—and went to work in the copper mines in Butte, Montana. Though most Irish Americans opposed U.S. entry into the war, Peter promptly enlisted to get on the fast track to U.S. citizenship and fought with the "Wild West" Division.

LEFT: Antonio Pierro was drafted four years after he emigrated from the south of Italy and served in the "All-American" Division's field artillery. Decades after the war, he was still angry about being called a wop at training camp in Georgia.

RIGHT: Samuel Goldberg, a product of the Russian Pale and the streets of Newark, New Jersey, enlisted in the cavalry at the urging of a recruiting officer. "Jesus Christ, a Jew in the cavalry," his Polish drill sergeant greeted him when he arrived at boot camp at Fort Oglethorpe.

Just months after emigrating from the Ukraine in 1899, Sam Dreben joined the U. S. Army to fight in the Philippines. Nicknamed the Fighting Jew, Dreben had an incredible career as a soldier of fortune before reenlisting in the army at the age of thirty-nine to fight in the Great War. General Pershing called him "the finest soldier and one of the bravest men I ever knew."

Born-again Christian and Tennessee marksman Alvin Cullum York was horrified at being "throwed" in with a bunch of foreigners at Camp Gordon training camp in Georgia. Awarded the Medal of Honor for bravery in the Argonne, York dedicated his memoirs to "the Greeks, Irish, Poles, Jews, and Italians who were in my platoon in the World War. . . . They were my buddies. I jes learned to love them."

The 27th Division, in whose 107th Infantry Regiment Epifanio Affatato and Michael Valente served, marching down New York's Fifth Avenue on August 30, 1917. The division's "send-off" parade was actually a send-off not to war but to boot camp in South Carolina.

Camp Wadsworth, outside Spartanburg, South Carolina, where the 27th Division trained. There was much resentment when the New York City bluebloods of the 7th "Silk Stocking" Regiment were merged with the upstate farmers and immigrant factory workers of the 1st Infantry Regiment to form the 107th Infantry.

FOOD WILL WIN THE WAR
You came here seeking Freedom
You must now help to preserve it
WHEAT is needed for the allies
Waste nothing

UNITED STATES FOOD ADMINISTRATION

Immigrants were targeted in a wide range of government propaganda during the war, including Liberty Bond drives, rousing patriotic speeches made by "four-minute men," and posters like this one exhorting the foreign-born to save food.

Eighteen thousand Doughboys, hundreds of immigrants among them, stood in formation on the parade ground of Camp Dodge in Des Moines, Iowa, on a stifling July day in 1918 to make this "Human Statue of Liberty." The image was intended for a Liberty Bond campaign.

When Tommaso Ottaviano wrote his mother of the *"sconvolgimento"*—devastation—he had witnessed in the Argonne in October 1918, this was what he was referring to. Such atrocities were part of daily life on the Western Front.

Those who died in battle were buried hastily near where they fell. Chaplains tried to ascertain the religion or national origin of the dead and say the appropriate prayers. After the war, thousands of corpses were disinterred and reburied in U.S. military cemeteries or, at the families' request, sent back home.

The *Leviathan*, the world's largest ship, entered New York harbor on March 6, 1919, with ten thousand returning men of the 27th Division. Epifanio Affatato, among those on board, received the British Military Medal that day from Britain's Prince of Wales (the future Edward VII).

Sergeant Matej Kocak (at right) with three buddies near Soissons, France. This photograph was taken on July 18, 1918, the day Kocak took out two German machine-gun positions unassisted—an act of heroism for which he was later awarded two Congressional Medals of Honor.

Michael Valente receiving the Congressional Medal of Honor from President Herbert Hoover in Washington, D.C., on September 27, 1929. "I did not forget that the president had decorated an American of Italian origin," he told a reporter after the ceremony.

which entered a heavily fortified tunnel just east of the Knoll, and advance to the town of Gouy. If they did all that and lived, they could tell their grandchildren that they had broken the supposedly impregnable Hindenburg Line.

At five thirty in the morning "a single gun barked," wrote Company C's historian; "then another; then a thousand. The din was indescribable." The men stood with jaws clenched and listened to the machine-gun barrage. That would be the only red carpet they would get in those first twelve hundred yards. Lieutenant Buell gave the command and they moved out at a fast walk. In a smooth steady wave the men rose against the side of the Knoll. There was no bunching up or panicked tossing of grenades. They were doing just what they were supposed to do. Twenty-five yards past the jump-off tape they encountered the first German outposts. The pop and zing of bullets. Every few seconds a soldier crumpled but the others kept going just as they had been told to do.

Company D jumped off right behind them. They had been assigned the "mopping up"—the job of dispatching or capturing any German stragglers left hiding in trenches or shell holes. This was essential in a rapid advance: if you failed to mop up, the stragglers would wait until your men had gone past, then rise up and shoot them in the back.

Half an hour into the assault, Company C hit serious trouble. At the dot of 6:00 A.M. the Germans launched a heavy counterbarrage of shells and machine-gun fire from their position on the Knoll. The tanks that were accompanying the American advance were blown up one after another by land mines or German antitank guns. The troops of Company C were now suffering badly—that exposed left flank was like a shooting gallery for the Germans. "It was a slaughter; we ran into a trap. The machine-gun fire was thicker than flies in summer," as one of the men put it. "Our men were being cut down rapidly," wrote Jacobson, "but still they pressed ahead." They had reached that critical point where the wave would either crest or dissipate. Lieutenant Buell, watching his men fall, understood that if they didn't surge ahead immediately they would be pinned down and picked off one by one. "Understood" is probably too rational a word for what was happening to Buell. With German fire converging on his company from three sides, the lieutenant snapped into action. He took off alone up the Knoll, shouting to his men to follow. Buell held an automatic pistol at the ready, like a party to a duel. There was a snarl of wire in front of him with a German machine-gun nest dug in just behind. While enemy machine gunners let fly with everything they had, the lieutenant slashed his way through the wire. Three Company C sergeants—Clark, O'Connor, and Schwegler—followed with their squads of men. There was now enough American firepower to distract the German gunners, and Buell seized the opportunity to gain the lip of the forward German trench. The lieutenant was standing over the trench with his pistol cocked when a burst of gunfire took him down. The men rushed ahead to destroy the machine-gun nest that had fired on their commander. All three sergeants were severely wounded in the charge.

Epifanio was among the Company C men still alive on the Knoll: leaderless, these survivors were now on their own in a firestorm of German bullets and shells. Epifanio had been in combat long enough to know that he was capable of remaining calm in the worst possible conditions. More than calm. But it was different fighting without an officer to back you up. The men hesitated. The Germans seemed to have an inexhaustible supply of ammunition and will. "How in God's name can anyone live in this?" kept running through the mind of one man on the Knoll. Epifanio looked around—there was practically no one in his squad left standing. How the hell could they get to the top of this hill? It was at that moment that the shell with his name on it exploded. A shrapnel shell. Hundreds of lead marbles poured down from the sky and one of them hit Epifanio's jaw and lodged in the bone. The force of the impact knocked him over. Blood gushed from the wound. At first he was too dazed to feel pain—but when the pain came a few heartbeats later it was bad. Then the pain eased off and an unbearable thirst took over. Blood flowed down his neck, through his fingers, soaked into his tunic. With his hand to the wound, Epifanio dragged himself on his elbows to the nearest trench—a scratch on the side of the Knoll that the Brits had named Lone Tree Trench. Four other Company C guys were already there, among them Sergeant Francis H. Doane, all of them wounded badly. The battle raged on over their heads. The air was so thick with smoke and fog they could barely see—but from the sound of things they figured they had gotten out ahead of the rest of the company. There they sat in no-man's-land, four guys with holes in their bodies, separated from their comrades, cut off from any possibility of help, trapped by enemy fire.

Suddenly, grenades starting hailing down on them. In the smoke of battle, it was hard to tell whether the grenades were being tossed by the Americans behind them and falling short, or whether they were coming from the Germans up the Knoll. Epifanio thought they were both. Again, thought is too lucid a word for what was going through his head. He really didn't care or stop to process who was throwing these grenades. He just had to get them out of Lone Tree Trench before they exploded. Epifanio was short—just five-feet-four—but very quick. His wound proved to be less serious than it looked. Even with the blood dripping from his jaw, he found he was able to get around with extraordinary speed. As fast as those grenades came in, Epifanio grabbed them and tossed them forward, into the enemy trench. He wasn't enraged or vengeful. He certainly wasn't doing it for God or country. It was just his job. The fact that it was a matter of life and death for him, and for the four other guys huddled too weak to get up, didn't change the nature of the job. As long as those live grenades came in, Epifanio kept tossing them out. If Germans were dying when the grenades finally exploded, that was part of the job. God knows Epifanio had nothing against Germans. He had nothing against anyone. He just wanted to keep himself and the four guys beside him from being blown to pieces.

At some point the grenades stopped falling. Company C, whatever was left of it, moved ahead. Now there was nothing to do but wait for help and pray they'd still be alive when it came. From the sound of things, it was going to be a long wait.

Mike Valente's moment came later that Sunday morning. With Company C decimated on the Knoll, Company D's task was considerably tougher than had been anticipated. Rather than simply mopping up, the men of Company D had to deal with fresh troops being fed in from underground passageways connected to the German stronghold in the St. Quentin Canal tunnel. If anything, the resistance was fiercer than during the initial attack—"murderous fire full in their faces" was how Jacobson described it. Pinned down in a shell hole, Mike was starting to find the murderous fire intolerable. As the bullets kept spraying from the German line, he was thinking that the only way out of this mess was to charge. They needed to move out, not hunker down. Why the hell didn't someone take those machine guns out? Mike lay there doing a slow burn until one of those German bullets hit his buddy. That was when he exploded. Rage born of frustration and adrenaline took over. Orders or no orders, he was moving out. Enlisting his Ogdensburg buddy Joseph Mastine, Mike grabbed his rifle and a bunch of grenades and left the shell hole. Anger gave him total focus. When he reached the offending nest, Mike set his sights on a German machine gunner, took careful aim, and shot the man through the head. When that man fell, he used the rest of the bullets in the magazine on the other occupants of the trench. Then he moved on to the next nest. No time to reload the rifle, so he reached for the grenades. At some point in his charge, a German bullet ripped into Mike's right wrist—not a serious wound, just enough to make him madder. His blue eyes blazing, he was now shouting at the top of his lungs. Our Italian boy, one of the officers present later reported, had blown his stack. A soldier in the company described the private's actions as "frenzied." With Joe right behind him, Mike worked his way deeper into the German line. Having taken out two machine-gun nests, the men now fell on a German trench, jumping down into it without hesitation and going to work. Enemy soldiers who had witnessed what the crazed Americans did to their comrades came out with their hands up. Those who tried to scramble to the rear got picked off by American snipers. Joe escorted the prisoners back to the American lines. According to soldiers witness to the fighting, Mike and Joe killed five Germans and took twenty-one prisoner, though Mike's grenades may well have killed more. There is no calculating how many American lives the two had spared or how far they advanced the 107th up the Knoll.

"Of course I should have asked permission of my commanding officer to do the job," Mike admitted later. "I was crazy then," he told his family. "I tried to fool the Germans into thinking they were in trouble. I didn't want to die. I had no choice about killing those men. And besides I never knew how many of them I killed." Some soldiers experience a terrible revulsion against the killing they do in the madness of war. "We feel that all this murdering is unworthy of the human race," a German soldier remarked after passing through a combined German-French cemetery with graves dated 1914, 1915, 1916, 1917, 1918. "I am ashamed not only of my deeds," wrote one English combat veteran after the next world war, "not only of my nation's deeds, but of human deeds as well. I am ashamed to be a man." But Mike Valente was not ashamed. He voiced no regrets for what he had done. Never would. Ordinarily he was the most genial of fellows—nice-looking, easygoing, even-tempered. Under normal circumstances, he was not a violent, impetuous, or even notably angry man. At the Hindenburg Line, he just snapped. Snapped in a way that the military recognizes as heroic.

Both he and Joe were eventually evacuated to the rear so their wounds could be treated.

"He is of Italian ancestry," the Ogdensburg newspaper wrote of Mike after the war, "and when he shouldered a rifle to take his place in the invincible Company D force he had two purposes in mind. One was to uphold the standards of honor cherished by his native country—Italy. The other was to prove that he was not only an American but was willing to prove it even unto death." Maybe.

Rabbi Lee J. Levinger, a chaplain with the 27th Division, recounted the story of a young Jewish soldier in the division who risked his life to rescue the wounded in the course of a terrible firefight: "I asked . . . why he had led a group of volunteers in bringing from an exposed position some wounded men of another regiment, an act in which the only other Jew in the company had been killed and for which my friend was later decorated. 'Well, chaplain, there were only two Jewish boys in the company and we'd been kidded about it a little. We just wanted to show those fellows what a Jew could do.'"

It's possible that one filament in the fuse that set off Mike Valente's rage that Michaelmas morning was spun from that same desire "to show those fellows." In war, the immigrant's delicate pride is a force to be reckoned with.

For his actions on the Knoll on September 29, Epifanio Affatato was awarded the Distinguished Service Cross for "extraordinary heroism." The same medal went to Lieutenant Buell, who survived the gunshot wound his men believed had killed him. The British gave four Military Medals to the American soldiers of Company D, among them Mike Valente. "By his courage and utter disregard for personal safety he set a fine example to his comrades," the British citation reads. An American medal was clearly in the offing for Mike as well—something big, maybe even the Medal of Honor.

Epifanio Affatato and Mike Valente were both disabled by their wounds and eventually both were moved to field hospitals behind the lines. But the slaughter on the Knoll and in the fields and thickets around the St. Quentin Canal went on unabated into the night. When the numbers were finally tallied, it emerged that the 107th Infantry lost more men in action that day than any other regiment in the war—a record of suffering that still stands. In Company C alone, 51 men died. As Private Ingalls, the company historian, wrote, "When, fifty-six hours later, having reached their objectives, the companies assembled and staggered down the road out of the lines, only 22 of the 151 who had sprung over the top with Company C answered roll call." In the 107th as a whole, 349 of the 2,500 officers and men who went into action on September 29 were either killed in action or died of wounds suffered that day; total regimental casualties (killed, wounded, captured, or gassed) came to 1,062, a rate of 42 percent. One of the men of the 107th killed in action that day was Corporal Alexander Kim of Company L—one of the few soldiers of Chinese descent in the regiment. Shot through the head by a German bullet, Corporal Kim died near the ruins of Guillemont Farm, yards away from the trench where Epifanio was pinned down.

The battle of the Hindenburg Line has one other grisly distinction: it marked the first time in the war that Allied forces used mustard gas. Of the three types of poison gas introduced during the war—chlorine and phosgene were the others—mustard was the most insidious and the most excruciating. Several hours after exposure to the gas's oily amber-colored clouds, human skin erupts in painful blisters, the eyes become sore and inflamed, the membranes coating the bronchial tubes are burned and flayed away. You vomit and bleed internally and externally; you choke and gag and gasp for breath; the pain is so intense that victims usually need to be strapped to their beds. Death, if it comes, takes a month or more to arrive. By the autumn of 1918, the Germans had been delivering mustard gas via artillery shells for a year, saturating entire battlefields with the poison that remained active for weeks. The Allies, after inaugurating mustard on September 29, would continue to use it for the duration of the war.

The 107th, or what was left of it, remained in the lines on September 30, and was finally relieved by Australian troops on October 1. It took four more days of ferocious combat, but at last the dread Hindenburg Line was broken, the tunnel of the St. Quentin Canal cleared, the sector around Le Catelet and Gouy secured.

Sixty-two years after the battle, Sergeant Francis H. Doane of Company C, one of the men who was in Lone Tree Trench with Epifanio that day, published an article on the op-ed page of the *New York Times* addressed to his buddy Jim who died in action on September 29: "When you fell, the gun landed on its tripod with you on top, with a bullet through your head. Whittle [Private First Class John R. Whittle] rolled you over, picked up the gun, and we kept going. Then Whittle fell and we lost both him and the gun and got pinned down in Lone Tree Trench where the Germans threw some potato-masher grenades at us. Affatato won the D.S.C. for throwing them back." Doane, then a retired lieutenant colonel, was eighty-three when he wrote this—but the memory clearly still burned in his mind.

In the days after the battle, the bodies of the dead were recovered from the trenches and shell holes and cratered farm fields where they had fallen and taken for burial in a gently sloping, relatively sheltered patch of ground near the ruined village of Bony. On September 29, this had been part of no-man's-land—the Knoll rising to the north, the earthworks marking the canal tunnel to the east. It was an obvious spot for a cemetery since so many had died on this ground or close by. When it was safe enough to pray here, three of the division's chaplains came to read the burial service for the dead. Rabbi Levinger recalled that a cold rain fell the day that he and the other chaplains stood over "those many graves, the result of the terrible battle at the Hindenburg Line. . . . There was none of the panoply of war, no bugle, firing party or parade, just the prayer uttered for each man in the faith to which he was born or to which he had clung. We did not even know the religion of every man buried there, but we knew that our prayers would serve for all. . . . We [chaplains] did our work together as parts of one church, the United States Army."

CHAPTER 13

BLANC MONT

The real man of the whole crowd is the ordinary garden variety of enlisted man, sgt., cpl., or private, the fellow who really does the work and the suffering," an officer with the 77th Division wrote to his mother back in North Carolina after the war ended. "In the American Army it didn't seem to make any difference whether he spoke Polish, Italian, Yiddish, Irish, or plain American, he was a wonderful soldier." Sergeant Matej Kocak was the epitome of this wonderful soldier. By the end of September 1918, the Slovak marine was going on his tenth month overseas. He had fought at Belleau Wood, Soissons, St. Mihiel—every major engagement that the AEF had undertaken in France. He had killed and captured enemy soldiers with extraordinary bravery, rallied his men in battle, kept his wits about him under fire, repeatedly risked his life by taking the lead in assaults. Officers and soldiers alike recognized his heroism. "The fellow who really does the work and the suffering"—no one deserved these words more than Sergeant Kocak. He had seen virtually the entire American war effort—and he would see it through to the end—the end of the war or his own end, whichever came first.

Kocak was not a letter writer—at least no letters of his have surfaced—and by all appearances he was not a sentimental man. Though
his parents were still living in his home village of Gbely (then part of
the Austro-Hungarian Empire and thus enemy territory), and though
he had family in Binghamton, New York—brother, sister-in-law,
nephew, cousins—Kocak had written "none" on the line of the military form asking for the name and address of "person to be notified
in case of emergency." He stated that he had no relatives or friends
and listed his former residence as U.S. Marine Corps. On the night
of October 1, when his unit was ordered into position for the assault
on Blanc Mont Ridge near the cathedral city of Rheims, Kocak was
750 miles away from his parents and his childhood home. But clearly,
the thirty-six-year-old marine had put his Slovak past behind him. His
home was the 66th Company, 5th Marine Regiment, 2nd Division.
In truth, since leaving Gbely twelve years before, the marines were the
only home he had known.

Even if he had been a letter writer, mail was terribly slow—letters
from the States took a month or more to reach soldiers at the front—so
Kocak would not have known that three days before his unit entered
the front lines at Blanc Mont, his cousin Paul had been discharged
from the army. Drafted in Binghamton the previous September, Paul
had been training at Camp Gordon in Georgia for a year while his
cousin Matej fought in France. A prolonged stretch in boot camp was
not uncommon, especially for foreign-born recruits. But when Paul
Kocak was finally deemed sufficiently schooled in the ways of the
American military to ship out to France, trouble arose. Somehow the
awkward fact came to light that Paul's father was serving in the Austrian army and thus Private Kocak was classified an enemy alien. This,
in itself, would not necessarily have disqualified him from staying with
his unit. Army regulations stipulated that loyal aliens, even those from
enemy nations, could fight; it was only when an alien soldier made it
clear to his officers that he "did not desire to serve" that he would be
dismissed from the service with a general discharge (and thus lose all

future benefits). Paul insisted that he never voiced this desire; he never petitioned his superiors to relieve him from service; he maintained that he was willing and able to fight for his adoptive country. But evidently there was some misunderstanding, because on September 27, while a million American soldiers, Matej Kocak among them, were massing for the assault on the Western Front, Paul Kocak was discharged from the army and put on a train back to Binghamton.

It was one of the ironies of this war that one cousin should be sent home in disgrace while the other marched heroically into battle. Such was the lot of the foreign-born in those times. With the massive Argonne campaign raging and the nation braced for a long winter of fighting, no one had the time or the stomach to ponder the irony of what befell the Kocak cousins or to rectify the misunderstanding that blotted Paul's military record. But the implications of the events and decisions made during those darkening September days would haunt the Kocaks for the rest of that century and into the next one. The story of war is also the story of families struggling to shine a light into the abyss of the past.

T he word that rippled down the line on the night of October 1 was that Blanc Mont looked like another bloodbath. As usual, the marines were being brought in to do a job that other soldiers had failed at. For four years the French had tried to dislodge the Germans from Blanc Mont—not really a mountain but a spinelike ridge, densely wooded at the top, that dominated the choppy white-clay Champagne countryside east of Rheims and west of the Argonne forest. As the most prominent rise south of the Aisne River, the ridge was key to the control of the Champagne sector. It was the same story as Mont-faucon and the Hindenburg Line: a position previously considered unassailable had now, by some magic of strategy or wishful thinking, been reclassified as a feasible objective, to be taken in a matter of hours no matter what the human cost. That human cost was to be paid over the next five days by the men of the 2nd Division, including, first and foremost, the marines.

What Kocak and the other marines in the trenches did not know was how freely and heedlessly their superiors had been gambling with their lives. Early in September, as plans were being forged for the great Allied offensive that was intended to win the war, the French command had asked Pershing to lend them two American divisions to beef up their forces in the Champagne region—an arrangement similar to the loan of the 27th and 30th divisions to the British for the assault on the Hindenburg Line. The battle-hardened 2nd Division, to which the 5th and 6th Marine regiments belonged, was one of the divisions tapped for service with the French; the other was the raw 36th, which had just shipped out from boot camp in Texas and had yet to suffer a crease in their uniforms. On the eve of the battle of Blanc Mont, Major General John A. Lejeune, the 2nd Division's commander, met with his French counterparts to iron out the details of how to coordinate their forces. Lejeune had been warned that the French intended to break up his division and feed the pieces into their own units—the American command's worst recurring nightmare. Bent on avoiding this at all costs, the general agreed to a hasty and ill-considered bargain: if the French left his division intact, the American forces would undertake the impossible mission of storming Blanc Mont alone.

This was how Sergeant Kocak and a thousand other men in the 1st Battalion of the 5th Marines found themselves on the night of October 1 massed at the base of an incline that rose steeply into what one historian called "some of the best defended and nearly impregnable positions on the western front." The men spent the following two days loafing on the region's famous chalky white soil, muttering about the "endless trenches, barbed wire obstacles and entanglements, traps, pill boxes, tunnels, and dugouts" that awaited them behind the German line, and sampling dehydrated potatoes and dehydrated soup for the first time (the verdict, as one marine officer put it, was that "the potatoes were not worth the water required to prepare them," but the soup "was delightful and quite interesting"). The attack finally broke on the morning of Thursday, October 3. The artillery barrage went off as planned starting at 5:30 A.M., and in the course of the day the infantry and the marines managed to storm the ridge. Kocak, with the 1st Battalion, was assigned to guard the battalion's highly vulnerable left flank with the understanding that French forces would move up and advance beside them. When the French utterly failed to appear, officers in the field started firing off frantic messages complaining that their left flank was "in the air." The ridge may have been in American hands, but the battle was far from over. By nightfall, with most battalion commanders hopelessly separated from their men and the exposed left flank an open wound, the advance was canceled. Major George W. Hamilton, the dashing young commander of the 1st Battalion, used the word "fiasco" to describe the state of things that night.

"The battalion had been in every engagement in which the regiment had participated up to this time, but the afternoon of the 4th of October, 1918, was by far the bloodiest and worst day of the entire war," wrote one of the 1st Battalion captains after it was all over. Every marine who fought and survived that day shared this opinion. Casualties—appalling as they were—do not explain the trauma of those hours. Marines had died at Belleau Wood and Soissons—but at least they died in those battles pushing their shoulders against the wall. At Blanc Mont on October 4 they were trapped on a sloping, forested tableland, surrounded, penned in, and slaughtered.

Orders that day called for the three battalions of the 5th Marines to move north toward the village of St. Etienne starting at six in the morning—3rd Battalion in the lead, 1st next, 2nd in the rear—with a distance of five hundred yards to be maintained between each battalion. As soon as the Americans came into the open, the Germans opened fire. The 3rd Battalion took the brunt of it. Nonetheless, the three battalions managed to push about a mile forward before orders came in late morning to dig in.

There is no mention of Kocak's name in the papers that survived the battle. Inevitably, the official documents—field messages and orders fired off in the heat of the battle, reports written in the months after the Armistice—concern the deeds, decisions, judgments, and fears of the superior officers. In the archived record, the hero of that gruesome day, if any man can be called the hero of a slaughter, was 1st Battalion commander Major George Hamilton. Handsome, athletic, worshipped by his men—one private in the battalion referred to him as "our motion picture captain"—Hamilton was an inspiring leader, and it's clear from the messages he wrote that day that he was extraordinarily brave and cool under fire. Kocak was one of about a thousand men under Hamilton's command—the 1st Battalion consisted of four 250-men companies, of which Kocak served with the 66th (the others were the 17th, the 49th, and the 67th). Since Kocak does not appear in the written record of that day, the scraps of paper that Hamilton covered with his blunt unhurried script will have to serve as a kind of proxy for Kocak's experience.

By early afternoon, the 1st Battalion was on the move again, advancing north into wooded terrain, but about one thirty they hit what Hamilton described as "heavy machine gun fire . . . laid down from the left flank." German artillery shells were also falling at a rapid rate on the advancing lines. A half hour of this punishment was all that the men could take. Marines had been through as bad or worse at Belleau Wood and Soissons and muscled through, so it's hard to explain why they cracked now. Maybe it was because so many of the old Leathernecks—the regular career marines like Kocak—had been killed or wounded. Maybe it was because so many of the men in the field now were wartime recruits with limited training and experience. Maybe it was something more ineffable—a subtle ebbing of morale, a pervasive weariness of war, a chill they caught from the nakedness of the sloping topography. For whatever reason, by 2:00 P.M., some of the marines broke ranks and started running to the rear. Hamilton somehow found the time to compose a lengthy, disgusted message describing seasoned warriors, officers among them, going to pieces: "In some instances officers were leading in what appeared to be a grand rout. Among those whom I noticed particularly was Capt. [DeWitt] Peck, Capt. [David T.] Jackson, (2nd Bn.) and Major Messersmith [3rd Battalion commander]. There were also several lieutenants whom I did not recognize. Major Messersmith explained that he had lost all his officers, but didn't show any initiative or leadership. Capt. Jackson was hopeless. When it became evident that the retirement had become a rout, Lieut. [James A.] Nelms ran out and endeavored to turn the men back. His task was a hard one and attempted at great personal exposure to machine gun fire and a violent artillery bombardment. We then were forced to draw our pistols, and it was only by this method that we were able to stop the retreat."

"Marines never retreat," insisted Elton E. Mackin, a private with the 5th Marines, in his memoir *Suddenly We Didn't Want to Die*; but Mackin conceded that at Blanc Mont "for the first time, panic had its way, and a few men ran as the attack closed in." *Panic had its way.* The fact that these veterans of Belleau Wood and Soissons had to be forced back into battle at gunpoint is a sign of how desperate the situation had become. The men could choose how to die—whether as coward or soldier—but for many of them death was certain and they knew it. Who was to blame? The men for refusing to do their duty—or Hamilton and his fellow officers for drawing pistols and forcing them into certain death? Or was General Lejeune at fault for letting the French manipulate him into this suicidal assault? It's impossible at this remove to draw clear lines or pass simple judgments. It was a war in which slaughter invariably overwhelmed strategy. After the battle, General Lejeune proudly told the survivors that "to be able to say when the war is finished, 'I belonged to the 2nd Division, I fought with it at the Battle of Blanc Mont Ridge,' will be the highest honor that can come to any man." Noble words. The panic that led some to choose the highest dishonor received no mention in the general's speech. Hamilton's message detailing the "grand rout" was slipped into a file and eventually deposited and forgotten in the National Archives. *Marines never retreat.*

At 2:15 Hamilton sent another, even more dire message back to regimental commander Lieutenant Colonel Logan Feland: "Absolutely impossible to carry out attack. Machine guns encircling us and reinforcements needed. It is doubtful if we can hold out here unless machine guns are cleaned out on flanks." Mackin wrote that "the men were stunned" by the viciousness of the German fire; "lashed down to earth by flailing whips of shrapnel, gas, and heavy stuff that came as drumfire, killing them. There was no place in all our little world for us to go. The fellows bunched against the fancied shelter of the larger trees in little close-packed knots, like storm-swept sheep, and died that way, in groups. . . . Never in my time had I seen such a deadly, killing fury." Hamilton, having said that to attack was "absolutely impossible," did precisely that. His 1st Battalion went out in advance of the other two and descended the slope in the face of relentless German fire. Those who made it found themselves so deep into enemy territory that bullets were now coming at them from three sides. Mackin called this terrifying belt of woods "The Box": "It was a place for men to die; a spearhead of out-flung battle line thrust deeply into the German front, exposed to fire from three sides, its line of communication cut off by enfilading Maxims firing from the flanks. It was a deadly place."

There were now only about one hundred men left of the thousand in Hamilton's 1st Battalion. The sole battalion officer still standing was a first lieutenant with the good Irish name of Francis J. Kelly Jr. Lieutenant Kelly took over the 66th Company, Kocak's unit, and made a rapid assessment of the situation. The Germans holed up near St. Etienne were starting to intensify their artillery barrage in preparation for a counterattack—a classic move when the enemy's initial attack surge flagged. Kelly had orders from Hamilton to pull his men back to a safer spot. But, in the words of marine historian George B. Clark, the lieutenant "did what any Marine would do in the same situation: he ordered his company to advance." Kocak was among those who went forward with Kelly. The only account of Kocak's role in this desperate push was written by Theodore Roosevelt Jr., who was then serving as the commander of the 1st Division's 26th Regiment. It's unclear how the president's son came to know the story of the Slovak marine—Roosevelt's division was nowhere near Blanc Mont and he was still recovering from being gassed and wounded at Soissons. But the details match up with the official documents, and his narrative has the ring of truth—so some eyewitness or buddy in the marines must have told him about it. Roosevelt wrote: "Sergeant Kocak's battalion, the 1st, was ordered to meet it [the counterattack]. They did not wait for it to reach them, but drove out at it, meeting the counter-attack with attack. With a shock the troops came together. For a few moments with bayonet and rifle they fought hand to hand. Then the Germans wavered and in an instant were in full retreat into the woods from which they had come. Flushed with victory, the marines followed hard on their heels. They burst into the woods. Kocak's company was held up by the sweep of a machine-gun. Without hesitation the sergeant started for it alone. He crawled, as he had at Soissons, to close quarters, then sprang to his feet and charged. Two Germans had seen him. Before he could reach them they fired and he fell dead."

Kocak's extensive military file contains a few additional details. After the body was recovered, a military clerk filled in information about the time, place, and cause of death on an army medical form; on the line marked "diagnosis" the clerk typed: "Wound, gunshot (multiple of body)", after which someone wrote in by hand, the letters KA (presumably "killed in action"). Another form indicates that his left leg was fractured below the hip and knee. The final entry in Kocak's service record book reads as follows: "While serving with 66 Co 5th Regt in France from July 1, 1918 to Oct 4, 1918. No offences. Oct 4, 1918. Killed in action by enemy gun fire in the Champagne Sector near Blanc Mont Ridge, France in line of duty. Date of burial, location of grave or grave number not known. Character: Excellent."

Sergeant Kocak was one of 292 marines killed in action at Blanc Mont. Another 1,893 were wounded in the fighting. In his eleven years of service with the marines, Kocak had but a single blot on his record—the time he went AWOL from the *Georgia* in Norfolk, Virginia, back in 1910. Since then his record had been impeccable. He consistently received the highest marks on his "professional and conduct record." His character was always rated "excellent." He had been promoted twice and was cited for heroism at Soissons. Given all of this, there is no way in hell that Kocak would have turned tail and joined in the "grand rout" earlier that day. It was simply not in his nature to run from danger. In the absence of letters or interviews with his comrades, one can only speculate—but it seems fair to say that Matej Kocak, Slovak American, died in the belief that he was doing his duty for the country he had embraced in 1906 and had come to love more than life itself.

No one but those present will ever know or fully appreciate what the battalion went through during the charge up this hill," wrote 1st Battalion Captain LeRoy P. Hunt of the afternoon's assault on Blanc Mont. "The rate of casualties was far above anything we had experienced, but the men kept on." The fighting and dying went on well into the night, but to little avail. St. Etienne, the 5th Marines' objective for the day, remained in German hands. The following morning, Saturday, October 5, after a harrowing night of enemy machine-gun fire and artillery, Hamilton sent a bitterly frank message to Colonel Feland: "It is hard to say 'can't,' but the Division Commander *should thoroughly understand the situation and realize that this regiment 'can't' advance as an attacking force. Such advance would sacrifice the regiment*." Feland understood. The fighting continued, but the 5th Marines were "finished for this campaign," in the words of historian George Clark. "No one had the gall to commit the 5th to any more engagements at Blanc Mont."

O n Sunday, October 6, the marines were finally relieved. Word came down through the ranks that a bunch of Texas and Oklahoma cowboys from the 36th Division's 141st Infantry were being brought in to finish the job at Blanc Mont. Mackin described the scene of the two long columns of men, one battle-ravaged, the other clean and eager, trading places on the narrow French road. The marines, Mackin wrote, were "a battered, filthy, ragged crew; they did not look like soldiers. Beards—week-old, bristly growth ringed around thin-lipped, silent mouths—helped to frame weary eyes that had a glare of madness in their depths." When the "fast-stepping column" of 141st infantry appeared coming toward them, the marines "fell out of ranks to rest and watch them pass by. They were tall, clean-cut fellows, walking rapidly toward the guns." Inevitably the two groups of men began trading insults and jokes:

> "Hey crumbs! Why don't you wash your dirty necks?"
> "—call them soldiers, son? Why that's a bunch of tramps!"
> "Why don't you get some uniforms—and use some soap?"
> They asked for it. We had answers by then for such as that.
> "Hey, you—the loud-mouthed bastard over there. You'll make a handsome-looking corpse, tomorrow!"
> One saw sudden dread spread over his face, below a sickly smile that masked his fear. . . .
> The effect of what was said went through the ranks and cut youthful banter short.
> The silent column paced away toward the guns. We took to our own road and cursed beneath our breath.

Not all the Texas infantrymen striding toward the guns were as brash and green as Mackin supposed. At the head of the 141st Regiment's Company A marched a squat, middle-aged first sergeant—eyes snapping, chest thrust out, every inch the professional soldier—who had been in combat of some kind nearly all his life. It was Sam Dreben, the Fighting Jew. Dreben and the boys of Company A may have been latecomers, but they were about to be scorched by one of the hottest baptisms by fire of the war.

The 141st Infantry had shipped out during the Aisne-Marne offensive at the end of July; they arrived at the port of Brest on August 6, and from there proceeded five hundred miles inland to a French training camp at Bar-sur-Aube, due south of the Argonne front. It was during those first weeks in France that the forty-year-old Dreben received a letter from his young wife in El Paso with the news that their infant daughter had died while he was crossing the Atlantic. Sam could have requested a hardship discharge and returned to Texas to comfort his wife, but he chose to remain with his men. He was Company A's top kick and these raw recruits needed a seasoned veteran to break them in, now more than ever. By this stage in the war, training periods were being compressed to the vanishing point. Sam's regiment had about six weeks to shape up before heading into action—and a lot of that time was spent dealing with an outbreak of influenza that ravaged the regiment. A new, intensely virulent strain of influenza had appeared at Brest just around the time that the 36th Division disembarked, and the men carried the infection to the training camp at Bar-sur-Aube. Twenty-three soldiers died of flu within a few weeks. The pandemic, which was exploding that summer and fall, would eventually kill upward of 50 million people worldwide, many more than the war that helped spread it.

In the first week of October, their training over, Dreben and the surviving men of the 141st Infantry were trucked to the Champagne sector and massed at Somme-Suippe, the same village where Kocak's unit had jumped off to storm Blanc Mont. On October 6, the 141st marched into the lines past the filthy, stubble-faced marines coming off the bloody battlefield.

Aside from the fact that thousands of men on both sides were dead, gassed, or maimed, not much had changed at Blanc Mont since Kocak had jumped off a week earlier. The Germans, though weakened, remained in possession of the town of St. Etienne, which had been the objective of Kocak's unit on October 4. The Americans had not altered their basic strategy of throwing fresh bodies at the enemy until they finally gave up. The one crucial difference was that those fresh bodies, with very few exceptions, had never been in battle before. Where the veterans of Belleau Wood and Soissons had failed, the callow cowboys of Camp Bowie and Bar-sur-Aube training camp were supposed to prevail, once their officers pried them out of their trenches. Top-kick Dreben had some kicking to do.

The new men got their first taste of fire when they were heavily shelled while marching north on Monday, October 6. Captain Richard F. Burges, the El Paso attorney in charge of Company A, said that the shell fire "became so hot that it was necessary to order [the men] to lie down on either side of the road until the bombardment stopped." It wasn't until well after midnight that Company A finally reached its position at the front; German shell fire remained so intense and relentless that the men did little for the next twenty-four hours aside from dig themselves shelter pits. The word was that the renewed attack on St. Etienne was supposed to start the next morning, but as of 3 A.M. that morning Captain Burges had yet to receive written orders about his objective, timing, or position (the captain learned later that the runner carrying his orders had been blown up by a German shell). The only confirmation Burges got that the attack was in fact going ahead came shortly before 5 A.M. when the American artillery barrage opened—at which point "Sergeant Dreben and I roused our men," wrote Burges, "and called them to get ready for action immediately as we were going over the top in a few minutes."

If Sam Dreben had not been an actual person, he might have been concocted by some Hollywood hack. From the moment he stepped off the dock in the United States in 1899, everything he did smacked of big-screen epic—enlisting in the army because of the promise of a free uniform and a free funeral, defying death in his first tour of duty in the Philippines, blasting his way out of jams as a soldier of fortune in Central America, gambling in dusty cantinas with Mexican banditos, supplying arms to Pancho Villa and then joining Pershing's Punitive Expedition against Villa, suffering the death of his only child while he was aboard a troopship to France. Thrill to the adventures of the Fighting Jew! Dreben's next few hours at Blanc Mont play like another scene in his movie—the Fighting Jew battles the Boche!—but incredible as it sounds, what Dreben did on the battlefield is substantiated by a load of documentary evidence. Shortly after the war ended, officers of the 141st Infantry were instructed to record everything they could remember about Blanc Mont. Captains, lieutenants, the regiment's Lieutenant Colonel Luther R. James wrote lengthy, vivid, hour-by-hour accounts of the fighting that occurred on October 8 and the days that followed. Everything in those documents, especially the account written by Company A's Captain Richard Burges, agrees detail for detail with the three single-spaced typed pages headed "Relating My Experiences in Going over the Top with 'A' Company 141st Infantry" and signed "Sam Dreben, 1st Sgt. Co. 'A'."

Dreben noted that Captain Burges gave the order to go over the top at "exactly 5:15 A.M." and within one hundred yards of jump-off his company encountered "strong machine-gun resistance." For the next ten hours Company A inched its way forward through heavy German fire: as the hours of that autumn day ticked by, they advanced, took cover when enemy fire became intense, and advanced again. At one point, said Dreben, "we saw about 25 or 30 of the enemy advancing across an open space in the woods and advancing to meet them, in the open, we killed wounded or captured the entire party." The climactic moment, at least for Dreben, came as dusk was gathering. His account of this skirmish offers not only a vivid soldier's eye view of combat, but also a rare opportunity to hear the voice and feel the spirit of this amazing man:

> About 4:00 P.M. that evening, one of our men on an outpost called my attention to some troops advancing in our direction and at the same time he fired a shot. I told him to stop firing as it was possible that the party advancing were French. However, after a close inspection with my glasses, (it was getting dark in the woods at this time) I satisfied myself absolutely that they were the enemy, and carrying several machine guns. I immediately called for volunteers to go out and attack them, for I realized that if they took up position on our right, they would command our position. From twenty to thirty responded to my call and we went on at double time to meet "Fritz", shooting and shouting all the time while we were charging them, as we were only a little over a hundred yards from them. Here is where I realized the value of the Browning Automatic Rifle, as in less than ten minutes about 58 big huskey [sic] well equipped Boches were stretched out nicely resting in peace. We took two prisoners for information. We gathered up four machine guns and one instrument that appeared to be used for signaling or telegraphing. About this time I saw a red signal or flare directly in front of our position and about 300 or 400 yards away. I immediately gave the order to double time back and not to mind the booty, and we returned without the loss of a man. No sooner had we reached our positions when they did open a terrific barrage on us.

"Back of that vaudeville exterior was a cool, calculating brain," Dreben's fellow soldier of fortune Tracy Richardson once said of him, "a courage that nothing daunted. He seemed to take thousands of wild chances with death and emerge by fool bull luck. But when you knew him, you learned how carefully he planned every detail and how little he left to chance." So it was at Blanc Mont the afternoon of October 8. That evening, during a lull in the German barrage, Captain Burges sent this message to regimental headquarters: "Enemy advance on our right repulsed. Sergeant Dreben has captured three machine guns, three prisoners, and killed fifty of the enemy." To which the colonel responded: "Great work. Keep it up. Percentage between killed and captured is about right." Soldiers on the scene attested that if it hadn't been for Dreben's raid, Company A would have been mowed down by those German machine guns.

The "great work" Dreben did on the right flank was about the only thing that went well for the 141st that day. It was a regiment of rookies and their inexperience showed. "Companies were disorganized, battalions were disorganized, everything was disorganized, and to add to our other pleasures the Boche proceeded to shell the woods at least three times daily," wrote Lieutenant Colonel James in his account of the chaos that prevailed throughout the attack. Two ranking officers were killed or wounded in the first hour; a major lost his mind under the pressure and shot one of his own runners after mistaking him for a German; communication and liaison degenerated to the point where "the fighting resolved itself into a series of independent fights conducted by detachments of varying sizes, composed of men from different companies"; no one in command had any idea where the dispersed companies were fighting or even where the front line lay. That night, the 141st dug in under a heavy and continuous artillery barrage. The following day, October 9, "we remained in the same position," wrote Dreben. "The night of the 9th we had a terrific machine gun barrage, from three directions, but we still held fast." The 141st and 142nd infantry suffered some 1,300 casualties on October 8 and 300 more the following day.

Despite the terror and madness that crippled the 36th Division on its first two days of combat, somehow they succeeded in turning the tide. The Germans, having been hammered for six days straight by two American divisions, did not surrender so much as melt away. They lost control of the village St. Etienne on the night of October 8 and by the morning of October 10 they had withdrawn all but token resistance in the sector—just a few machine guns left behind to harass the American infantry. The following day they retreated rapidly thirteen miles north to the Aisne River. Finally the Texas and Oklahoma cowboys got their wish for a full-throttle war-whooping sprint—"a wild open field dash," in the words of one historian, though in fact the cowboys were rushing through a gate that the marines had blasted open for them.

In retrospect, the capture of Blanc Mont stands as one of the great turning points in the war, a crucial victory that precipitated the general German withdrawal. But at the time, the cowboys, and the marines before them, were hurting too badly to savor the moment much. The accounts written by the officers of the 141st were remarkable for their absence of back-slapping or high-flown rhetoric. Blanc Mont Ridge had fallen, the "sacred city of Rheims" had been freed, as one soldier put it, the Germans were on the run in this sector, but still the war dragged on. Some 7,800 American soldiers died or were wounded on this chalky rise. "They had paid a price hideous even for this war," wrote a marine who had survived. *They*, meaning the men on the ground. As for the generals who had sent the men up the ridge, all but the most vainglorious ultimately decided that the less said the better.

The burial of the dead began on October 11, the day the Germans started pulling back. Fifty men from the 36th Division were assigned to the burial detail and it took them four days to get all the bodies on the battlefield underground. Captain W. S. Montgomery of the 141st Regiment who was in charge of the burial detail reported that the corpses had been stripped of their valuables—no watches remained on wrists, no money in pockets, even the shoes and socks had been removed from some of the bodies. Sergeant Matej Kocak, killed eight days earlier, was among the soldiers that Montgomery's men laid to rest on October 12. The remains of the fallen marine were wrapped in canvas, placed in a wooden coffin and buried in a field cemetery near the village of St. Etienne. T. B. Lugg, a chaplain with the 6th Marines, presided over the ceremony. When the prayers had been said, a plaque with the words "Kock. M. Sgt. M.C." was attached to a peg and placed over the body. The grave was numbered 172.

Sergeant Kocak's file was not closed yet. Grave Registration Service Form No. 8 circulated through the military bureaucracy with the lines marked "relative, relationship, and address" checked and this message typed in: "This man told that he had no relatives or friends whose names he might use." As a result, no one would be notified that their heroic son, brother, uncle, cousin, friend had died in France.

It's rare in a modern war that one soldier can be said to make a difference—but Kocak and Dreben did each make a difference at Blanc Mont Ridge. Their bravery and their sacrifice alone did not win the battle—but the battle could not have been won without them and others like them. Others who fought for an adopted country, who fought all the harder because it was a country they had chosen. "These men fought for America to be Americans," wrote career soldier Peter I. Pellicoro (U.S. Army 11th Special Forces Group Airborne) of his Italian-born father and grandfather, both of whom served with the U.S. Army during World War I. The same could be said of Kocak and Dreben and thousands of others who fought and died alongside them in France and Belgium.

Without exception, these guys were having a hell of a time during those dark days of October 1918. The United States had been fighting with ever-increasing force since the late spring—long enough for individual battles, advances, victories, and retreats to lose their crisp edges and blur into one long nightmare of gas and bullets and blood. The Hindenburg Line had been broken, Blanc Mont Ridge seized and declawed—but to the men on the ground it seemed to make little difference. There was no end in sight to the violence. No matter how cunning the battle plan or how noble the name of the push, German bullets still killed, shell fragments still tore off limbs, gas still burned their throats and shriveled their lungs. All along the Western Front soldiers were muttering darkly about digging in for the winter.

By October 12, the day Sergeant Kocak was buried, the Argonne campaign that had opened sixteen days earlier with such an overwhelming display of American firepower was stalled just about everywhere. Meyer Epstein, having gone out on the first wave of the offensive with the 58th Infantry, was now stuck in a foul wood called Bois de Fays deep in the Argonne forest. No advance, no retreat—just day after day of being shelled and gassed from every side. One officer with the regiment wrote of the black despair that settled over his men in this quagmire: "It is said that the Bois de Fays means 'Woods of the Fairies.' Were I to name it, I would call it 'The center of Hell.' Any man who ever spent any time in those woods, from the 4th to the 17th of October, knows that even that term does not adequately express the true situation. The shell torn woods were wet and muddy; everything was wet and damp, raw, cold and clammy. Not a breeze blew to clear the gas laden air. The sun never shone, it was always dark and murky. Down the sides of our fox holes, water trickled or seeped through the walls. From all sides came the odor of death and decay. Mangled bodies of men were everywhere to be seen. Our bodies ached from the cold and wet. The foul surroundings made one sick at heart. We were hungry, yet unable to eat but little of the food which came up. For hours at a time we were forced to be without water, for to go after it was to gamble with death. The mental strain was maddening, the physical strain exhausted us. . . . Sleep was impossible." Meyer was in even worse physical shape than most of the men because he was starving from lack of protein. Jewish dietary laws forbade all but kosher meat, so Meyer traded his meat rations with the non-Jewish soldiers— meat for bread. But he was steadily losing weight and strength. How was he going to fight when the order finally came for Company H to blast its way out of the "center of Hell"?

Tony Pierro was also stuck in hell. The All-American Division had been held in reserve at the start of the Argonne campaign, but on October 5 they were ordered into the line in the western Argonne where the Aire River wound through dense enemy-infested woods. The All-Americans were mostly city kids, half of them immigrants, all of them green and unblooded. The couple of hours of blasting that Tony's field artillery unit had done at St. Mihiel barely counted. This nasty dripping jungle in the Argonne was their first time in the real show, and it was a bad time and bad place to enter.

Since soldiers at the front didn't see American newspapers, Tony may not have realized that he was going in just a few miles away from where the so-called Lost Battalion was trapped behind enemy lines. The story of how this group of soldiers got separated from their division and pinned down in the Argonne forest—and the question of what would happen to them—was one of those cliffhangers journalists live for. On October 2, parts of nine companies from the 77th "Melting Pot" Division—554 men in all, many of them Jewish and Italian immigrants from the Lower East Side of New York—advanced so rapidly into enemy territory that they had become surrounded by German forces and entirely cut off from the rest of their division. By October 7, when the All-Americans entered the sector, the Lost Battalion had been subjected to five days of incessant shelling and sniping. Food, ammunition, and medical supplies were nearly exhausted—and casualties were mounting. The situation was becoming more desperate by the hour.

The All-Americans were enlisted to take part in a massive coordinated assault just east of the "pocket" where the Lost Battalion was lodged. Tony's artillery battery opened fire at 5:40 the morning of October 7, and Battery E kept up intermittent barrages and harassing fire over the next eighteen hours. In the course of the day, the All-American infantry managed to advance half a mile—not far enough to free the Lost Battalion, but far enough to put severe pressure on the German forces in the sector. Later that night, men from the 77th Division's 307th Infantry finally broke through. Of the 554 men of the Lost Battalion who had been caught behind enemy lines four days earlier, only 194 were able to walk out on their own; another 144 had to be evacuated on stretchers. The other 216 were dead.

Retribution of a kind came swiftly. The following morning, Corporal Alvin York and a group of his buddies from the All-American's 328th Infantry walked into a firefight with some Germans manning a machine gun. York watched six of his comrades from Company G, including his best friend Corporal Murray Savage, get shot to pieces. Armed with a rifle and pistol, the superb marksman from the hills of Tennessee exacted his revenge. York chose his targets carefully from among the surrounding German soldiers and began picking them off one by one. When York was done shooting, he began taking prisoners. Eventually, with the help of a captured German lieutenant who had lived in the United States and spoke English, York rounded up 132 prisoners and won himself a Medal of Honor.

York was a deeply religious Christian, and the bloodshed he had witnessed and caused on October 8 troubled him sorely. The following day, he asked to be excused from the line so he could return to that scrap of woods and search for survivors—American, German, anyone who might have been wounded in the exchange of fire. But there was no one. York would not leave the site before bowing his head in prayer. "I prayed for the Greeks and Italians and the Poles and the Jews and the others. I done prayed for the Germans too. They were all brother men of mine. Maybe their religion was different, but I reckon we all believed in the same God and I wanted to pray for all of them." York prayed, but what he had seen and done on the field of death had nothing to do with God. "God," he wrote afterward, "would never be cruel enough to create a cyclone as terrible as that Argonne battle."

Over the next week, as Tony inched his way with the All-Americans through the fetid woods astride the Aire River, he came to regret that he had ever been tapped to work with horses. One of his jobs now was to drive the horse-drawn supply wagons to and from the front: going up to the line, Tony carried supplies and ammunition, going out

he carried bodies. Tony never talked much about what he was thinking when he drove that cart with a load of corpses behind him; but it
was probably no different from what another soldier told a reporter
almost ninety years later when the country was at war again, this time
in Iraq: "This is not easy to talk about. Part of our job, our duty, was
that we loaded, you know, bodies. We were in charge of the airfield,
and we would load these heroes into the aircraft. My platoon sergeant
had a policy. He didn't want lower-ranking soldiers involved. He told
us, 'You guys are going to carry this with you, whether you realize it
or not, for the rest of your lives. If I can protect the privates, I will.' I
don't know if I could ever explain what that was really like. I loaded
these guys—and I know all their names—onto a plane. And you don't
know how heavy a guy in a body bag can be. It's not just his weight.
He may be 180 pounds, but it's a lot more than just a 180-pound guy.
You're loading his entire life." That was what Tony was doing with his
horse and cart in the Argonne forest; like the soldier interviewed in
Iraq, he carried it with him all his life.

On one of Tony's trips to the front an artillery shell landed directly
in front of his cart. The horse took the impact of the explosion and
died instantly. Tony realized that his horse had saved his life—and he
grieved as if he'd lost a buddy. He'd gotten to love that horse in all
those trips back and forth to the line. Other guys felt the same about
their horses, their dogs, any innocent living thing that kept them company through this horror. It might seem strange to mourn for a horse
when men were dying all around, but that's how it was.

Peter Thompson, Leonardo Costantino, and the rest of the Wild
West Division were in no better shape. Though the Wild West
boys had acquitted themselves admirably in their first couple of days
in the Argonne, by October they were slogging miserably along with
everyone else. Leonardo's log entries, brief though they are, capture the
sense of sinking hopelessly, blindly into a never-ending battle:

Sept. 30 One more long night spend in a trench seem like a year. All of us are tired and sick. No sleep, no eats, no water for 84 hours.

Oct. 3 No relieve for us, so Had to go back on the line at 6:30 last night. Can't advance, for the boys are all in and sick. Our clothes are all wet and torn. I have now a blanket thank God. I'm freezen. Had some stew and Bread, that safe my life.

Oct. 4 Feal very tired and Hungry. Try to locate Archi but was impossible. War is worst than "Hell." That is all I have to say

Oct. 5 Had my shoe off first time in two weeks. Poor feet of mine are certainly sore. Had a hot meal, first time in two weecks but I can't even eat that now.

Wet clothes, sore feet, long nights freezing out in the rain, never enough to eat: that was the reality of the soldier's life in war. If there was progress on any given day, Leonardo was mostly oblivious to it. He made no mention of objectives or strategy, battle plans or liaison—but no mention of fear either. For Leonardo the war was the immediate sensation of his living flesh and the fate of buddies. He might not ever win a medal, but he was going to make damn sure he survived the war. San Diego suited him fine and he had every intention of going back there.

Andrew Christofferson was also hungry that October, even hungrier than he'd been as a boy growing up poor in Norway. He was slight of build—five feet, seven inches, never broke 140 pounds—and so he had less in the way of reserves than the huskier guys. Andrew had also had less time to get broken into the army. Drafted on June 25 from his Montana homestead and sent down to South Carolina to train with the 81st "Wildcat" Division, he was in France by August and in the trenches by the middle of September. Andrew's regiment—the 321st Infantry—had been assigned to a relatively static sector in the Vosges Mountains not far from the German border, about a hundred miles southeast of the main action in the Argonne. It was a cushy berth compared to the worst of the Argonne; nonetheless, for the month he was stationed there, Andrew endured sporadic shelling and machine-gun fire from the enemy and chronic short rations from his own side. Things got so bad he took to following the bread wagon that supplied the front and scrounging for whatever fell off. If a loaf landed in a pile of horse manure, he would just brush it off as best he could and eat it. Drinking water was scarce enough that on occasion Andrew had to get down on his stomach in the mud and sip from the puddles that collected in the horses' hoofprints.

There was a guy from Chicago in Andrew's company, a banker's son who never quit boasting about how rich his family was and how luxurious his life had been before he got drafted. One day a bunch of the men sat together under a tree sharing a single loaf of bread that one of them had scavenged. The guy with the loaf tore it into chunks and passed them around. It tickled Andrew no end when the banker's son accepted his shred of dirty stale bread and murmured ruefully, "Oh, if my family could see me now!" He ate the bread all the same, Andrew noted, and was happy to get it—same as the freckled, sandy-haired country boys from the Carolina hills, same as the bantam-weight immigrant from the west coast of Norway who had always thanked God for every blessed morsel of bread that came his way.

No regiment stayed in a quiet sector forever. By the middle of October, new orders came down for the Wildcat Division. Andrew, the banker's son, and the rest of them packed up and started on the long march to the Argonne.

A mong the infinite variety of suffering on the battlefield, there was a special torment reserved for the soldiers of German descent: the fear that they might be fighting one of their relatives. Charles Minder, a New Yorker of German blood serving with the Melting Pot Division, gave voice to this anxiety in a letter to his mother written not long after he arrived in France: "I wonder about the enemy. They are the same as me, I have German blood in me, I never know when I might be shooting at one of my own cousins or uncles. We don't want to shoot each other, but we are forced to."

By the time Minder was fighting in the Argonne, this fear had become a kind of mania. In another letter, he unburdened himself of an encounter with a dying German soldier that had driven him to the brink of insanity:

I walked up to the machine-gun nest and there were the two Germans stretched out on their backs. One of them was unconscious and the other opened his eyes very weakly as I came up to him and when I looked into his face, I felt like dying. I had a ghastly fear that he was Uncle Franz, for he looked like him. I lifted his head, and blood spurted from the wounds on the side of his neck. I asked him, Sint sie nicht Franz Barg, von Bremen aus? He opened his eyes very slowly again and looked at me; then his eyes closed, and he was still. He gave one slight gasp and passed on. I knelt there for a little while just dumb. I couldn't think or do anything. Of course, he was not Uncle Franz, but I kept thinking of him in the German lines and my cousins there, too.

An officer came along and started hollering at me: "What the hell are you doing there? This is no time for souvenir hunting! Don't you know we are advancing? Leave those dead Germans alone! Come on!" I gathered up the ammunition boxes and followed on. I was so heartsick I couldn't talk. . . .

This is about the worst place that we have ever been in. This day has been like a dream to me, a terrible nightmare. I can't write any more tonight.

"If there is a God, why doesn't he put a stop to this?" Minder asked his mother. "Is this evil force, War, more powerful than God?" How many soldiers of every nationality and religion must have wondered the same in those darkening October days.

CHAPTER 14

WHY SHOULD I
SHOOT THEM?

Tommaso Ottaviano, dutiful Italian son that he was, spared his mother the bloody truth when he wrote home to Rhode Island. He talked about how slow the mail was, how good his health was, and how hopeful he remained that the war would end soon; he wrote to ask whether his sister had received the handkerchiefs he sent as a gift; he inquired about the relatives in Italy; he talked about his buddies, especially his best pal Antonio Giosi, another Italian from Rhode Island. He wrote at length about money—how he had instructed the army to send half his monthly pay of $30 home to Rhode Island, and how he was paying $6 of his own $15 share each month into his army life insurance. Three days into the Meuse-Argonne offensive Tommaso wrote to ask his mother if she had received a document indicating that he had taken out a $10,000 insurance policy—"you must have the document to be sure," he cautioned her.

It wasn't until the middle of October that the violence began to shadow Tommaso's letters home. On October 13, he told his mother that he had been at the front for fifteen days, though only six of those days were in the "first line"—the rest of the time they had been in reserve. Now, however, they were en route to "another front, whose name I can't reveal to you."

That new front, a bleak, dangerous place, provoked the one shudder of horror that Tommaso let slip in his letters home to his mother.

> *War zone, 18 October, 1918*
> *Dear Mother,*
>   *With a little time that I've been indisposed I'm going to resume telling you my news, which by the grace of God is good aside from what we've been through this past week. I really can't say anything at length, but it would be hard enough just to describe the devastation. . . . Now we are at the heels of the Germans—what I really mean is that we still need good luck and the grace of our Lord: this is what I pray for always.*

The full charge of horror was loaded into a single word—*sconvolgimento* which can be translated as devastation, convulsion, or personal tragedy. But it was a word that spoke volumes about everything Tommaso was holding back. Having given his mother this glimpse, Tommaso dropped the war and went on to inquire after family and friends—Uncle Pietro, Uncle Nicandro from Italy, Uncle Giovanni Giuseppe, Carluccio, Patrino—and to insist that he lacked for nothing, though if they did want to get him a Christmas present he could use a pair of gloves lined with fur inside that would be good in the rain. He went on for a couple of pages more, another of his warmhearted upbeat letters from "somewhere in France." But *sconvolgimento* burned like a coal in the first lines.

What Tommaso did not tell his mother on October 18 was that the previous two days had been the worst of the war for Company I, 310th Infantry—and that there was nothing ahead but darkness and more fighting.

Tommaso's regiment had been rotated into the Argonne only a few days earlier. When the great American battle opened on September 26, the 78th Division had been holding a trench system near

Thiaucourt, well east of Verdun—one of those so-called quiet sectors where instead of being slaughtered wholesale men were picked off daily by snipers and shells. By the time they headed north for the Argonne on October 5, the 310th had suffered losses of 121 killed, 446 wounded, and 95 gassed. Not such a quiet sector after all. It took them a week to reach their new position—the "other front" whose name Tommaso could not disclose. Their new home was a bluff just south of the Aire River commanding the villages of St. Juvin and Chevieres, with tiny Champigneulle beyond folded behind a rise. Lush productive country, at least it had been productive in peacetime, with fields sloping up to the feet of compact villages, a big boxy church rising like a fortress in the center of St. Juvin, a neatly squared-off block of woods covering the crest to the west of Champigneulle. Bois des Loges the wood was labeled on the maps—a name the men would come to hate in the weeks to come.

The attack on Bois des Loges, the first of several as it would prove, began on the dim, rainy morning of October 16. The 310th was assigned the job of advancing to the northern edge of the wood—a couple of kilometers from their jump-off point—by the following morning. This in itself was ambitious, but the orders came with an ominous coda: "This is most important," read the communiqué from Corps headquarters. "Whole military situation demands success of push through the Bois des Loges tonight. Go to it. Good luck. Your work has received much praise." Given the fact that the "whole military situation" hung on this advance, it's remarkable how rushed and poorly planned it was.

Tommaso and the other men of Company I equipped themselves with two bandoliers of ammunition and a hundred extra rounds in their belts and jumped off "with all possible vigor," as their orders demanded. Casualties were heavy as soon as they emerged from the ravine of the Agron River into the open fields. From here to the woods there was no cover at all—just a wide open expanse that rose gradually to the edge of the woods, uphill all the way. The Germans timed their artillery barrage precisely to catch the first waves of American infantry. "The 310th was learning how to fight," wrote the regimental historian. They were also learning how to die. "Third Battalion [to which Tommaso's Company I belonged] held up on line approximately S.W. corner Bois des Loges to village Champigneulle," read an urgent field message sent by runner that afternoon. "MG [machine-gun] fire from all directions makes it impossible to advance beyond this line. Meantime enemy artillery pounding both Battalions. Casualties extremely heavy caused entirely by artillery and MGs. What shall we do." By some miracle, a forward party made it all the way to the objective line at the northern edge of the woods—but the unit to their right had failed to advance. It would have been suicidal to remain in place with their right flank entirely exposed, so at the end of day one the 310th retreated.

The attack resumed the following day, this time with Tommaso's company in support. The woods were dense, the trees that had not been shelled were spindly and close together, the ground, pitted and spongy, was crumpled in a series of corrugated ridges that made rapid forward movement impossible. The stagnant air held captive by those trees was so permeated with gas that the men had to fight wearing masks. Not all the gas had been released by the Germans: the All-American Division's field artillery, which had been stationed just east of the 78th for the past week, had been directed to put down 2,600 rounds of phosgene on the village of Champigneulle (at the eastern fringe of the wood) at 6:10 that morning, and another 1,200 gas rounds at 11:30. Very likely Tony Pierro carried some of these gas shells to the big guns, since his artillery unit was involved in the fighting. In the event, the only thing the All-American's gas barrage accomplished was to drive German soldiers out of the village and into the woods where they fell on the struggling 310th infantry.

Again the casualties were horrific. In the course of the day the aid stations set up in St. Juvin were overwhelmed. The next day, October 18, Tommaso's company was pulled back, which explains why he had time to write his mother about the *sconvolgimento* he had witnessed. But other elements of the 310th kept fighting—to no avail. The men reported that the central ridge of the woods was infested with thirty German machine guns arranged in an impenetrable interlocking network. On October 20, after another costly push, the entire regiment withdrew to a position straddling the east-west road between Grandpré and St. Juvin and there, in the valley below the cursed Bois des Loges, they hunkered down and absorbed whatever shells the German fired down on them.

There followed ten days of waiting, wondering, and trying to remain sane and alive through the intermittent German artillery barrages.

On Monday, October 23, burial details were organized to collect the bodies of the men who had been killed in the first failed Bois des Loges attack. The corpses of thirty-five soldiers from Tommaso's regiment were taken off the fields between the Aire River and the Grandpré–St. Juvin road—there were many more in the woods, but those would have to wait until the Americans dared to set foot in that toxic forest again. On October 29, Tommaso and the rest of Company I were marched five miles to the rear to the village of Cornay. Rumors were rife, as always, that they were being relieved—but those rumors proved false. The men were ordered to strip off the filthy uniforms they been wearing for most of October, bathe, delouse, and pick up fresh uniforms. By six o'clock that night they were back in the line. Now the rumor was that another assault on the Bois des Loges was to be mounted any day, though why the men should have been prettied up only to be sent into action again no one could figure. Maybe it was supposed to be good for morale.

Two days later, on a turnip field in Belgium, scrappy little Peter Thompson had his moment of glory, bittersweet though it was. After suffering through some of the worst fighting in the first weeks of the Argonne, the Wild West Division had been pulled off the line on October 12. The division as a whole had lost a quarter of its men in the Argonne—five thousand killed or wounded. The Montana boys in Peter's 362nd infantry regiment had fared even worse—more than half were casualties by the time they were relieved.

The respite from fighting was welcome but brief. "The Capt went out this evening," Leonardo Costantino wrote in his log on October 15, "am Keeping the fire going at his Room. Good feed to-day. Had beans twice to-day." Two days later the Wild West men were in trains bound for Belgium. "Oct. 18. Arrived at some place in Belgium at 6:30 P.M.," recorded Leonardo. "Pitched tent at 8 P.M. While doing it, saw many rats, large AS CATS. We are some distance from the lines. The country is full of shell holes." In fact, they were near Ypres, scene of some of the worst and most prolonged fighting in the war. Forty-eight hours later they were on the move again, marching through a lunar landscape of mud, rubble, and leaden skies. Leonardo's log: "Oct. 20. In hiking, saw nothing but shell holes and dead trees. All muddy hard hike. Saw many tanks but they were out of order on account of being shelled." "Oct 21. While waiting for orders to move to the front line, saw a Belgium lady, middle age, take off socks from a dead French soldier."

The division took its position at the front outside the Belgian town of Waereghem on October 30, and at 5:35 A.M. the next morning, Peter's regiment attacked (Leonardo's regiment, the 364th, was held in reserve that day). The fighting went relatively well, and when they dug in that night on the flat expanses of western Flanders, the men felt pretty good about things. They all agreed that the mud of Flanders was preferable to the Argonne forest. Peter spent the long night in a trench watching the sky light up with the occasional exploding shell—German shells. Unable to sleep through the noise and stench, Peter sat up and scanned the flat expanse before him. At some point his eyes fell on Arnold Pratt, a sergeant in the regiment he had never had any use for. Peter's granddaughter, Christy Leskovar, describes what happened next in her riveting family memoir *One Night in a Bad Inn*:

Peter watched [Sergeant Pratt], and then suddenly, the sergeant lurched
and fell to the ground. He'd been shot and was lying there helpless, fully
exposed to enemy fire. Without thinking, Peter scrambled out of the
trench and ran to him, and with shell after shell exploding all around
and machine gun bullets whizzing by, Peter knelt down and bandaged
his wounds. Then he got up, grabbed Sergeant Pratt, and though Pratt
was nearly twice his size, Peter singlehandedly dragged or carried him
back to the trench. Once at the trench, Winks Brown [a comrade in the
company] helped Peter pull Sergeant Pratt down into it. He was very
badly wounded and had to be evacuated to a hospital, but he was alive.
Peter had saved his life and at grave risk to his own. He would later tell
his sister Nellie, "I only did it because I hated him."

The fact that Peter hated Pratt didn't matter. Rescuing him from
certain death was enough to make Peter Thompson a hero.

Tommaso went back into the Bois des Loges the following day.
The word going around was that this was meant to be the knock-
out punch—not only in the Bois des Loges but along a wide swath
of the front stretching from Grandpré in the west to Cunel in the
east, territory divided between seven American divisions. If the punch
landed right, the American First Army would finally succeed in breaking through to the strategic transport hub of Sedan and severing the
German supply line. Any doubt that something big was brewing was
laid to rest by the magnitude of the advance artillery barrage: for two
days, the Americans relentlessly pounded the woods with explosives
and poison—50,000 rounds of mustard gas, 10,000 shells. It was the
most massive single release of poison gas in the history of the U.S.
Armed Forces. "Day and night was continuous pandemonium," as one
of the officers put it.

The 310th was a heavily Italian regiment—lots of immigrants like
Tommaso who been drafted from the big East Coast cities. Some of the

guys spoke only rudimentary English, just enough to understand orders and gripe when the going got rough. Pasquale Marcone, serving with Company A, remembered that the night before the attack, the Italians sang songs they remembered from when they were kids and swapped family stories. To keep old memories alive for one more night.

On the morning of November 1, Tommaso and the men of Company I were rousted out in the pitch black and given breakfast at 4 A.M.—coffee and stale bread. They were instructed to roll their packs and leave them behind—officers didn't want anything to slow the men down. Four companies, fewer than nine hundred men in all, were assigned to the front line, Company I placed second from the right. Everything was the same as in the previous attempt to clear the woods—it would be another frontal assault into the teeth of the German machine guns—the exact same strategy that had failed before.

Whistles were blown to alert the men to get ready; the final preparatory barrage opened at 4:30, reportedly the largest high-explosive barrage the American army had ever seen—16,000 rounds of HE and phosphorus crashing down on a scrap of woods that measured maybe a half-mile square. Then at 5:30 came the whistle to charge. Things went all right for the first couple of hours—the men advanced in pace with the rolling barrage; they took cover from German fire in shell craters; the first wave managed to infiltrate the southern fringe of the woods. But at 8:00 A.M. they hit heavy machine-gun fire along the western edge of the woods and the advance halted. The men discovered that their massive artillery bombardment had accomplished almost nothing: the Germans had simply waited it out under cover and then popped back up unscathed with their machine guns.

At 10:40 the 3rd Battalion commander sent an urgent message back to the command post: "We are completely held up by many m.g. nests in left half of Bois des Loges. . . . We are using T. M's [trench mortars], 37 mm's and M.G's on few known locations. Many E M G's [enemy machine guns] are on crest of hill which we cannot locate"—the same crest fortified with interlocking nests of machine guns that had repulsed the earlier attacks. "Am trying infiltration to the right and left. We need further artillery on northern and western sections of woods. His artillery is shelling our positions heavily with gas in woods." The 10:45 message: "Am held up at all points by m/g nests. . . . Trying to knock out m/g nests while infantry pushes forward around left flank. Casualties heavy. One Company probably wiped out."

Tommaso never wrote to his mother about the fighting that morning, and even if he had, he would not have told her the truth about what really happened. But other soldiers left accounts that capture something of what Tommaso must have seen and smelled and felt in his gut in the Bois des Loges. Leslie Allen McPherson, a private from St. Paul, Minnesota, serving with Company E of the 360th Infantry, 90th Division, was fighting at the exact same time about ten miles away from Tommaso's regiment. This is what McPherson wrote in his diary about the morning of November 1:

> As I looked at the faces of my comrades I saw no fear, but every face was set for they knew what was expected from each one of them and they as well as myself [knew] that some that went out into that hell of bursting shells and poisoned gas would never live to see sundown. Our Sgt. got us together and started with us but we only got a few yards when a gas shell bursted and one of our Lieutenants fell to the ground gassed. . . . Every step we took some poor boy fell either dead or wounded and believe it would unnerve the bravest of men to go thru a place like that was that morning it was not a surprise to see your pal or comrade with his head blown off and in lots of cases a shell would kill 6 or 8 men and there they would lay in a pile torn beyond recognition. I had a friend who was Scotch by birth who I had soldiered with ever since I came to the army he was attached to the machine gun co and had just stooped over to pick up his box of ammunition when a shell bursted and tore his head off. It was an awful sight to see that morning the ground was almost covered with dead and wounded from both sides.

Precisely when the bullet hit Tommaso was never recorded—but odds are it was shortly before 11:00 A.M. when his company was being mowed down by the machine guns that held the crest of the Bois des Loges. Nor was the place where the bullet entered his body ever specified. His medical records carry only the three letters GSW—gunshot wound. There was no need to indicate that the wound was serious—before antibiotics, any gunshot wound sustained in those conditions of filth, gas, mud, and rampant infection was serious. There is no knowing how many hours Tommaso lay bleeding on the damp forest floor, but it's likely he was not evacuated until there was a break in the action after dark.

The German machine guns and artillery finally, mercifully, fell quiet around 7:00 P.M., and teams went in with stretchers to bring out the wounded. Regimental casualties for that single day in the wood ran to 519—18 officers and 501 enlisted men, "a deplorable waste of life," the division's commander, Major General James H. McRae, conceded afterward.

Did Tommaso know that his best friend, Antonio Giosi, had been killed in the Bois des Loges that day? The friends no doubt fought and fell side by side, so he probably did know.

Nothing survives about Tommaso's 370-mile journey from the Bois des Loges to Base Hospital 34, located in a seminary in Nantes near the Atlantic coast in western France, but it was certainly agonizing. From this point there are only feeble, contradictory glimmers. A paper filed with the state of Rhode Island suggests that Tommaso may have contracted influenza in the hospital: this paper states that Tommaso died of purulent meningitis—an inflammation of the membranes covering the brain and spinal cord often associated with influenza. But the chaplain who filled in the grave location blank indicated the cause of death as GSW. And on yet another form in Tommaso's burial file, the cause of death is given as DWRIA—died of wounds received in action. All that is known for certain is that Tommaso Ottaviano died at Base Hospital 34 on November 22, 1918, twenty-one days after he was wounded in battle and eleven days after the Armistice, and that he was buried the next day in Grave 188, Section B, of American Cemetery 88 near Nantes. He was twenty-two years old.

Tommaso's death was all the more bitter for his family because he fell on the last day, indeed during the last hours of serious fighting in the Bois des Loges. When the German guns fell silent that night at 7:00 P.M., they fell silent for good. Over the next five hours German troops quietly and efficiently cleared out of the woods and beat a hasty retreat to the north. For nineteen days they had fought the 78th Division to a standstill, and now they just turned and walked away as if the place had a curse on it, just as they had done at Blanc Mont. It wasn't an American victory; it was a German abandonment. But it amounted to the same thing.

One more day, and Tommaso and his buddy Antonio would have lived to chase the Germans—and eventually return home to America.

On the morning of November 2, what was left of the 310th Infantry strolled incredulously through the Bois des Loges past German bodies and elaborate observation posts still equipped with gleaming modern instruments. When it finally sank in that the enemy had packed up and left, the American infantry lit out after them.

The retreat from the Bois des Loges was, in the words of one historian, "the first stage of the great *Kiregsmarsch*, the withdrawal of the German armies to the Meuse." All along the Argonne line from the Bois des Loges in the west to Cunel in the east, American divisions were giving chase as the Germans pulled back. The military breakthrough precipitated a political crisis for Germany and its allies. Turkey surrendered on October 31, Austria-Hungary capitulated on November 4, and as the German government tottered, the kaiser fled Berlin to take refuge in the Netherlands. With Germany's navy on the brink of mutiny and German workers rioting in the city streets, the Fatherland seemed headed for a Bolshevik-style revolution. But still the German army refused to surrender. The outcome of the war was now all but certain, but the killing and the dying continued.

Indeed, for some American soldiers the war was just beginning. Joseph Chmielewski was one of them. Though the Polish coal miner from Fifficktown, Pennsylvania, had enlisted in the army only two months after the United States declared war on Germany, he had yet to fight on the front line when the Americans broke through on November 2. Joe had dutifully trained for the better part of a year with the 6th Division's 16th Machine Gun Battalion, and he shipped out of New York with tens of thousands of other Doughboys on July 6, 1918. But since their arrival in France on July 23, Joe's unit had done nothing but drill, march, and camp. On September 24, two days before the jump-off in the Argonne, the 16th Machine Gun Battalion took up a position in a quiet sector of the Vosges Mountains (near where Andrew Christofferson was stationed with the Wildcat Division), and here they remained, four miles back from the front lines, until October 12. The Vosges was beautiful, conceded divisional historian Captain Lloyd C. Parsons, but the men found it depressing on account of "all their traveling by hiking." Finally, on October 23, orders came to load their guns and ammunition on trains in preparation for a move to the Argonne. "Cars were scarce," wrote Parsons, "so the property was spread over the floors of the boxcars and the men crawled up on top." Cramped though they were, the men were nonetheless glad to be riding rather than marching in the rain—which had been their customary mode of transport since arriving in France. In fact, the incessant marching had become something of a joke, giving rise to the division's nickname, "the Sightseeing Sixth."

After a twenty-six-hour train ride to the city of Sainte-Ménehould south of the Argonne, the division was back on the march again. For four nights they camped fifteen miles from the front—close enough to hear the boom of the big guns—and then on November 2, they were ordered to take up a position above the Aire River south of Grandpré and St. Juvin, the same spot that Tommaso and the 78th had occupied before the first assault on the Bois des Loges. Bivouacked in the ruins of Grandpré on the night of November 4, Joe and his comrades had a brush with fire when a German plane flew over and, spotting their campfires, began to bomb them. A major was killed in the attack and several men were wounded. "Within half a minute after the explosion of the first bomb every fire was smothered," wrote the division's historian.

The next day they marched through Authe, a village nine miles north of the Bois des Loges that the 310th infantry had secured three days earlier. The marching Americans were greeted by villagers "amazed and appreciative" to be free after four years of German occupation. There was some hugging and waving of handkerchiefs, but the joy of the moment was dampened for the Sightseeing Sixth by the fact that their packs had doubled in weight from absorbing so much water in the incessant rains. Since the battalion lacked for horses, Joe and the other machine-gunner privates had to drag the loaded carts themselves. From Authe the march proceeded on to Oches. "In a valley just beyond Oches 2 shells screamed overhead, bringing gas masks to the alert and a more satisfied look to the faces," wrote the historian of Joe's battalion. "At last we were near it." The following day, they assumed, they would be *in* it: the orders on November 6 called for the Sixth Division to take a position on the front line. "How did the news affect them? After all those long and weary days and nights of continuous hiking; after all these months of training; not for a moment were they dissatisfied with their lot. They sat down coolly, even smiling, got out their oil cans and gun rags and cleaned their rifles."

But those weapons were destined never to be fired. On November 9, the division was ordered to turn around, retrace their steps to Grandpré, and continue east to Metz. With the 16th Machine Gun Battalion in the lead, the Sightseeing Sixth hit the road and marched to the rear.

And that was Joe Chmielewski's war. The Polish immigrant who had joined the army a year and a half earlier in order to become a U.S. citizen never did much fighting, but he learned one heck of a lot of marching songs.

D uring the first week of November, the now seasoned Texas and Oklahoma fighters of the 36th Division, having chased the Germans from Blanc Mont as far north as they could in the Champagne sector, were transferred east to Lorraine. The story goes that when Sam Dreben's outfit, the 141st Infantry, reached the post marking the border between Champagne and Lorraine, the march was halted. Since Germany had annexed part of Lorraine after the Franco-Prussian War, this line could be taken as the symbolic boundary between France and Germany, and so the commander of the 141st chose to take it. While the column of soldiers stood in place, an order got passed from front to rear—"Sergeant Dreben, front and center!" Dreben dutifully made his way to the front of the column, and when he got there the regiment's commander saluted smartly and boomed out for all to hear: "Sergeant Dreben, we are entering German territory. You've earned the right to set the first foot on enemy soil. Take the point." The regimental band leader raised his baton and led his musicians in a rousing rendition of "My Old Kentucky Home," and the Fighting Jew strode into the Fatherland.

F or Andrew Christofferson, combat began in earnest on the first hours of November 11. Ten days prior, his Wildcat Division had been ordered to proceed to the Woevre Valley bordering the Meuse River east of Verdun to relieve the 139th Infantry. They traveled by train as far as Sampigny, about thirty miles due south of Verdun, and proceeded by foot along muddy roads until they reached the front lines near Vaux-devant-Damloup, the scene of intense fighting during the ten-month battle of Verdun in 1916. Here on November 5 they took up position in extremely muddy trenches and awaited orders. By now wild rumors were circulating about an imminent armistice—but nonetheless the business of war went on as usual. Artillery barrages were launched, infantry went over the top, machine guns mowed them down. Andrew's turn to experience all of this for the first time came at 6:00 A.M. on November 11. The orders that day were to move on a tiny hamlet called Moranville—to "make the assault vigorous and keep pushing ahead." It began as just another grueling day of war. The Germans laid down a "terrific bombardment" at 5:55 A.M.; despite the German bombs, the men of the 1st and 3rd Battalions moved out promptly at 6:00. Their officers soon realized that a wide gap had opened between the two battalions (Andrew, in Company M, was with the 3rd Battalion), so the position of the advance had to be adjusted. At 7:30 Andrew and Company M passed through the southern part of Moranville. By 10:00 A.M. both companies hit a snag as a heavy German artillery barrage and machine-gun fire held them up. At 10:35, the 2nd Battalion commander scribbled this message— "Commander of F Co attack vigorously to the left of H Co move north east, cut off woods"—and handed it to a runner to take to the line. The message was never delivered. Major Louis E. Schucker of the 321st Infantry explained why in an account written after the war: "The reason this message was not delivered: just as the runner was starting we received the order to cease hostilities which we sent immediately to all Co and unit Commanders at 10:55 a. m."

Major Schucker could not resist adding that in the last seconds, "H. Co pushed one automatic under the wire and had taken hold in the enemy trenches, killing two machine gun crews of seven men. G Co also had men through the last wire and in 15 minutes the enemies [sic] trenches would have been in our possession."

If only the damn war hadn't ended.

Private Christofferson had a different reaction to the "cessation of hostilities." As his daughter, Nellie C. Neumann, said nearly ninety years later, "I remember Dad saying, 'I trew dat gun as far as I could trow it.' He told me that men were shouting and embracing their comrades. Suddenly, what uniform you were wearing made no difference as Americans rushed forward to hug Germans. Dad made it all quite personal as he told me one time, 'I had nothing against those German fellows. Why should I shoot them? A lot of them were farm boys just like me, just doing what we were told to do.'"

Major Schucker's desire to keep killing right up to the bitter end was shared by many. Taking all sides together, some 10,900 men were killed and wounded on this last day of the Great War—a higher number of casualties than during the invasion of Normandy on D-day less than twenty-six years later.

CHAPTER 15

POSTWAR

What did they die for? The enormity of the killing made it almost a sacrilege to ask such a question, let alone attempt to answer it. More than 9.5 million soldiers died in battle—and for what? A botched treaty that set the stage for an even bloodier war twenty years later and for conflicts that still bedevil the world? No great principles, ideals, or even dreams remained intact to glorify or justify the Great War after it was over. Looking back, it's impossible to say with confidence which side was worse. It was not the war to end all wars. It did not, as President Wilson promised, make the world safe for democracy. It ended nothing, resolved nothing, made the world safe for nothing except more war. The only great thing about the Great War was the scale.

So why did they fight?

The question was especially fraught for America's immigrant soldiers. To fight for your own country is an inescapable part of the social contract. In exchange for the benefits of a secure civil society we offer our bodies, and if need be, our lives in time of war. But the foreign-born were asked—indeed, forced—to serve without having executed the social contract in full. In the streets of America they were aliens—but in no-man's-land they were expected to fight as fervently as native-born Americans. And, for the most part, they did. It was that loyalty in action that changed everything. They righted the imbalance of the social contract not by protesting but, paradoxically, by submitting. Their pride in serving won them, and their families, the status they could never have gained without the war. God knows they weren't all heroes. But the fact of their service was heroic. In a war remembered more for senseless slaughter than personal courage, the service of the foreign-born shines. Nearly a hundred years later, it's one of the few things about the Great War that still does.

The Armistice took effect with eerie numerical alignment on the eleventh hour of the eleventh day of the eleventh month of 1918—but the war didn't end when the guns stopped firing. Indeed, the aftershocks of the four years of bloodshed and the treaties that formalized its termination are still rippling through central Europe and the Middle East. Less obvious are the tremors that run through families. For many immigrant families, the memories and emotions remain fresh and accessible, even though they have long since assimilated to the mainstream culture. "I can speak only for my family but service in the military transformed them into Americans with an Italian heritage," writes Patricia A. Valente, whose Italian-born ancestors served with the U.S. Army in both world wars. "July 4th picnics may have had lasagna on the menu but they were our holidays. I attended a funeral of an older family friend who had the U.S. Army insignia in the coffin. When I asked why this was chosen, his wife replied he felt that this gave him the most pride. He too was born in Italy but he was proud to be an American." Pride is a word that comes up again and again—even among those whose sons and fathers never came back. Not pride in the war, not even necessarily pride in the United States— but pride in their ancestor for having fought. "Service in the military transformed them into Americans"—in one way or another, many say the same. Some say it with a touch of bitterness over how much their family was asked to sacrifice; some with amazement at what an ancestor endured; some with an abiding sense of connection—a path blazed, footsteps followed ever since. In many immigrant families, the military service that began with the Great War continued into the following war, and the one after that, and the next, and the next, and continues still in the wars we are fighting today.

This is not, however, a chorus that sings in unison. There are those who look back in shame and anger at the waste of so many lives. Those who proudly remember not their ancestor's service but his refusal to serve because of his faith, principles, or political beliefs. The price these men and their families paid for refusing is another strand in the immigrant experience of the war.

It's astonishing, given the volume of propaganda the Wilson administration pumped out in 1917 and 1918, how quickly the nation turned its back on the war and how reluctant it has been to turn around to look again. The Great War was too painful, too irrelevant, too quickly superseded by the next and even greater war. The attempt to find meaning and make sense of the carnage proved futile—if anything, the war and its aftermath consumed and destroyed the very idea of meaning in art, in politics, in civic life and rhetoric.

For the foreign-born that retreat from meaning was especially troubling because of the hostility to all things foreign that gripped the nation in the years immediately following the war. The Wilson administration had spent the war years hammering at Huns, alien agents, disloyal hyphenates, German speakers or sympathizers, slackers of all stripes—and the public, once infected, remained feverish with hate well into the next decade. In the hysterical xenophobia of the Red Scare period, Jews, Bolsheviks, and immigrant workers were lumped together as enemies of the American way. Medals, military honors, and loyal service counted for nothing if you spoke with an accent, held a union card, dared to advocate the brotherhood of man.

The United States' experience in the Great War is commonly derogated by the word "only": the country was in the war for *only* nineteen months and really fought for *only* six of those; *only* 53,513 American soldiers died in combat (another 63,195 succumbed to disease, accidents, and privation), compared to 1.3 million Frenchmen, 900,000 Britons, 1.6 million Germans, 1.7 million Russians. For the United States, the Great War was *only* a small war—the other war, the forgotten war. But many have not forgotten. For immigrant soldiers, their families and descendants, the First World War marked their watershed experience as Americans. Whatever happened to those men in their months of hell redounds still, privately but nonetheless vibrantly, in the lives and experiences of those who came after.

Motorized traffic was a rare thing in the south of Italy in 1918, so rare that the arrival of a bus in the main piazza of Forenza high in the stony hills of Basilicata was something to make the townspeople stop and stare. There was more staring than usual the day a young American soldier stepped off the bus. Polished and gleaming from his jaunty brown cap down to the puttees wrapped around his calves, the youth looked like he had just come from a military parade. But what on earth was he doing in Forenza? The war was over, *finita*, so why were the Americans sending a soldier to their village?

Eyes and whispers followed the American as he made his way through the narrow streets, stopped in front of the Pierro house, craned his neck to look up at the windows, and then opened the door and walked in without even knocking. Nine-year-old Nicola, one of the three Pierro boys still living at home, gaped in amazement as *il soldato americano* strode into the house. Even the grin on the clean-shaven unlined face failed to tip the boy off. When the first flush of shouting and crying and embracing played itself out, everyone had a good laugh. How could Nicola possibly fail to recognize his own brother Antonio on leave from the American army that had fought so valiantly, shoulder to shoulder with the Italians, to win the Great War?

Nicola never got over it. His brother Tony had left Forenza five years earlier, a teenage peasant in homespun with a pasteboard suitcase—and he came back a man, a soldier, a hero, an *American*. No wonder Nicola didn't recognize him. It was nothing short of a miracle.

Tony visited with the family in Forenza for four days; then he kissed his mother good-bye, got back on the bus, went to Naples, caught a train, and returned to the All-American Division in France. The war was over, *finita*, but that didn't mean they got to go home. Tony's regiment was now stationed in Bordeaux, and all through the long dreary postwar winter the officers racked their brains trying to keep the restless, bored, demoralized veterans from going crazy. Officially, there were wholesome organized activities like theatricals, ballgames, classes (the most popular were auto repair, accounting, and bookkeeping), and the inevitable and by now utterly useless drilling. Unofficially, there was a lot of drinking and fraternizing with the French villagers. Tony had been careful to stay away from French girls while the war was going on—too many stories about disease—but that winter in Bordeaux he fell in love. Tony met Magdalena when the lieutenant he served sent him into town to find comfortable billets. By this point Tony spoke passable French, and he'd always had a way with the ladies—so he had no trouble chatting up the pretty French girl and arranging a date. Many more dates followed. The couple went to dances and spent a lot of time walking. Tony curried favor with Magdalena's father by bringing him American cigarettes—not much of a sacrifice since Tony had never been a smoker. It seemed like a promising match—the good-looking, clean-living American soldier; the pretty French girl; the father's blessing.

But it was not to be. Tony told Magdalena's family that he was an American now and when his regiment sailed for America he would be sailing with them. He didn't tell them that he had a girlfriend back in the States, Maria Pierro, a distant cousin living in Boston.

Tony's field artillery unit left for the United States on May 9, 1919, and Tony received his honorable discharge six days later. He married

Maria the following year, took a job in a shoe factory in Lynn, Massachusetts, not far from his home in Swampscott, and settled into civilian life. There had never been any question in his mind about returning to the States, rejoining his father and brothers, becoming an American citizen, getting an American job. The older he grew, the more patriotic he became.

When Tony's mother and three younger brothers arrived in the United States in 1920, the last tie with Italy was severed. Tony never went back to Forenza. As the years passed, his memories of the war faded and blurred. In time he would even forget how to speak Italian. But he never forgot Magdalena.

W
hy would I want to go back to Ireland?" Peter Thompson once brusquely asked a friend. "All I ever got in Ireland was black bread and black tea." The United States had always suited Peter better—in the army, in the mines in Butte, it didn't matter as long as he didn't have to go back to black tea and bread.

But he did go back to Ireland once, right after the war. The Wild West Division was still in France in March 1919 when Peter managed to get a furlough to visit his ma and siblings in Belfast. In a photograph taken during the visit, Peter and his brother Denis stand side by side in their uniforms—two victorious Irish soldiers, one in the BEF uniform the other in the AEF khaki, both of them looking pale, stunned, and serious.

Like Tony Pierro, Peter, now Sergeant Thompson, arrived back in the States in May 1919. When he strode into his aunt's house in Butte and flashed his melting blue-eyed grin and boomed his hearty hello, the woman cried out in amazement, "Peter, you sound like an American." To which he replied, "I am an American—I've my papers to prove it!" On the back of a photograph taken at a family reunion to honor his return, the aunt wrote, "Peter just returned from France, helped save democracy."

Butte was roiled in yet another round of labor unrest, but Peter settled back to working in the mines without any trouble—or without any more trouble than usually followed him. Even more restless than most war veterans, he moved from mine to mine, working for a few months and then quitting. He drifted out to Seattle, worked there for a spell, and drifted back to Butte.

It was on his return that he learned he had won a military honor: the Republic of France had awarded him the Croix de Guerre for dragging that damned sergeant to safety in the Flanders mud in October 1918. An American recruiting officer stationed in Butte had the French medal and certificate and was eager to hand it over. But with typical cussedness, Peter refused to accept it. He insisted that he wouldn't take the award until his comrade Winks Brown, who had participated in the rescue, had also been decorated. A tussle ensued in which the officer kept proffering and Peter kept refusing the Croix de Guerre. Finally, the officer threw up his hands, left the medal with Peter's Uncle John, and closed the case.

Peter married in 1921, fathered a family, and divided his time between mining and gambling, with the occasional brawl thrown in. He loved his kids dearly, but he was not cut out to be a steady, reliable father. Peter moved his family often—Idaho, Oregon—in search of work, a change of scene, the hope of something better; but for some reason the lure of Butte was too powerful to resist, and they always came back to the rowdy old mining town. Peter was working at the Tramway Mine in Butte when he was crushed deep underground by a trolley tram motor. He died of his injuries at the age of forty-two on October 31, 1937, nineteen years to the day after the act of heroism that won him the Croix de Guerre that he refused to accept.

Peter Thompson was buried in Butte's Holy Cross Cemetery with full military honors. His casket was brought to the cemetery draped in an American flag, and an honor guard fired three volleys into the air. A bugler sounded taps as Peter's body was lowered into the ground.

Jacob Wipf and the three Hofer brothers, Joseph, Michael, and David, had been in Alcatraz for four months when the Armistice was declared. For obeying what they believed was the word of God and for clinging to the language of their people, the four young Hutterites had been subjected to treatment that can only be described as torture. Four months of isolation, nakedness, near starvation, cold, and darkness had ruined their health, but it had failed to weaken their beliefs or break their spirits. "He promised us, his children, that he will be with us till the end," David Hofer wrote home to South Dakota. The others believed the same.

Two weeks after the Armistice went into effect, the military police transferred the four men from Alcatraz to Fort Leavenworth in Kansas. They arrived in Kansas on the night of November 23, exhausted from the three-day train ride and debilitated after the months in solitary confinement. At the train station, the prisoners were chained, handed their luggage, and then, prodded by bayonets, marched to the military prison. By the time they reached the gate they were wet through with sweat—and when they stopped moving they began to shiver uncontrollably in the raw night air. The torture they had endured in San Francisco began anew. Military police ordered them to strip to their underwear and stand outside for two hours until prison uniforms were brought out to them. At one in the morning they were finally allowed to collapse in bed, but four hours later they were routed out and again made to stand outside in the cold. Joseph, age twenty-four, and Michael, twenty-five, complained to the guards of pains in their chest—probably pneumonia—and were placed in the prison hospital. Jacob Wipf and David Hofer were returned to their solitary cells. When they refused to work, Wipf and Hofer were made to stand for nine hours a day with their hands thrust through the prison bars and chained together.

Somehow Wipf learned that Joseph and Michael Hofer had fallen dangerously ill in the prison infirmary, and he telegraphed their families in South Dakota to come quickly. Joseph and Michael's wives, both of whom were named Maria, set out for Kansas, but due to a mix-up the women ended up at Fort Riley instead of Fort Leavenworth, and a full day was lost before they reached the prison that held their husbands. The wives did succeed in seeing Joseph and Michael briefly on the night of November 28, a Thursday, but by then the men were so far gone they could barely speak.

When the women returned to the prison the following morning, Maria was told that her husband Joseph was dead. She pleaded with the guards to be allowed to see his body one last time but was rebuffed. Finally the colonel in charge relented, and Maria was ushered into the room where the open coffin stood. To her horror she saw that Joseph was laid out not in the homespun clothing that he had worn when he left home six months before, but in the khaki uniform of the U.S. Army that he had refused to touch so long as he had breath to refuse.

Michael died three days later. His father and brother David were with him when he died—and as a result, Michael Hofer was buried wearing his own clothes. "I stood there all day and cried," David wrote afterward; "but I could not even wipe my tears away, since my hands were chained to the bars of my prison cell."

The Hofer wives returned to the Rockport Hutterite Colony in South Dakota with the bodies. Joseph's Maria, seven months pregnant, had two small children waiting for her at home. Michael's Maria had an eight-month-old daughter. On December 4, the brothers were laid to rest under the Dakota prairie grass. Their small unadorned grave markers were inscribed with the years of their births and deaths and the word "martyr" after their names.

The army at last relented. The war was over and there was nothing to be gained by keeping two ailing Hutterites in prison and risking more embarrassment if they too perished. David Hofer was released from Leavenworth on January 2, 1919; Jacob Wipf's release came on April 13. The men returned to their wives and children in South Dakota and told and retold the story of their ordeal.

On January 28, 1919, Joseph's widow, Maria, gave birth to his child—a son she named Jacob—but the baby died six weeks later.

The Hutterites of the upper Midwest were traumatized. As a direct result of the deaths of Michael and Joseph Hofer, nearly all of the Hutterite colonies in the Dakotas sold their holdings and relocated to Canada, mostly Alberta. Better to move on, as they had done so many times in the past, than face the martyrdom of more of their young men. Many colonies, however, chose to return to the States after World War II. In that war, unlike the first, provision had been made—made clearly and with the cooperation of the so-called peace churches—for conscientious objectors. The government now offered two options to men whose faith forbade them to fight: either they could serve in the medical corps or some other noncombatant unit of the military; or they could remain in the States and perform "alternative service" through the Civilian Public Service. Those who chose the latter were interned in CPS camps for the duration of the war and put to work doing manual labor for nine hours a day, six days a week. It was grueling and deadening work, and for the privilege of performing it each man was assessed $35 a month for room and board. But it was better than what Jacob Wipf and the Hofer brothers had been forced to endure.

Today large Hutterite colonies flourish once again on the prairie of the Dakotas and eastern Montana. Their members still speak and worship in German. They still refuse to wear the uniform of the armed forces or fight in the nation's wars. They still remember the ordeal of Joseph, Michael, and David Hofer and Jacob Wipf. When Hutterites travel to San Francisco nowadays, it is for one purpose only: to pay their respects at the dungeon at Alcatraz where four of their brethren were imprisoned and tortured in 1918. Someday Hutterites hope to see a plaque placed at Alcatraz commemorating what their ancestors suffered while the nation was fighting to make the world safe for democracy.

Epifanio Affatato arrived back in New York on March 6, 1919, on board the world's largest ship, the *Leviathan* (known as the *Vaterland* until the U.S. Navy seized it from Germany when the country went to war). While the immense three-funnel steamer was anchored offshore waiting to disembark thousands of eager veterans of the 27th Division, the Prince of Wales (the future Edward VIII and then, following his abdication, the Duke of Windsor) came on board to bestow British medals. Epifanio was awarded the British Military Medal for his heroic acts of September 29—an honor he would cherish all his life.

The division's welcome-home parade took place on New York's Fifth Avenue on March 25, and it was an epic event even by New York standards. Crowds estimated at 3 to 5 million people turned out to watch the soldiers parade up Fifth Avenue from Washington Square Park. Two airplanes kicked off the gala by buzzing low through the canyon of skyscrapers, and nine black horses, one of which was riderless with stirrups reversed, moved out in solemn procession escorting a caisson decked with memorial wreaths. Then in a mighty surge came the soldiers of the 27th Division, marching up the avenue company by company. The men of the 107th Infantry had painted their battle helmets a darker shade, so Epifanio and his fellow survivors of the Hindenburg Line stood out like a dark current in the river of bobbing helmets.

Epifanio wasn't quite done with the army. After the parade he had to go out to Camp Upton on Long Island to await his discharge, which came through on April 2. An honorable discharge, of course. At Camp Upton, he also took care of one other piece of official business: he became a citizen of the United States of America. No proof of residency or declaration of intention was needed—you fought for Uncle Sam, you were a citizen, bang, just like that. Hundreds of thousands of other guys did the same as Epifanio. In total, over 280,000 immigrant soldiers became citizens by virtue of their service in the war.

For Epifanio the high point of the postwar euphoria came on April 14, twelve days after his discharge, when he learned that he had been awarded the military's second highest award, the Distinguished Service Cross, for his heroic actions at the Hindenburg Line.

So now he was a war hero, a citizen, a buddy, a veteran—in a word, an American. All he needed was a job. Family memory is a little vague about what Epifanio did in the years immediately following the war—trade school, night classes to perfect his English, and whatever jobs he could pick up. He had a case made to store and display his war medals—the British Military Medal, the French Croix de Guerre, and the DSC. When the Great Depression hit and jobs were hard to come by, Epifanio took the case with him from one potential employer to the next. The bosses looked at the medals and shook their heads. Nobody was hiring.

Finally, in 1935 Epifanio landed a good steady job as a machinist with the New York City Department of Sanitation. That same year he bought a house and married Filomena Mancuso, a young Calabrian immigrant from a coastal village only a few miles away from Scala Coeli where Epifanio had grown up. The difference in age—he was forty, she was twenty-six—didn't matter to them in the slightest. They made a fine couple. They had three children, all boys—Domenick in 1936, Edward in 1939, Charles in 1941. Inevitably, as the boys grew up, they asked their dad about the war and the medals he had won—but invariably, Epifanio refused to talk about it. Occasionally, he had nightmares about the fighting, even years later, but he never told the boys what the nightmares were about. He had a small scar on his face where the shell fragment had hit, and he kept the bit of exploded steel that the doctors extracted as a souvenir. The other souvenir he showed the boys was the German officer's belt inscribed with GOTT MIT UNS that he had traded his warm hat for. But Epifanio kept the reality of the war to himself.

Despite his reticence, Epifanio was a proud and active veteran. For a while he served as president of the local Veterans of Foreign Wars chapter, and he remained buddies with a couple of guys from the regiment. Every year, he put on his medals and marched in the big Memorial Day parade on Eastern Parkway in Brooklyn. "My father was a shy, conservative man—not a flag waver," recalled Ed, the middle son; "but he was always very proud of his service and of his country." His country, meaning America. As for Italy, he had put the land of his birth behind him. Epifanio and his wife spoke Calabrese dialect around the house, and when the boys got to high school he taught them to speak proper Italian. But he never had the slightest desire to return to Italy.

Today, the two surviving Affatato brothers speak of their father with devotion and reverence that go beyond filial piety. "Everyone loved him," said Ed (who died of leukemia in 2009). "They thought he was like a saint—he would settle family problems, he took care of his parents and in-laws, he gave money to the poor, he was very patient and kind as a father. A kind, gentle, and loving man." A family story has come down that says a lot about Epifanio's fundamental humanity. The details differ a bit with the teller, but the gist of it is this: At the Department of Sanitation, Epifanio befriended a German who had emigrated to the United States after the war. The German fellow was missing one eye, and in time he let it be known that he had lost the eye while fighting in the German army—in fact, he had lost the eye in the battle of the Hindenburg Line in which Epifanio won his DSC. The German took to ribbing Epifanio, saying, "Maybe you're the one who shot my eye out." To which Epifanio responded, "Yeah, maybe you're the one who threw that grenade." The men became buddies and Epifanio introduced the German guy to his brothers and sons. Strangely, or maybe not so strangely, being enemies in the war sealed their friendship in civilian life.

Another family story speaks to the depth of Epifanio's love of his adopted country. He was forty-six years old in 1941 when the United States entered World War II. Nonetheless, when he heard the news on the radio about the bombing of Pearl Harbor and the declaration of war on Japan, he told his family he was going to reenlist. "What, are you crazy? You have three children," everyone said. "They'll never let you in the army again." Crazy or not, Epifanio believed it was the right thing to do. But the army turned him down.

Epifanio died in 1959 at the age of sixty-four, just shy of his twenty-fifth anniversary with the Sanitation Department. He had been hospitalized for a hernia operation and his sons believe he died of the pneumonia he contracted after the operation. Epifanio Affatato was buried at Long Island National Cemetery, a military cemetery in Farmingdale, Long Island, about thirty miles west of Camp Upton. His gravestone is inscribed with his name, his military unit, and the initials of his two American military honors—PH (Purple Heart) and DSC (Distinguished Service Cross).

A couple of weeks before the grand 27th Division parade in New York City, the *Ogdensburg News* ran a short article about local "celebrated hero of the great war" Mike Valente. Mike, the article stated, was "the type of youth that every person is glad to shake hands with not because he is about to receive distinguished laurels when the 27th Division parades in New York City but because he is very human and sincere." In one way or another, everybody who knew Mike said the same thing. Mike Valente was a great guy and the role of war hero fit him like a glove.

Though Mike never again did anything as spectacular as his rampage at the Hindenburg Line on September 29, 1918, his day of heroism was enough to keep him in the public eye, in a modest way, for the rest of his life. When things settled down after the postwar festivities, Mike moved down to Newark, New Jersey, to study electrical engineering at the Newark Technical School. He married a Sicilian woman named Margarita he had met in Newark, and the couple settled in Long Beach on the south shore of Long Island. Mike ran a building and electrical contracting business, and he sold real estate during the boom years of the 1920s, when property prices kept shooting skyward.

On September 27, 1929, eleven years after his actions at the Hindenburg Line, Mike was given the nation's highest military award, the Congressional Medal of Honor. Whether the long delay was due to a bureaucratic foul-up or whether it was because he was Italian never became clear. "It's the proudest moment of my life," Mike told President Hoover when he accepted the medal at a ceremony in Washington, D.C. One hundred and twenty-four men received Medals of Honor for valor in the Great War, but Mike Valente was the only man of Italian descent among them. The Italian American press went crazy with pride. One dazzled reporter, begging Mike for a message to convey to the Italian American community, got this response: "Tell them this in my name: I'm happy that the duty I carried out in the war gained such high recognition in Washington yesterday. Say that I did not forget, while the president of the republic was conferring the Congressional Medal, that he had decorated an American of Italian origin. Proud of these origins, happy that through him honor can come to the entire mass of Italians who emigrated here, of which I am a humble part."

The stock market crash a month later hit Mike and his family hard. His real estate business failed and the contracting work dried up, but Mike managed to hold onto the Long Beach house—and the house remains in the Valente family to this day. In the 1930s and '40s, he became involved in local politics, serving as a Long Beach city marshal and a committeeman for the Democratic Party. He was active in the American Legion and VFW—but he always maintained that his favorite veterans group was the Jewish War Veterans. Even though he wasn't Jewish, he felt at home with these guys.

Like Epifanio Affatato, Mike had pretty much put Italy out of his mind. But history has a way of reminding people of their origins. In the next war, the United States fought a prolonged and costly battle to wrest Italy from the Germans, and at a critical juncture in this battle American planes bombed and destroyed the ancient monastery of Monte Cassino on February 15, 1944. Sant'Apollinare, the satellite village of the great Benedictine monastery where Mike was born and raised, was bombed in the attack and some of his relatives and family friends were killed.

When Dwight David Eisenhower was inaugurated president on January 20, 1953, Mike was invited to Washington to attend the ceremony with a squad of World War I MOH winners. Five years later, on Memorial Day 1958, he attended the internment of the Unknown Members of the Armed Forces of World War II and Korea at Arlington National Cemetery. By nature affable and mild-mannered, Mike made it a point of pride never to use the medal to advance himself or gain special privileges. As he got older, he spoke with a touch of wonder, even incredulity, of the twenty-three-year-old soldier who had blasted through the German trenches all those years ago in Picardy. It was hard for his grandkids to picture the raging, adrenaline-pumped war hero in the kindly, courtly gent who rode his bike along the Long Beach oceanfront and grew tomatoes in his backyard. But the soldier's steel never went to flab. "He was like a bull," recalled grandson Ralph Madalena. "He kept himself in fantastic shape all his life."

Had his country called on him to serve again, Mike Valente would have answered the call proudly and unhesitatingly. He died at the age of eighty on January 10, 1976, and was buried at the Long Island National Cemetery, a few paces from the grave of his exact contemporary Epifanio Affatato, fellow Italian American, fellow veteran of the 107th Regiment, fellow hero of the breaking of the Hindenburg Line.

N ot all the foreign-born soldiers ended up as rooted and happy as Mike Valente and Epifanio Affatato. Some drifted. Some lived on the margins. Some never recovered from war wounds, whether physical or psychic.

Alexander Raskin, a smart, good-looking Jewish immigrant fight-

ing with the 78th Division—the one who wrote in horror to his girlfriend about recruits rioting on the train in New Jersey—was never
the same after being gassed in the Argonne. He returned home, married his sweetheart, got his law degree, had children—but he never
fully regained his health or vitality. For the rest of his life, Alexander
suffered from lung disease. He never really felt whole after the war,
and despite his law practice, he never made enough money to support his family the way he wanted to. It was a fate that many other
victims of gas attacks shared. In fact, medical studies have established
that the long-term effects of poison gas are both physical and psychological: men exposed to gas were susceptible for the rest of their lives
to chronic ailments ranging from bronchitis, emphysema, laryngitis,
and cancer; adverse psychological effects included mood and anxiety
disorders, sexual dysfunction, and post-traumatic stress syndrome. The
excruciating external blisters raised by mustard gas eventually healed,
but the damage to the respiratory system and the mind of the victim
was permanent.

The expatriate artists and writers in Paris were not the only lost
generation.

Polish Americans had a double stake in the war once the Wilson
administration committed itself to the cause of Polish independence. Whether they had fought in the semi-autonomous Polish army
in France or with the AEF, America's Poles fervently believed that the
Great War was *their* war. Yet when it ended and an independent Polish
Republic was wrested from Russia, Prussia, and Austria, Polish Americans were faced with a dilemma: return to the homeland they had
been dreaming of for so long, or remain in their adopted country?
Most chose to stay, realizing as never before how thoroughly, ardently
American they had become. They might fly the Polish flag side by
side with the American flag on Memorial Day; they might teach their
children to speak Polish and worship at the Polish church—but after

the war most Polish Americans acknowledged that their future was in America and only in America.

J oseph Chmielewski was one of the drifters, one of the Polish vets who came home not to hang flags on his porch or pay tribute to the dead, but to struggle and search and finally sink beneath the surface of American life until he disappeared without a trace.

Though he had been among the first to enlist when the nation went to war, Joe never saw any action with the 16th Machine Gun Battalion; still, he ended up staying in the army for a full two years, from June 17, 1917, until his honorable discharge at Camp Dix in New Jersey on June 26, 1919. At a loss for where to go or what to do next, Joe returned to Fifficktown, Pennsylvania, and moved in with his brother Frank and his sister-in-law Mary and their growing family. Inevitably, he went back to work in the coal mine. "The war was the best thing that ever happened to him," one of Frank and Mary's sons later said about his uncle Joe. "He learned a lot about America by being a soldier, learned to speak English, connected more closely with this country." But somehow the connection was never strong enough to make Joe really take hold. He had never liked being a coal miner, and since that's the only work there was in South Fork, he left a few years after the war—the family thinks it might have been 1925—and tried his luck in Michigan. According to the census, in 1930 Joe was lodging in Detroit with a Macedonian immigrant chef and his German American bookkeeper wife and working in an auto factory.

Joe apparently returned to Fifficktown in 1932; he lived for a year or so with Frank and his family and went back to work in the mine. According to one of Frank's grandsons, Joe's return had something to do with the higher wages won for coal miners by John L. Lewis, the heroic leader of the United Mine Workers of America. In any case, by 1934 he was back in Detroit. That year he claimed $200 in veteran's compensation under the terms of a bill enacted on January 5, 1934;

when his application was approved the following year, he was living at 303 Highland Avenue in Saginaw, Michigan. After that, the traces grow fainter. One of Joe's nephews thinks he moved to Duluth, Minnesota, and went to work as a deckhand for the Grantland Steamship Company. At some point he served with or was employed by the U.S. Coast Guard. One day in the 1940s a package turned up at the South Fork post office addressed to Joseph Chmielewski—and it was delivered to Frank's son who was named for his uncle. The package contained some of Uncle Joe's clothing and an uncashed check made out to Joseph Chmieliewski dated July 4, 1945, and drawn on the Continental Illinois National Bank. Frank's family assumed Joe had died—but they never learned any details.

Compared to his younger brother, Frank Chmielewski led a happy, robust life, but it wasn't a long one. Frank fathered seven children—four sons and three daughters—and he could barely contain his pride when three of his boys went to fight for their country in World War II. But Frank was never the same after one of them died in a training operation while still in the States: it was the third son, Joseph, named for the uncle who had fought in the Great War. Frank died in 1949, at the age of sixty-five, of silicosis—the lung disease that shortened the lives of so many miners.

A month after the Armistice, the Ivy Division was ordered to march into Germany as part of the Army of Occupation, and Meyer Epstein and his comrades in the 58th Infantry took up residence in the city of Koblenz, southeast of Bonn at the confluence of the Moselle and the Rhine rivers. "The area now occupied by the Regiment was a very pleasant one," according to the regiment's historian. "The German people were kind, courteous, and very amendable to the restrictions placed upon them by the military authorities." But some of the shine wore off as the months dragged on with no prospect of returning home. Training and drilling, which seemed so pointless now that the war was over, continued with maddening monotony. In their free time, the guys played basketball, watched movies, and took boat trips on the Rhine. Baseball teams were organized with the first hint of spring weather. The regiment ran mandatory classes for men who were unable to speak or read English—but by this point Meyer's English was quite proficient, so he was probably excused.

Meyer and his comrades finally sailed back to the United States on board the *Mount Vernon* on July 24, 1919, arriving in New York Harbor at midnight of July 31. Six days later, Meyer received his honorable discharge and went back to the Lower East Side—and back to work.

Meyer was lucky in the years that followed. Though he lived through tumultuous times in an uproarious city, he managed to lead a steady, rewarding, honorable life. He never lacked for what he valued most—love, respect, faith, and hard work. In 1922, he married a fellow Jewish immigrant from the Pale named Ida Rubinstein—a vivacious twenty-two-year-old woman who had been working in a garment factory since the age of twelve and who could recite long passages of Wordsworth and Kipling from memory—and the couple moved to Brooklyn and started a family. Their first son, Julius—named for Meyer's deceased father Yehuda—was born in 1923, and two more sons followed—Harold in 1925 and Leonard in 1931. In 1926, Meyer received his license as a master plumber, and in time he also became a licensed steamfitter and oil burner mechanic and installer. For the next forty years, he worked steadily and with scrupulous honesty and reliability to support his family. Every Saturday he went to pray at the Glory of Israel Synagogue in the East New York section of Brooklyn, and for a time he served as the synagogue's vice president.

But it was his year of service with the Ivy Division that remained Meyer's proudest moment outside of his family. "To him, November 11 was like Yom Kippur—always a very important day," said his youngest son, Len. Many other Jewish war veterans felt the same. American Jews were quick to point out that not only had they served their country loyally, they had also served in sizable numbers. Though only 3.27 percent of the U.S. population in 1917, Jews made up 5.73 percent of the army; 72 percent of Jews in uniform served in combat units, compared with 60 percent of all military personnel. Two thousand American Jews were killed in action, and total Jewish American casualties topped ten thousand. "When the time came to serve their country under arms," General Pershing wrote in gratitude, "no class of people served with more patriotism or with higher motives than the young Jews who volunteered or were drafted and went overseas with our other young Americans to fight the enemy."

Meyer kept his AEF uniform, gas mask, and steel helmet all his life and proudly displayed his victory medal and service ribbon with four Bronze Stars, one star for each of the battles he had fought in France. He told his sons stories of mustard gas, trench foot, living in mud and water, and eating stale bread. He was an active member of the VFW and the Jewish War Veterans.

Meyer wept the day his oldest son, Julius, was drafted in 1942—not because he opposed the nation's entry into World War II, nor because he wanted his boy to sit it out, but because he knew what war was and what his son would have to face. Julius fought in an antiaircraft unit in the Coast Artillery and saw action in New Guinea and the Philippines. Harold, the second son, enlisted in the navy and served from 1943 to 1946 as a radio operator on a destroyer escort. Len was too young for World War II, but he was drafted during the Korean War and served from 1953 to 1955.

To his dying day, Meyer Epstein remained fiercely proud of his military service. Len recalls visiting his father in the hospital in the last week of his life as he struggled with colon cancer. "My father felt that the hospital staff was not attentive enough (and they weren't!)," says Len today. "As we approached his room, we heard him complain indignantly, 'Is this how you treat a war veteran?'"

Meyer passed away on July 23, 1976. "When my parents died," wrote Len, "several strangers came up to us at the funerals and told us of how they brought bags of groceries to them each week, without which they would have starved. My parents could not pass a beggar without opening their purse. They also gave substantially to religious institutions."

The tradition of *tzedakah* that had saved him from homelessness in the Pale of Settlement and his military service in the Great War remain Meyer's chief legacy to his family.

M eyer Epstein was still in Germany with the Army of Occupation on May 21, 1919, when an estimated half a million people took to the streets of New York City to protest the pogroms that were sweeping through Poland, Galicia, and Rumania. Ten thousand Jewish American war veterans, most of them from the 77th "Melting Pot" Division, led what became known as the Mourners' Parade—and in fact, it was the presence of veterans that kept the huge demonstration orderly. A second protest rally, dubbed a Day of Sorrow, was organized in the city on November 24 to protest the mass killings of Jews in the Ukraine. Meyer by now was back in the States, and he may well have been among the estimated twenty-five thousand Jewish war veterans who marched that day from the Lower East Side to Carnegie Hall.

These postwar attacks on Jews in eastern Europe—attacks which the struggling new governments of Poland and Ukraine refused to stop or cynically condoned—presaged an era of murderous anti-Semitism. The infection of hatred spread through eastern and western Europe, ultimately becoming most virulent in Germany, but no Western nation was immune, certainly not the United States. Anti-Semitism was high on the agenda of the Ku Klux Klan, the post–Civil War white-supremacist brotherhood that came roaring back to life in the 1920s. Jews, blacks, Bolsheviks, Catholics, and foreigners—all became public targets during the Red Scare period. Congress, responding to the ugly national mood, voted overwhelmingly to bar the door to "undesirable" immigrants. The Emergency Quota Act of 1921 and the even more stringent Immigration Act of

1924 imposed tight quotas based on national origin. The 1924 law restricted new immigration to 2 percent of the population of a given nationality resident in the United States in 1890 (with the total yearly number of immigrants limited to 164,667 individuals) and banned Asian immigration altogether. The effect of these new laws was to award the relatively small number of slots to "Nordic" immigrants (since the nation's population in 1890, even with the mass migration from eastern and southern Europe, had been predominantly British and German), and to keep out the "inferior strains." Immigration from Italy plunged from about 200,000 in 1900 to about 4,000 in 1924—and the figures for Poland, Russia, and the regions that had been part of the Pale were similar. Never mind that aliens had "done their bit" in France. In the Red Scare, everything alien was suspect.

It was in this poisonous atmosphere that Sam Dreben, the Fighting Jew, returned home to El Paso, Texas. Dreben had had a glorious war—the capstone to his glorious military career. He was awarded the DSC for bravery at Blanc Mont, and after the Armistice, General Pershing made Dreben his personal guest at the American headquarters in Paris. Pershing, whose acquaintance with Dreben dated back to the Punitive Expedition in Mexico in 1916, went on record stating that the Fighting Jew was "the finest soldier and one of the bravest men I ever knew." Dreben even got to fulfill a lifelong dream one night in a Paris restaurant when he encountered Nikolai Nikolaevich, the exiled Russian grand duke who had been commander in chief of the Russian army during the war, and socked him in the nose in revenge for generations of pogroms.

When the 141st Infantry pulled into El Paso in April 1919, Dreben and his buddy, Captain (now Major) Burges, were welcomed home as war heroes with the usual rounds of parades, toasts, and flowery editorials. But for Dreben the postwar euphoria quickly turned sour. His young wife had never forgiven him for failing to come home after the death of their infant daughter, and there were rumors that she had consoled herself with other men. Two months after the hero's welcome, Dreben filed for divorce.

The Fighting Jew picked up the pieces of his life, swallowed his pride, got a job selling insurance, and became active in the American Legion, the veterans organization founded by returning soldiers in 1919. This became the site of Dreben's last great battle. When Dreben learned that the KKK was trying to win control of the El Paso post of the American Legion with its own slate of candidates, he took to the field again—not with weapons but with words. He proposed a resolution barring Klansmen from holding office in the El Paso post. In a rousing speech defending himself, and by extension all immigrant soldiers, from the calumny of the Klan, Dreben declared: "I am a loyal member of the Legion and a loyal American citizen. These men [indicating the Klansmen], oath-bound to secrecy, hide behind their masks and say that because I am a foreign-born Jew I am not good enough to be an American. Every time America has called for volunteers, I have put on the uniform. They did not ask me at the recruiting office if I was a Jew, and they did not ask me on the battlefield what my race or religion was. The soldiers didn't wear masks in France, other than gas masks, and they don't need them now."

In the heated debate that followed, the chairman ruled that Dreben was out of order; but Dreben pressed for a vote and his resolution carried unanimously. "It was the first major defeat for the Klan in El Paso," wrote Dreben's biographer.

On the third anniversary of the Armistice, General Pershing invited Dreben to Washington, D.C., to be part of the honor guard at the burial of the Unknown Soldier at Arlington National Cemetery. Among the other war heroes chosen by Pershing to accompany the casket of the Unknown from the Capitol Rotunda to Arlington were Sergeant Alvin York of the All-American Division and Major Charles Whittlesey, commander of the Lost Battalion.

A crowd of a hundred thousand lined the five-mile route of the procession. At the stroke of noon, the artillery barrage that had been booming all morning held fire as the nation observed two minutes of the Great Silence in honor of the fallen. But even on this reverent occasion there were mutterings of disillusionment and disgust. John Dos Passos ended his brilliant novel *1919*, the second volume of the U.S.A. Trilogy, with a scathing reverie on the interment of the Unknown. Dos Passos did a kind of mordant jazz riff on how the "body of an American" had been chosen from the rows of reeking coffins stacked in the "tarpaper morgue at Chalons-sur-Marne":

> *Only one can go. How did they pick John Doe?*
> *Make sure he aint a dinge, boys,*
> *make sure he aint a guinea or a kike,*
> *how can you tell a guy's a hundred percent when all you've got's a*
> *gunnysack full of bones, bronze buttons stamped with the screaming eagle*
> *and a pair of roll puttees?*

Dreben would have been horrified.

In 1923, the Fighting Jew married a young widow from Dallas and moved to California. He died two years later at the age of forty-seven as the result of a bizarre medical error. Ailing and drinking heavily, Dreben was being treated by a Los Angeles doctor; in the course of one of his treatments, a nurse accidentally gave him a syringe filled with a toxic substance. The *New York Times* reported on March 18, 1925, that "the internationally famous soldier of fortune and hero of the World War" died of "accidental poisoning." "He fought on scores of battlefields," stated the *Times* in its obituary, "sometimes as an American soldier, sometimes as a freelance. He won every possible medal for bravery in the World War."

Glowing tributes appeared in the two El Paso newspapers. "Here's to Sam Dreben," toasted the *El Paso Herald*, "Jewish immigrant, of old world peasant stock, a fine upstanding American who loved the United States with a passionate devotion of which many of us are incapable." The *El Paso Times* ran a more sharply pointed memorial reflecting the climate of the times: "It is with some bitterness that one realizes that had the 'alien laws' now in force in this country been in effect when Sam Dreben reached our shores, he would have been barred. We would have no place for the immigrant boy who proved one of the greatest heroes of the American Army."

"He had one great, tremendous love, and that was love of his adopted country and his country's flag," wrote Dreben's friend Damon Runyon in his newspaper column. "It was his religion. I have known many thousands of men who have worn the uniform of the American Army but I have never known a man who held it in such absolutely devout love as Sam Dreben."

M any an immigrant family sent their son to war in ignorance of the laws of the land. Did a foreign-born boy have to serve if he hadn't yet declared his intention to become a citizen? On what grounds could an alien request an exemption? If he did serve, what would he be paid? If he died, what would happen to his body?

When their sons failed to return from France, these same families learned the answers to all of these questions to the letter and the decimal point. The War Department rose splendidly to the occasion of death in war. Bereaved families were, eventually, informed in exhaustive, repetitive detail of everything the army knew about their sons' earthly remains and the financial arrangements they had made before they died. For many immigrant families, the official response to the death of a son profoundly affected their futures in the United States, both emotionally and financially. For some it determined their very survival.

Private Max Cieminski died in action in the Aisne-Marne offensive on July 22, 1918, and was buried the following day near where he fell outside the village of Trugny, France. His death and burial set in motion a series of actions and reactions in the Cieminski family that continues to this day.

On June 10, 1919, almost a year after he was killed, Max's body was disinterred and reburied with thousands of other American soldiers at a large military cemetery outside the French hamlet of Seringes-et-Nesles. Two years after his reburial, Congress authorized this ground as a permanent American military cemetery and designated it the Oise-Aisne Cemetery—the second largest of the eight permanent World War I military cemeteries in Europe. At this point the army got in touch with the Cieminski family to inquire about their wishes in regard to the final resting place of Max's body. The Cieminskis responded that they wanted him returned to Wisconsin. Nearly three years after his death, the process of bringing Max home began.

A small folder stored at the National Archives documents exactly how and when Max's body got from Seringes-et-Nesles in France to Polonia, Wisconsin. This folder is the most thorough and detailed record of Private Cieminski's war service, all of it, unfortunately, posthumous. A "report of disinterment and reburial," dated May 17, 1921, notes that the body was originally buried in a "5 ft. earthen grave, uniform, burlap, disintegrated, unrecognizable"; the report indicates that it was impossible to determine his height or weight; under "wounds or missing parts" is noted: "left shoulder blade shattered, G.S.W. in base of skull, left upper arm missing." The only intact identifying marks were the ornaments on his collar: Max's buttons indicated that he served with Company B, 345th Regiment, United States National Army. His remains were exhumed from the Oise-Aisne Cemetery, transported to Antwerp, placed on board the SS *Wheaton*, and shipped to Hoboken, New Jersey. On July 20, in the care of one Sergeant James V. Hendrix, the coffin was placed on a car of the Soo Line Railroad and sent west via Minneapolis and Sault Ste. Marie to Stevens Point, Wisconsin. Here, Sergeant Hendrix surrendered the coffin to Max's brother, Boleslaw. On July 22, Boleslaw signed a receipt stating that he had received the body "in satisfactory manner and condition," and Sergeant Hendrix returned east to his post.

The official record ends here and family lore takes over. At some point between the arrival of Max's body in Stevens Point on July 22 and his funeral at the Sacred Heart Catholic Church in Polonia on July 26, Boleslaw opened the coffin and looked inside. What he saw devastated him. Not only did the pile of bones and rotting flesh wrapped in burlap bear no resemblance to the smiling blue-eyed young man he had last seen three and a half years earlier—but the brass buttons on the corpse's uniform convinced Boleslaw that this was not in fact his brother. Max had died while fighting with Company C, 102nd Infantry, 26th Division—but the brass collar ornaments on these remains were stamped Company B, 345th Infantry Regiment. Not Max's unit—not even close.

Nonetheless, the funeral went ahead as planned on July 26. Friends of Max who had fought in the war put on their uniforms and carried the coffin to the church cemetery; a volley was fired in tribute as the body—*someone's* body—was lowered into the ground. The Cieminskis wept in grief and bitterness that their young man had given his life for his country, but his country did not even have the decency to send the proper body home for burial.

Before he went to war, Max had told his sister Mary that he was making her the beneficiary of his war risk insurance policy. He promised that no matter what happened to him, she and her daughter Marguerite, his favorite niece, would be taken care of for the rest of their lives. Mary had wondered at the time what Max was talking about. Now she found out. Every month, Mary received a check for $57.50—not a huge sum of money even back then, but enough, as it turned out, to keep Mary and her family going through the Great Depression. For the rest of her life, Mary remembered Max in her prayers and grieved for him.

In 1984, Marguerite, who had been eight years old when her uncle died, petitioned the army to award Max the Purple Heart. Sixty-six years after he died in a wheat field in central France, Max Cieminski received his medal.

Marguerite's son John Riggs, himself a decorated Vietnam War veteran, says that his mother and grandmother never got over the ordeal of burying the wrong body. The psychic wound never healed. Now Riggs believes that his relatives may have suffered needlessly. Reopening the files after I contacted him, Riggs realized that the brass ornaments on his great-uncle's tunic collar—345th Infantry, Company B—were in fact the insignia of the unit he had trained with at Camp Pike in Arkansas. "When I left Vietnam," Riggs wrote me, "I was not issued a dress uniform until I got to Oakland. Then they gave it to me, and everything to put on it to go home in." Riggs speculates that the outdated brass ornaments on Max's collar may have been the result of a similar army oversight. He has now concluded that the body shipped to Stevens Point in 1921 was in fact his great-uncle, Private Max Cieminski.

The body of Tommaso Ottaviano was also brought back to the United States for burial at the request of the family. Tommaso's mother, Antonia, had made contact with the army on April 26, 1920, stating her desire to have "the remains of my son sent to my home," and the process of exhuming and shipping the body was carried out that September. Unlike the Cieminskis, the Ottaviano family at least had the solace of paying their final respects to a body they believed had been their son and brother. Tommaso's headstone in the North Providence cemetery bears his photograph in his Doughboy uniform and this inscription:

TOMMASO OTTAVIANO

310TH INF. CO. I

FOUGHT AND FELL IN FRANCE

NOV. 22, 1918

AGE 22 YRS 6 MOS

BORN IN ITALY

In his last letters to his mother, Tommaso had written anxiously and repeatedly about his army life insurance—and his foresight in this matter ended up providing his family with a sizable legacy. According to his great-nephew John Ricci, Tommaso had actually taken out two insurance policies totaling $15,000—a considerable sum in the 1920s. With the proceeds of her son's insurance, paid out over a period of twenty years, Antonia built a two-family home in North Providence. The ground floor was rented out, but the upper floor and the big vegetable garden in the backyard served as the Ottaviano family anchor for the next two generations. Without her son's sacrifice, Antonia would never have been able to afford this house.

But Tommaso's legacy went beyond the material. "He was the family hero and martyr," says his great-niece, Pamela Ottaviano Rhodes. "He was the oldest son and in the family he was revered like a saint. We still talk about him when the family gets together. He remains a touchstone for us." Pamela notes that her father, Otto, son of Tommaso's brother Ascanio, was obsessed all his life with his uncle's military service. In 1989, after extensive research in regimental histories and army records, Otto brought his family to France to seek out the place where Tommaso fought and was fatally wounded in the autumn of 1918. "We walked on the road leading over the hill," Pamela recalls, "and we entered the Bois des Loges where Tommaso was shot. I was surprised at how little had changed. My father found artillery shell casings in the woods. To me that ground is soaked in blood."

According to John Ricci, Tommaso inspired his family's continuing tradition of service in the U.S. Armed Forces. John's father, Basil, fought with the army air force during World War II, and after the war he put in twenty-eight years in the army reserve—as did Basil's brother John. "Tommaso did what his country asked him to do," John Ricci says today. "These immigrant families worked hard, they sent money back to the old country, they prospered if they were lucky. That's the American way. Military service is part of paying your dues in this country. They learned to be Americans, and their sons went to war and sometimes didn't come back. We all look to Tommaso—he was the first."

Pamela Rhodes has a different perspective. "The family did buy into the American dream. Others did serve in the wars that came after. Yes, there was pride in Tommaso's sacrifice. But there is also a lingering feeling that this was a lot to ask—the life of a young man, an oldest son with everything ahead of him. It was a lot to ask of my family."

Matej Kocak had told his superior officers that he had "no relatives or friends" to be notified of his death—but in time the army tracked down his family both in Binghamton, New York, and in the newly created nation of Czechoslovakia. There were two pressing matters to take up with the Kocaks: first, where to send the two Congressional Medals of Honor (one each from the army and the navy) that Sergeant Kocak had been awarded posthumously for his "extraordinary heroism" at Soissons on July 18, 1918; second, where to bury the hero's body.

Kocak had dedicated his entire American life to the marines, and in the end, only the marines turned out to be interested in the final disposition of his remains and medals. Though military authorities made numerous attempts to contact Kocak's sister-in-law, Julia, in Binghamton, she failed to reply or acknowledge receipt of the correspondence. Julia's life in America had not been easy—her husband was institutionalized for mental illness before the war, and on November 4, 1918, just a week before the Armistice, their seven-year-old son John died. Perhaps the burial of a brother-in-law she had rarely seen and never heard from was of little interest to her. On May 21, 1921, Kocak's case was closed and the final decision was made to leave the body in France. On October 13, 1921, the remains—"badly decomposed, features unrecognizable"—were placed in a casket and reburied at the Meuse-Argonne Cemetery in the shadow of Montfaucon.

Sergeant Kocak's medals—not only the two MOHs but also the French Croix de Guerre with Palm and the Italian Croce al Merito di Guerra—continued to generate official correspondence. Finally, on August 20, 1925, they were sent to Julia Kocak. Presumably Julia took them with her to Czechoslovakia when she returned to go to work as a housekeeper for her parish priest.

Word of Kocak's heroism and death eventually made its way to his parents in Czechoslovakia. In July 1937, his mother, Roza, now a widow, applied to the army for "adjusted compensation" for her son's pay, and her application was approved. At the height of the Depression and on the eve of the next war, Roza Kocak received a check for $563.75.

There is one final item of interest in the file: in 1943, twenty-five years after the marine's death, the city of Binghamton decreed the day of his death—October 4—Sergeant Kocak's Memorial Day. The stated purpose of this holiday was to encourage "increased participation in the American war effort" on the part of Slovak Americans. General Pershing, Theodore Roosevelt Jr., Josephus Daniels (Wilson's secretary of the navy), and Governor Thomas Dewey were among those who contributed testimonials to the Slovak Marine's bravery—and a "liberty ship" was named for him. At some point, the city of Binghamton quietly forgot about the holiday, and October 4 came and went without any further mention of Kocak.

The military has been more faithful. Over his grave in the Meuse-Argonne Cemetery stands a white marble cross with an inscription capped by a five-pointed star. Among the 14,246 crosses and Jewish stars planted in perfect rows in this beautiful garden of the dead—the largest American military cemetery in Europe—Kocak's marker is one of ten in which the chiseled letters have been etched in gold. Nine crosses and one Star of David at Meuse-Argonne gleam when the sun strikes them, signifying the burial place of a Medal of Honor winner.

Andrew Christofferson, having tossed his gun in the air on the final day of war, was discharged from the army at Fort D. A. Russell in Cheyenne, Wyoming, on June 29, 1919, and went back to his homestead north of Chinook, Montana. It was the source of some pleasure to Andrew that he now owned the land free and clear (the Land Office at Havre had issued the patent to the claim on March 5, 1919) but not much. By now Andrew knew all too well how hard it was to make this dry land yield a crop he could live on. He lasted four winters shivering out there by himself, and then, in 1923, he packed up his few possessions, moved into Havre, and went to work as a carpenter. It was the right move. Andrew proved to be a gifted carpenter—precise, patient, meticulous, painstaking—and he rarely lacked for work. He built his own house and the houses of many folks throughout the county; but the building he was always proudest of was his church. When a group of Christians in Havre got together to start an Assembly of God church, Andrew was among them. Since he was the only member of the fledgling congregation to own an electric saw, he practically constructed the new church building himself. Christian faith had been Andrew's rock and his anchor ever since his conversion at that Nazarene camp meeting back in North Dakota before the war; when he joined Havre's Assembly of God, he dropped his anchor for good. "He was in church every time the doors were open," as one who loved him put it.

In 1925, Andrew married Juline Ostrem, the sweetheart he had left behind in Norway fifteen years earlier and had faithfully loved and waited for ever since. They had three sons, but their happiness together did not last long. Juline fell sick and died in 1938. Three years later, Andrew married again and started a new family. He fathered four more children, the last one, Norman, born when he was sixty-five.

Andrew's sons, like Epifanio Affatato's, were always after their dad to talk about the war—but he wouldn't. "My father was not an emotional man," says Norman today, "but he would tear up whenever I tried to talk to him about the war. I don't know what he saw, but it must have been horrendous."

One legacy of the war that impressed itself on all the Christofferson children was their father's absolute ban on guns. As his daughter Nellie wrote, "Dad made a vow when he tossed his gun in the air on Armistice Day—he never again would pick up another gun, and he would not allow one in his home. I recall as a little girl, a friend of one of my older brothers brought a BB gun into the house to show it to my brother. Dad happened to meet the young man at the door, and in no uncertain terms, he told the boy, 'Get that thing out of my house! I saw enough guns in the war, and I will not allow one in my home!' Such forceful language was rare for Dad." Andrew become a U.S. citizen soon after the war, and he never skipped an election and always admonished his children to vote.

When the United States entered World War II in December 1941, this gentle Christian carpenter—a man who rarely raised his voice in pain or anger—threw himself on his bed and wept for all his family to see. "But I fought the war to end all wars!" Andrew cried. "Why am I sending my sons to war?" He had believed President Wilson's vow implicitly and it broke his heart to see his sons go off to fight another war.

On November 15, 1988, the *Havre Daily News* ran a photo of Andrew and his second wife, Ruth, with a caption noting that at ninety-eight years old he was Havre's oldest living World War I veteran. In the last months of his life, Andrew's Norwegian past resurfaced. One of his sons recollects that the week before his death Andrew quit speaking English and reverted to Norwegian. He had never told his children much about the old country—but now he described in great detail the place where he had grown up and the people he had known. His daughter Nellie remembers being roused by song at three in the morning when she was staying with her father in the last days of his life: Andrew was standing in his bedroom by himself and singing the Norwegian national anthem.

Andrew Christofferson read his Bible so often that he had to bind the volume with electrician's tape to hold it together. That beloved book was beside him when he died on December 12, 1988, at the age of ninety-nine.

Whoever advised Samuel Goldberg to enlist in the cavalry was a wise fellow. For sixteen months, while his compatriots fought and died and inhaled poison gas in the trenches of Europe, Sam patrolled the Mexico–New Mexico border on the lookout for German attacks that failed to materialize. He never made it to France, never saw any action. The war gave Sam nothing to brag about, but nothing to complain about either. If anything, when he was discharged on September 27, 1919, he was healthier than when he went in. The New Mexico climate agreed with him. No more scrawny Jewish kid from the ghetto. Sam's mother commented that he ate like a truck driver when he stopped off to visit her after his discharge.

I met Sam in the summer of 2006—ninety-nine years after he had landed on Ellis Island, eighty-eight years after he had joined the cavalry. At 106, he was the last living Jewish immigrant to have served in the American military during World War I—a scrappy, witty, outspoken, fearlessly opinionated man. Sam's eyes, though watery and weak, were still bright blue, his grip firm, his memory unbelievably sharp. Over the course of two hours, he told me about the blood he had seen during pogroms in Lodz, the stench of the immigrant ship that had brought his family to America, his fierce anger toward his bully of a father, still festering all these years later. He told me about the jobs he had held in the automobile business, his Irish pals in the old neighborhood in Newark, his occasional brushes with anti-Semitism. He said little about his wife and kids. Sam refused to draw any deep meaning from his experience as a Jew in the cavalry during the war. He was still angry at the bastard Polish sergeant who had tried to humiliate him by threatening to issue him a baby blanket. The part of his military service that seemed to bring Sam most pleasure was the memory of his best buddy, a fellow named Edward Moellering. Moellering was a German American from St. Louis who had grown up next door to a Jewish family and picked up Yiddish from hanging around with their kids. When the Jew and the German discovered that they both spoke Yiddish, a friendship sprang up. "That guy could do anything," Sam told me of his pal, "play the bugle, the drums, run the fastest hundred-yard dash. He was amazing." It pleased Sam no end when he overheard another soldier say, "Whenever you see Moellering, you see Goldie." They were that tight.

When I got up to leave, I was practically trembling from the intensity of having talked at length with someone who had been alive for so long and remembered so much. "I'll send you a copy of my book when it's finished," I promised. To which Sam replied, "I hope I don't live that long"—said not in bitterness but with the inexpressible weariness of having lived beyond his time. Samuel Goldberg died five months later on December 10, 2006.

You get in the army in wartime, your life is not yours," Tony Pierro told William Everett, the producer of a radio documentary about the last surviving World War I veterans, in the summer of 2006. "You're there to be used. You have an enemy, you want to kill him—get a hold of the enemy and kill him so he won't shoot at you. Shoot to kill." Tony's words were fierce, but if you listen to the interview, his voice comes across as kindly and relaxed. A sweet old gent summoning up another world, another time. Tony went on to say how happy he had been when the war ended and he got to resume his civilian life: "It's nice to be in peace now and do what you want. You do it your way—do what you want—nobody tells you what to do. I want to forget all those bad days. Thank God I came out alive."

Everett had spoken with Tony shortly before my own interview was scheduled, and I phoned him to ask how it went and get some advice. Tony was lucid, Everett said, but hard to get through to: it wasn't that his memory was gone, he just didn't care to say much. The one thing he still seemed to respond to was women, and Everett advised me that my interview would go better if I had a pretty young woman with me. So I pressed my daughter Emily, who fits the bill, into service when I visited Tony in Swampscott, Massachusetts, on July 8, 2006. Tony did indeed light up a bit when he saw Emily and pressed her hand warmly—but in the course of the interview he spoke little, and only after being pumped by his brother and his nephew. The only time Tony truly became animated was when the conversation turned to Magdalena. Then the ghost of a smile played on his face, his large faded eyes flickered inside their wrinkles, and he crooned deep in his throat, "Ahh, Magdalena." Somehow it was heartening to me that love was more enduring than war.

As we sat in the sun in the immaculate backyard garden of the Pierro home in Swampscott that summer day, I felt a great sense of gentle calm coming from Tony. His brother and nephew talked about how fanatical he used to be about keeping his car perfect and his house spotless, how guys used to call him the Duke because he dressed so neatly for his job as manager of an auto body shop and later when he worked on small aircraft engines at a General Electric plant, how sad he and his wife Mary had been that they couldn't have children. But all of that seemed long behind him. Most of the time, Tony appeared to be dreaming—dreaming deep beneath the calm clouded surface of consciousness. His nephew Rick showed me a photograph of Tony in his Doughboy uniform—spruce, back straight, dark brows nearly meeting over his nose, bright eyes fixed intently, boldly on the camera—and I tried to imagine that youth inside the old man in the baseball cap. As we got up to leave, Tony bent over and kissed my daughter's hand.

Antonio Pierro died on February 8, 2007, just days before his 111th birthday. He had been the oldest living American and the oldest man born in Italy, and he was the last foreign-born American soldier to have fought in the Great War.

IN MEMORY OF

COLONEL ROBERT ARMSTRONG,

UNITED STATES ARMY RESERVE

BORN NOVEMBER 12, 1948; DIED APRIL 3, 2008

SOLDIER, SCIENTIST, AND GREAT FRIEND

ACKNOWLEDGMENTS

I could not have written this book without the help and generosity of the families of the men I wrote about. The children, grandchildren, nephews, nieces, and cousins of the veterans were unfailingly patient through multiple interviews and follow-up questions. In particular, I'd like to thank:

Epifanio Affatato's sons Charles, Domenick, and especially Edward, who was truly heroic and unfailingly informative in answering emails and putting up with my efforts at corresponding in Italian. Ed's son Michael supplied additional anecdotes and enthusiasm. I was deeply saddened to learn of Ed Affatato's death shortly after I finished the book—just weeks after he fielded my final round of questions. Ed was a wonderful man and a devoted son. I would like to express my sympathy to Ed's wife, Gloria, and my thanks to their sons Michael and Jonathan for picking up where their dad left off. It has been my privilege to get to know and write about this fine family.

Meyer Epstein's sons Harold, Julius, and most of all Len, who repeatedly combed through family records and memories at my behest.

Joseph Chmielewski's nephew John Chimelewski; his niece, the late Josephine Rolincik; and Josephine's daughter Dorothy Vancheri and son Paul Rolincik.

Andrew Christofferson's daughter Nellie Neumann, his sons Clarence Christofferson and Norm Christofferson, and his widow Ruth Christofferson.

Maximilian Cieminski's great-nephew John Riggs.

Matej Kocak's cousin's daughter Marion Pekar.

Tommaso Ottaviano's great-niece Pamela Rhodes and great-nephew John Ricci. Pamela became a friend in the course of our many interviews, and that friendship has been among the many pleasures of working on this book.

Peter Thompson's granddaughter Christy Leskovar. I owe multiple debts of gratitude to Christy—first for granting me permission to draw on her extraordinary memoir *One Night in a Bad Inn* in creating my portrait of her grandfather, then for reading and correcting my passages about Peter, next for answering numerous questions along the way, and finally for helping me penetrate the labyrinth of the National Archives. Christy is one of the most brilliant and dogged researchers I have ever met—and surely the most generous.

Michael Valente's niece Michelina Rizzo and grandson Ralph Madalena. Special thanks to Ralph for sending photos of his grandfather.

It was my great privilege to meet with Tony Pierro and Samuel Goldberg shortly before their deaths. I'd like to offer my heartfelt thanks to both of them and to their families for facilitating the interviews. Mr. Pierro's brother Nicholas and nephew Rick were both extremely helpful and hospitable, as was his nephew Robert. Mr. Goldberg's son Philip Dorian was invaluable—indeed, the interview never would have happened without him.

Many other families shared stories, memories, journals, letters, scrapbooks, photos, pride, and love. I'd like to thank Diane Lopez for sending me the diary of her grandfather Leonardo Costantino, and my mother's cousin, Barbara Weisenfeld, for talking with me about the World War I experiences of her veteran father, Hyman Cohen, and for putting me in touch with *her* cousin, Barbara Siegel, who shared stories of her own war veteran father, Alexander Raskin, as well as the war letters he wrote to his future wife. Among the others who came forward with stories of ancestors in the war were Richard Warmowski, Rebecca Herzfeld, Stanley Shapiro, Bernard Kessler, Robert Angelo, Eric Boe, John Polomis, Bernadette Wiermanski, Sophie Thom, Elizabeth Parthun, Genevieve Karpowicz, Frank Milewski, Phyllis Hazecamp, Karen Majewski, Mary Bartus-Sidick, Lucia Morelli, Peter Pellicoro, Barb Alvord, Sidney Levin, and Robert Snoozy.

I'm grateful to Duane Schrag of Freeman, South Dakota, for helping me research the wartime persecution of Mennonites and Hutterites. Lynell Hofer told me about the torture and deaths of the two Hofer brothers, who were kin of his, and he put me in touch with Norman Hofer, who supplied books, articles, and many useful details. I am truly indebted to Norman for putting me in touch with Susan E. Cohn, who helped me in more ways than I can describe. Susan has devoted a huge amount of her time, energy, and imagination to bringing to light the story of the Hofer brothers and Jacob Wipf, and she generously shared with me material she has found. I am deeply touched and inspired by her moral passion and commitment to justice.

Thanks to Jim Skurdall and Vidar Warness for translating from Norwegian; Agnes Migiel for Polish translation; Myra Mniewski for Yiddish translation; and thanks to my Italian teachers Cecilia Strettoi and Virginia Agostinelli for helping me correct and refine Italian translations. Thanks to Manning Bookstaff for help with research in Milwaukee.

I received help and encouragement at all stages of the book from a number of writers, university professors, and researchers. Thanks in particular to Jonathan D. Sarna, Daniel Soyer, Nancy Gentile Ford, Nancy Foner, Donna Gabaccia, Deborah Moore, Glennys Young, Mary Wyman, Elliott Barkan, Alan Kraut, Brad Baltensperger, Rich Loftus, Joseph T. Hapak, Stanley Cloud, General Edward Rowny, Paul Va-

lasek, Anthony Bajdek, Ewa Wolynska, James Pula, and Jan Lorys. Tom
Simons, Ewa Sledziewski, and Dorothy Pula Strohecker helped me get
going on my research into Polish immigrants. Stan Ingersoll, David
Aune, and Priscilla Pope-Levison answered theological questions.

I'd like to thank Christina Holstein and Tony Noyes of Verdun Tours
for expertly guiding me through the battlefields in the Argonne region
and my daughter Alice for help with translation in France and for her
patience through three days of all Great War all the time. My sister-in-
law Jane Cowles and my nephew Alex flawlessly guided us from Paris to
Picardy to Rheims and shared adventures along the way. Thanks also to
Mureille Defrenne and Craig Rahanian at the Somme Cemetery, Phil
Rivers at the Meuse-Argonne Cemetery, and Jean-Paul de Vries of the
Romagne '14-'18 Museum in Romagne-sous-Montfaucon.

I'm grateful to the staffs of the following libraries and archives
for help with research and for valuable resources: Suzzallo and Allen
Library at the University of Washington; the Interlibrary Loan Division of the King County Library System (Washington State); Frederic
Krome at the Jacob Rader Marcus Center of the American Jewish Archives; Wes Anderson, curator of the Barnes County Historical Society,
Valley City, North Dakota; Joel Wurl, Donna Gabaccia, and Daniel
Necas at the Immigration History Research Center at the University
of Minnesota; the Polish Museum of America in Chicago; Department
of Special Collections, Young Research Library, University of California, Los Angeles; Kim Holland at the Norwegian-American Historical
Association in Northfield, Minnesota; Cambria County (Pennsylvania)
Historical Society; Heinz History Center in Pittsburgh; David Keogh at
the U.S. Army Military History Institute at Carlisle Barracks, Pennsylvania; Marlene Welliever of the Wibaux County (Montana) Museum;
Ogdensburg, New York, Public Library; El Paso County Historial Society; Patrice Kane, Fordham University Archives; the American Jewish
Historical Society at the YIVO Institute for Jewish Research in New
York City; Susan Lemke at the National Defense University in Washington, D.C.; Dee Anna Grimsrud at the Wisconsin Historical Society;

Steve Daily at the Milwaukee County Historical Society; Mitchell A. Yockelson at the National Archives and Records Administration in College Park, Maryland; Brian Shovers at the Montana Historical Society; Douglas Murdock at the Sioux Falls (South Dakota) Public Library; Cheryl Waldman at the National Museum of Jewish American Military History; the Liberty Memorial in Kansas City, Missouri; Kara R. Newcomer of the United States Marine Corps History Division in Quantico, Virginia; Victoria Leslie and Carol Cepregi at the Medal of Honor Society.

Peter Kacur has done a fantastic job of researching the life and military career of Matej Kocak; I'm deeply grateful to him for sharing his knowledge and for sending me a copy of Sergeant Kocak's complete military records.

Thanks also to Bill Rogers, Richard Rhodes, Lee Soper, Mike Knudson, and Lori Ann Lahlum for steering me in the right direction early in my research and, in Bill's case, for sharing his superb photos. Thanks to Chris Scheer and Robert Erickson for help in tracking down surviving veterans. Thanks to Eunice Unger for help at the start and for her patience during the long years I held onto volumes from her collection. Thanks to Jim Witkin, Barbara Roberts, Avice Meehan, Darlene Sadler, and Leslie Armstrong for hospitality in Washington. D.C. Thanks to Mary Ann Gwinn and Michael Upchurch of the *Seattle Times* for sending me so many Great War books.

Mark H. Feldbin has done a wonderful job putting together and maintaining the O'Ryan's Roughnecks website and I'd like to thank him for sharing his knowledge and for poring over trench maps of the Hindenburg Line on my behalf.

My thanks to Tim Kuzma, Frank Tapek, and Stan Gurzynski for help with the Polish Falcons; to Dona de Sanctis, director of Order Sons of Italy, for helping me find descendants of Italian American veterans; to Will Everett of Treehouse Productions for sharing tapes from his interview with Tony Pierro; and to Stephen Harris for invaluable research tips. Thanks also to Major William R. McKern and Angie Owens.

My gratitude also to Chester Kozik, Eugene J. Dreger, John E. Skarzenska, Mary J. Gaynor, Fred Cyran, Irena Szewiola, Edward Sobczak, and Michel Morel for responding to my requests for information.

Thanks to my brother-in-law, Dr. Donald O'Neill, for helping me with questions on the physiology of combat wounds; and thanks to Dr. Peter Esselman, Dr. Stanley Feldman, and my mother, Dr. Leona Laskin, for additional help with medical questions.

I can't thank Tom Gudmestad enough for his help, generosity, and astonishing expertise. Tom is a walking encyclopedia of the Great War and he has a world-class collection of books, photos, and memorabilia, all of which he made available to me—along with hours of his time. Tom made many corrections and pointed out many mistakes in the manuscript, for which I am deeply grateful. What an incredible stroke of luck to find this amazing resource a short drive from my home. I'm also grateful to Major Stephen C. McGeorge (retired), deputy chief historian, U.S. Army Training and Doctrine Command, for answering numerous questions and passing along book suggestions, photos, and references to many helpful individuals.

I'd like to thank the staff of the American Battle Monuments Commission in Washington, D.C., for invaluable help with research and contacts: Brigadier-General John Nicholson, Martha Sell, Charles Krohn, and Michael G. Conley. Mr. Conley kindly gave me a copy of the ABMC World War I guidebook *American Armies and Battlefields in Europe* at the start of my research—it proved to be a treasure I came to appreciate more and more.

I'm blessed with wonderful friends and never have I appreciated them more than during the long haul of bringing this book to completion. I want to single out my old buddy Robert Armstrong, to whose memory this book is dedicated. Bob helped me clarify my theme when this project was just the glimmer of an idea, and he facilitated my first research efforts at the library of the National Defense University where he worked. During the research and writing, Bob was my indispensable guide to all things military, happily fielding my most inane questions and putting me in touch with colleagues in all branches of the armed forces. His enthusiasm for the book never flagged during his fatal illness—in fact Bob and I had a great talk about military strategy, the wartime influenza epidemic, and the Treaty of Versailles just weeks before his death. I can't express how sad I am that he is not here to accept my thanks.

I think Bob would have been pleased that I did a little networking at his funeral service, enlisting the help of his friends and colleagues Colonel Dave Oaks, Bill Doyle, and Timothy Snider and his former wife Susan. These friends in turn led me to Philip W. Hill, Chaplain (Colonel) U.S. Army, and Marcia McManus, director of the Army Chaplain Museum, who fielded my questions about prayers and last rites on the battlefields of World War I.

My friend Jack Levison has heard so much about this book that he could practically write it himself—thanks for listening, advising, cheering, suggesting titles, and so much else. Thanks to Jim Moran for keeping me (relatively) sane and calm through three decades of friendship. Thanks to Ivan Doig for help with all things Montana, for neighborly companionship, and for setting an inspiring example of the writer's life. Thanks to Mary Whisner for help with all things legal and for her incomparable research acumen. And thanks to Erik Larson, David Williams, and Pat Dobel for friendship and writerly advice/ commiseration. My friend Kevin Francis, professor of the history of science at Evergreen State College, read the manuscript and provided insights, suggestions, comments, and corrections—and gave depth and nuance to my discussion of eugenics.

Jill Kneerim is a magician who, with a twitch (or in this case many twitches) of her wand, turns foggy ideas into viable book projects. Agent does not begin to describe Jill's role in my life—co-brainstormer, ingenious problem solver, cheerleader in chief, ardent fan with an eagle eye for infelicity, and above all dear friend. This is the second book of mine that Tim Duggan has signed up and edited—and I could not ask for a finer editor or a stronger advocate inside HarperCollins. Tim's assistant, Allison Lorentzen, has been unfailingly helpful, organized, and enthusiastic; I am truly grateful for her collaboration on the photo insert. I'd like to thank Martha Cameron for a careful and sensitive copyediting job—she has a light touch and a fine ear.

Finally, I want to acknowledge the support of my family. My brothers Bob, Dan, and Jon; my sisters-in-law Sue and Jane; and my parents Meyer and Leona Laskin probably have no idea of all the ways they have helped me. Thanks to my daughters Emily, Sarah, and Alice for tolerating long spells of paternal abstraction and distraction. An extra thank-you to Emily for coming up with a key word that clicked the title into place. My wife, Kate O'Neill, has given me more than I can ever thank her for—her love and her belief in me make it all worthwhile.

SOURCES

The individual stories of the twelve foreign-born soldiers I follow in this book are based primarily on family interviews, letters, scrapbooks, and military records. The one exception is Sam Dreben, the Fighting Jew, whose life and adventures were recorded in a biography by Art Leibson as well as articles written by friends, comrades, and admirers.

In describing what these men did and experienced on the battlefield, I have relied whenever possible on personal accounts—family stories and memories, letters and diaries. Where personal records do not exist, I have drawn on unit histories, battle diaries, and recollections and letters written by those serving alongside "my" men. I have found these recollections and unit histories (some published in book form, others in typescript) archived at the National Archives and Records Administration (hereafter NARA) in College Park, Maryland, and the U.S. Army Military History Institute at Carlisle Barracks, Pennsylvania, and in the personal collection of Tom Gudmestad in Seattle, Washington. I have gleaned additional details from field orders and memos written during battles held at NARA in Record Group 120, as well as accounts of individual battles or moments of battle written by officers after the war and archived at NARA in Record Group 117. The holdings at NARA are often deep and detailed enough that I was able to find reports written by company captains or even platoon lieutenants who led the men I wrote about; in some cases I have located diaries of privates or noncommissioned officers serving in their platoons or fighting nearby on the same day. In creating narratives of what my soldiers went through on a given day or a given hour of combat, I have made composites of these detailed reports and battle diaries and the accounts left by my men or recounted later to their families. When I write, for example, of Meyer Epstein's suffering in the miserable Bois de Fays in October 1918, I found the details of this misery—the cold and wet and stench of death—in a report written by one of his battalion commanders. Even though I had no documentary evidence that Meyer was equally miserable in this wretched wood, I thought it safe to proceed on that assumption since he was present at the same time and in the same place where the vivid report of the conditions was written. The detail about Meyer starving for lack of protein because of his refusal to eat meat that was not kosher came from an interview with his son.

This has been my method throughout: to weave together (and corroborate) eyewitness accounts with details taken from the memories, letters, and anecdotes passed down through families. All dialogue I quote either comes from eyewitnesses, letters, and diaries or from interviews with family members.

In the notes that follow, I indicate the most important source materials I used in each chapter. I have provided notes for quotations, statistics, and events for which I have found conflicting or improbable accounts or claims and thorny issues that have stirred up debate. These are not strict "academic" notes, but they should give the interested reader ample opportunity to pursue subjects in greater depth and the historian or scholar the references he or she needs to track down the source of crucial facts, figures, assumptions, and details.

INTRODUCTION

I learned of Tony Pierro's war and immigrant experience in an interview conducted with him and his brother Nicholas Pierro and nephew Rick Pierro at their home in Swampscott, Massachusetts, on July 8, 2006.

xvi *"Some half a million other immigrants"*: Nancy Gentile Ford, *Americans All! Foreign-Born Soldiers in World War I* (College Station: Texas A&M University Press, 2001), 3. I found Ford's book, which offers a detailed account of the drafting, training, and military experiences of immigrant soldiers, extremely useful.

xix *"Our minds were becoming warped"*: Joseph N. Rizzi, *Joe's War* (Huntington, WV: Der Angriff, 1983), unpaginated.

xix *"Combatants live only for their herd"*: Chris Hedges, *War Is a Force That Gives Us Meaning* (New York: Public Affairs, 2002), 38, 40.

xxi *"Their service is steeped"*: Patrik Jonsson, "Noncitizen Soldiers: The Quandaries of Foreign-Born Troops," *Christian Science Monitor*, July 5, 2005; http://www.csmonitor.com/2005/0705/p01s03-usmi.html.

xxi *The largest group of immigrant soldiers*: Though thousands of men of Asian, Pacific Island, South and Central American birth and ancestry served with the U.S. military in World War I, I have chosen to focus on those of European ancestry. My primary reason is that the stories of the European men fit together into a coherent narrative: theirs was the classic immigrant experience of the late nineteenth and early twentieth centuries, and even though they came from different countries, they had a great deal in common—most important, from my point of view, the fact that they were returning in military uniform to the continent from which they had recently emigrated. Many of these European immigrants had family members still living in the war zone or fighting in the armies of the European powers, which was not true for the Asians and Latinos. The men I write about represented the major immigrant groups of the period. The great waves of Asian and South and Central American immigration came later—in fact, Mexican Americans and Filipino Americans are the two largest immigrant groups serving in the U.S. Armed Forces today.

The contributions to World War I made by Mexican Americans is discussed in the excellent article by Carole E. Christian, "Joining the American Mainstream: Texas's Mexican Americans During World War I," in *Southwestern Historical Quarterly*, April 1989, pages 559–595. Christian calls the war "a crucial stage in the assimilation of Hispanics into the political and social life of Texas and of the nation" and argues that the war was even more powerful than the Great Depression in shaping the identities of Hispanics in the United States. The war "marked the first concerted effort by the American government and Anglo society to promote the involvement of Hispanics in national life," notes Christian. Even though most Mexican American soldiers were illiterate and few spoke English, they "fought bravely, though they did not understand for what they were fighting." Fighting in the war led "returning veterans . . . [to] spread American ideals and values as well as an awareness of being Mexican Americans, rather than Mexicans, to large segments of Texas Hispanics." I believe—and argue throughout this book—that the same held true for all ethnic groups.

CHAPTER 1: OLD COUNTRIES

The books that I relied on most heavily in portraying the conditions that led Jews to leave the Russian Pale include: *The Promised City: New York's Jews 1870–1914*, by Moses Rischin (Cambridge, MA: Harvard University Press, 1962); *Hands Across the Sea: Jewish Immigrants and World War I*, by Joseph Rappaport (Lanham, MD: Hamilton Books, 2005); *World of Our Fathers*, by Irving Howe (New York: Harcourt Brace Jovanovich, 1976); *Shores of Refuge: A Hundred Years of Jewish Emigration*, by Ronald Sanders (New York: Holt, 1988); *My Future Is in America: Autobiographies of Eastern European Jewish Immigrants*, edited and translated by Jocelyn Cohen and Daniel Soyer (New York: New York University Press, 2005); and *The Rise of David Levinsky*, by Abraham Cahan (New York: Harper, 1960).

I drew on the following books for background in describing the Italian immigrant experience: *La Storia: Five Centuries of the Italian American Experience*, by Jerre Mangione and Ben Morreale (New York: HarperCollins, 1992); *Blood of My Blood: The Dilemma of the Italian-Americans*, by Richard Gambino (Garden City, NY: Doubleday, 1974); *Unto the Sons*, by Gay Talese (New York: Knopf, 1992); *The Story of the Italians in America*, by Michael Angelo Musmanno (Garden City, NY: Doubleday, 1965); *Passage to Liberty: The Story of Italian Immigration and the Rebirth of America*, by A. Kenneth Ciongoli and Jay Parini (New York: ReganBooks, 2002); *The Italian Americans*, by J. Philip di Franco (New York: T. Doherty, 1988); *Americans by Choice*, by Angelo M. Pellegrini (New York: Macmillan, 1956); *Immigrant's Return*, by Angelo M. Pellegrini (New York: Macmillan, 1951); *Pascal D'Angelo: Son of Italy*, by Pascal D'Angelo (New York: Macmillan, 1924); *The Italian-Americans*, by Luciano J. Iorizzo and Salvatore Mondello (New York: Twayne, 1971); *From Immigrants to Ethnics: The Italian Americans*, by Humbert S. Nelli (New York: Oxford University Press, 1983); *Gli Stati Uniti d'America e L'Emigrazione Italiana*, by Luigi Villari (Milano, Italy: Fratelli Treves, 1912); *Christ in Concrete*, by Pietro di Donato (Indianapolis, NY: Bobbs-Merrill, 1939).

The books I found most useful in understanding the experience of Polish immigrants include: *The Polish Press in America*, by Jan Kowalik (San Francisco: R and E Research Associates, 1978); *And My Children Did Not Know Me*, by John J. Bukowczyk (Bloomington: Indiana University Press, 1987); *Behold! The Polish-Americans*, by Joseph A. Wytrwal (Detroit: Endurance Press, 1977); *Polish Americans: An Ethnic Community*, by James S. Pula (New York: Twayne, 1995).

The story of Meyer Epstein came primarily from my telephone interviews with his sons Harold, Julius, and especially my personal interview with Leonard Epstein, in Monroe Township, New Jersey, on October 22, 2006, and several subsequent telephone interviews. The story of Magnus Andreas Brattestø (later Andrew Christofferson) came primarily from interviews and letters from his daughter Nellie Neumann, and especially from our interview in Washington, D.C., on January 3, 2007. The story of Frank and Joseph Chmielewski came from my initial telephone interview with Frank's son, John Chimelewski, on March 3, 2007, and from personal interviews with Frank's daughter, the late Josephine Rolincik, and with Josephine's daughter Dorothy Vancheri in Levittown, Pennsylvania, on April 27, 2007.

1   *Sons as young as twelve*: Details on Jews in Russian army from Louis Greenberg, *The Jews in Russia*, vol. 1, *The Struggle for Emancipation* (New Haven: Yale University Press, 1944–1951), 49. Greenberg calls Russian conscription a "military martyrdom on Jewish youth."

1   *To avoid the virtual death sentence*: The detail about draft-age Jewish men slicing off their trigger fingers comes from my family history. My grandfather, Samuel Laskin, chopped off his own trigger finger as a young man in the Polish sector of the Russian Pale in order to avoid being drafted into the Russian army. Samuel emigrated to the United States in the 1920s and lived a long and robust life as a carpenter and as the father of four children, including my father, Meyer.

3   *"Feeding the hungry"*: Greenberg, *The Jews in Russia*, 63.

7   *"We plant and we reap wheat but never do we eat white bread"*: Mangione and Morreale, *La Storia*, 33.

7   *In December 1908, a 7.5 magnitude earthquake*: Details on the 1908 earthquake in Italy from http://www.pbs.org/wgbh/amex/rescue/peopleevents/pandeAMEX99.html.

10   *Two years later.* Tony Pierro's birth certificate says he was born February 22, 1896, though Tony maintained the date was February 15.

12   *That summer.* The Pierro family believed that Tony came to America in 1914, but the Ellis Island website clearly indicates that an eighteen-year-old named Antonio Pierro sailed out of Naples in the summer of 1913; see http://www.ellisisland.org (search under Antonio Pierro).

13   *Magnus Andreas Brattestø was born*: Nellie Neumann told me that her father was born near Haugesund, though his military papers and obituary show the town as Skjold, which is in the same region but 19 miles from the coast. Since Brattestø, the name of the large farm where his family lived, refers to a landing place, and since Andrew worked on a fishing boat as a young man, I have chosen to follow Ms. Neumann and go with the coastal town of Haugesund.

15   *"It was pretty well taken for granted"*: Nels Quam, *Birth of a Nation* (Northfield, MN: privately published, 1977), 3–4.

17   *To the extent that the idea*: Details on the Polish Catholic Church, from James Pula, *Polish Americans*, 20.

19   *Peter Thompson, born in County Antrim*: Details on Peter Thompson's childhood from Christy Leskovar, *One Night in a Bad Inn* (Missoula, MT: Pictorial Histories, 2006), 146ff. Leskovar's compelling and brilliantly researched book has been my source for all details about her grandfather Peter Thompson, with follow-up phone interviews and e-mail correspondence.

20   *"I was standing in front of our house"*: Samuel Goldberg's quotes here and his quotes and stories throughout the book are from my interview with him near Providence, Rhode Island, on July 9, 2006.

21   *Between 1880 and the 1920s*: Immigration statistics from Nancy Foner and Richard Alba, "The Second Generation from the Last Great Wave of Immigration: Setting the Record Straight," Migration Information Source, October 2006, http://www.migrationinformation.org/Feature/display.cfm?ID=439; and Maldwyn Allen Jones, *American Immigration* (Chicago: University of Chicago Press, 1992), 179.

CHAPTER 2: JOURNEYS

24   *The Affatato boys had descended*: Details on the journey of Epifanio Affatato come from telephone interviews with his sons Charles, Domenick, and Edward on October 25, 2007, and a series of follow-up e-mails and conversations with Ed.

24   *The long bumpy cart ride*: Ed Affatato told me that his father "probably" went from Scala Coeli to Reggio di Calabria by cart and then boarded a train to Naples, though he couldn't be sure. On a map this route appears rather circuitous.

26   *"What makes the emigrant so meek"*: Broughton Brandenberg, *Imported Americans: The Story of the Experiences of a Disguised American and His Wife Studying the Immigration Question* (New York: F. A. Stokes, 1904), 142.

29   *"For such quarters"*: Ibid., 176. I also consulted http://www.ohranger.com/ellis-island/immigration-journey for information on immigrant ships and the profits of the shipping companies.

31   *The Cunard Line built the* Lusitania: Patrick O'Sullivan, *The Lusitania: Unravelling the Mysteries* (Cork, Ireland: Collins Press, 1998), 37–38.

33   *"Hundreds of people had vomiting fits"*: Quoted in Sanders, *Shores of Refuge*, 67–68.

34   *"What dirty little imps"*: Brandenberg, *Imported Americans*, 192.

34   *"Who can depict the feeling"*: from Cahan, *David Levinsky*, 85.

40   *"Where was I to go?"*: from Mangione and Morreale, *La Storia*, 109.

CHAPTER 3: STREETS OF GOLD

45   *"the crying evil of race prejudice"*: Quote from W. E. B. DuBois, statistics on the cost of food, and information on the Wobblies are taken from Meirion Harries and Susie Harries, *The Last Days of Innocence: America at War, 1917–1918* (New York: Random House, 1997), 16, 19, 20. *The Last Days* contains an excellent discussion of American society during the war years, as well as highly vivid and detailed descriptions of battle scenes. Additional background on American society and culture in the first decades of the twentieth century are from David M. Kennedy, *Over Here: The First World War and American Society* (New York: Oxford University Press, 1980).

46   *"Through this metal wicket drips"*: H. G. Wells, *The Future in America: A Search After Realities* (New York: Harper, 1906), 36.

46   *By 1914*: Harries and Harries, *The Last Days of Innocence*, 22.

47   *"Noise is everywhere"*: Mangione and Morreale, *La Storia*, 120.

48   *"Ugly wooden houses"*: Interview with Maria Valiani, "Italians in Chicago—Oral History Project," June 10, 1980, on file with the Immigration History Research Center, University of Minnesota.

48   *The typical immigrant left Ellis Island*: Ann Novotny, *Strangers at the Door: Ellis Island, Castle Garden and the Great Migration to America* (Riverside, CT: Chatham Press, 1971), 76.

49   *It was "a gray, stone world"*: Leon Kobrin quoted in Howe, *World of Our Fathers*, 72.

49   *"America was . . . noise"*: from Cohen and Soyer, *My Future Is in America*, 141.

53   *"Everywhere was toil"*: from Gambino, *Blood of My Blood*, 93.

56   *Tommaso Ottaviano was seventeen years old*: The story of Tommaso Ottaviano and the Ottaviano family's immigration to the United States are from a series of interviews with Pamela Rhodes and John Ricci conducted in Seattle, Washington, in 2007 and 2008 and from follow-up phone interviews.

61   *"They have their own churches"*: Quoted in Pula, *Polish Americans*, 23.

62   *"My fellow workers are Polish"*: Quoted ibid., 24.

CHAPTER 4: THE WEAK, THE BROKEN, AND THE MENTALLY CRIPPLED

65   *One day, soon after he got to the United States*: For background on anti-immigrant sentiment, I drew on John Higham's seminal work *Strangers in the Land: Patterns of American Nativism, 1860–1925* (New Brunswick, NJ: Rutgers University Press, 1988). Also helpful were Hans P. Vought, *The Bully Pulpit and the Melting Pot: American Presidents and the Immigrant, 1897–1933* (Macon, GA: Mercer University Press, 2004), and Harries and Harries and Kennedy, cited above.

66   *"There is no swarming"*: Henry James, *The American Scene* (New York: Penguin Books, 1994), 100.

66   *Harvard zoologist Charles Davenport*: For information on Charles Davenport, see "Race— the Power of an Illusion," Public Broadcasting System, http://www.pbs.org/race/000_ About/002_04-about-03-01.htm. I am indebted to Kevin Francis, a professor of the history and philosophy of science at Evergreen State College, for refining and deepening my understanding of the early twentieth century eugenics movement. Francis writes that "the trend among historians has been to view eugenics as part of the Progressive Movement aimed at social reform and increased scientific management of society. Some eugenicists were early and influential environmentalists."

67   *"the weak, the broken and mentally crippled"*: quoted in Jones, *American Immigration*, 268. Other quotes are from Madison Grant, *The Passing of the Great Race* (New York: Charles Scribner's Sons, 1916), xxxi, 91.

68   *"When the test of actual battle"*: Grant, *The Passing of the Great Race*, xxxi, 91.

69   *Fair-haired, blue-eyed, strapping*: Material on Matej Kocak's youth, emigration, and early experiences in the United States is taken from a chapter in Theodore Roosevelt's book *Rank and File: True Stories of the Great War* (New York: Charles Scribner's Sons, 1928) and from Kocak's military file, copied and given to me by Peter Kacur.

73   *"Shoot de woiks"*: My information on Sam Dreben comes from the following sources: a series of articles entitled "A Soldier of Fortune's Story" written by Dreben's fellow soldier of fortune and sometime comrade in arms Tracy Richardson that appeared in *Liberty Magazine* on October 10, 17, 24, 31, and November 21, 1925; Dreben's obituary in the *New York Times*, March 18, 1925; "Sam Dreben—Warrior, Patriot, Hero," an article by Hymer E. Rosen, available online at http://www.jewish-history.com/WildWest/dreben.html; "Fighting Jew—Forgotten Hero" by Gerard Meister, at the Doughboy Center website http://www.worldwar1.com/dbc/dreben.htm; "The Fighting Jew" by Rabbi Martin Zielonka, Publications of the American Jewish Historical Society, no. 31 (1928): 211–217; and Art Leibson's biography, *Sam Dreben: The Fighting Jew* (Tucson, AZ: Westernlore Press, 1996).

73   *"He was the last man"*: Richardson, "A Soldier of Fortune's Story," October 10, 1925.

74   *While still in his teens Sam ran away*: There are some minor discrepancies in facts and dates between the Leibson biography and the Rosen article—for example, Leibson asserts that Dreben left home for good at age seventeen and arrived in New York in February 1899.

75   *"Do they give the uniform too?"*: Rosen, "Sam Dreben—Warrior, Patriot, Hero."

77   *"Handling a machine gun"*: Leibson, *Sam Dreben*, 44.

CHAPTER 5: THE WORLD AT WAR

79   *Four and a half years later*: Statistics on the numbers killed in the war vary widely by source and by how the statistician breaks down the numbers—for example, whether or not to include among civilian deaths those who died in the Armenian genocide and those who perished in the war-related influenza epidemic. Rough figures for military and civilian deaths combined, including Armenian civilians and excluding influenza victims, run from 9 million to 16 million. So the figure I give of nearly 10 million is on the conservative side.

79   *But instead the ultimatum*: Barbara W. Tuchman, *The Guns of August* (New York: Ballantine Books, 1962), 71.

85   *What Andrew found most disturbing*: For a description of the rape of Belgium, see John Keegan, *The First World War* (New York: Knopf, 1999), 82.

86   *Butte, Montana, is 237 miles west*: The description of Butte in 1914 draws on Leskovar, *One Night in a Bad Inn*, and Clemens P. Work, *Darkest Before Dawn: Sedition and Free Speech in the American West* (Albuquerque: University of New Mexico Press, 2005), 66.

87   *Maximilian Cieminski in Bessemer, Michigan*: The story of Maximilian Cieminski and his family comes from a series of interviews, letters, and emails I received from his great-nephew John A. Riggs.

91   *"Three elderly, frightened folk"*: from *Forverts*, September 6, 1910, translated by Myra Mniewski.

92   *"Seven million Jews are involved"*: Mordecai Soltes, *The Yiddish Press: An Americanizing Agency* (New York: Arno Press, 1969), 193.

92   *A story circulated*: S. Ansky, *The Enemy at His Pleasure: A Journey Through the Jewish Pale of Settlement During World War I* (New York: Metropolitan Books, 2002), 31.

94   *"A German victory"*: Quoted in Rappaport, *Hands Across the Sea*, 39.

95   *"No work could have been more perfectly calculated"*: Pellegrini, *Americans by Choice*, 81.

96   *The death toll all along the Western Front:* Statistics on deaths in first months of war from Keegan, *The First World War,* 136.

97   *In that same tender 1915 spring week:* Information on gas attacks are from Joel Vilensky, *Dew of Death: The Story of Lewisite, America's World War I Weapon of Mass Destruction* (Bloomington: University of Indiana Press, 2005), 13–15; James W. Hammond Jr., *Poison Gas: The Myths Versus the Reality* (Westport, CT: Greenwood Press, 1999), 10–11; and Keegan, *The First World War,* 197–198.

98   *When Meyer Epstein picked up his copy:* Quotes on *Lusitania* from *Forverts,* May 9, 1915, translated by Myra Mniewski.

99   *The Cincinnati Volksblatt:* Quoted in Carl Wittke, *German-Americans and the World War* (Columbus: Ohio State Archaeological and Historical Society, 1936), 72.

100   *Some 8 million individuals:* Statistics on numbers of German Americans are taken from Bureau of the Census, *Thirteenth Census of the United States,* vol. 1: *Population* (1910), 875. Don Heinrich Tolzmann puts the number at over 18 million, 25 percent of the population in 1900; see Tolzmann's *The German-American Experience* (Amherst, New York: Humanities Books, 2000), 268. Tolzmann may be including all Americans of German ancestry, rather than only those born in Germany or the offspring of at least one German parent.

100   *"We are Germans, of course":* Quoted in Wittke, *German-Americans and the World War,* 74.

100   *In the early days of the war:* Information on Germans returning to fight from Harries and Harries, *The Last Days of Innocence,* 30. But Chad Millman notes that relatively few German Americans actually returned to fight because German submarine attacks made it too dangerous to cross the ocean; see *The Detonators: The Secret Plot to Destroy America and an Epic Hunt for Justice* (New York: Little, Brown, 2006), 6–7.

102   *"Each of the belligerent nations":* Harries and Harries, *The Last Days of Innocence,* 30.

103   *In the coming months:* Musmanno, *Italians in America,* 141.

105   *By 1915, he was pulling in $3.83 a day:* Information on Peter's pay and wages in the mines is from Leskovar, *One Night in a Bad Inn,* 161; and Work, *Darkest Before Dawn,* 63.

105   *"Miners with families did not make enough":* from Work, *Darkest Before Dawn,* 63.

107   *One historian wrote:* Quoted in Keegan, *The First World War,* 285.

108   *Villa's men opened fire:* The exact numbers that died in the Villa raid are disputed.

110   *When Pershing pulled out in February 1917:* Mitchell Yockleson argues that the expedition was successful in displaying American power and that it "provided military training experience for the eleven thousand regular soldiers who made up the expedition" and set up Pershing to be commander of the AEF in the war; see Yockleson, "The United States Armed Forces and the Mexican Punitive Expedition," *Prologue Magazine* 29, nos. 3–4 (Fall–Winter 1997).

110   *Michele Valente had come to America:* Background on Michael Valente comes from an interview with his grandson Ralph Madalena in Rockville Center, New York, on October 23, 2006, and from several telephone interviews with Valente's niece, Michelina Rizzo.

113   *The German press was quick to point out:* See Wittke, *German-Americans and the World War,* 87.

113   *Grimmest of all:* Facts and figures on the Somme from Keegan, *The First World War,* 295, and Martin Gilbert, *The Somme* (New York: Henry Holt, 2006), xvii.

115   *President Wilson dismissed the explosion:* Wilson quote on Black Tom from Millman, *The Detonators,* 97.

120   *On the last day of the debate:* Kennedy, *Over Here,* 17–18.

CHAPTER 6: THE ARMY OF FORTY-THREE LANGUAGES

123   *As one newspaper informed its readers:* Leskovar, *One Night in a Bad Inn,* 203.

123   *For all the bluster about preparedness:* Statistics on the U.S. army in 1917 from Christopher M. Sterba, *Good Americans: Italian and Jewish Immigrants During the First World War* (New York: Oxford University Press, 2003), 32. Sterba's book contains much excellent material on how the

war affected the Italian "colonia" in New Haven and the Jewish community of the Lower East Side. I relied heavily on Sterba's section "Training the New Immigrant Soldier" for details and background. Another source of statistics on the army in 1917 was http://www.history.army .mil/books/Lineage/mi/ch2.htm.

123   *In strength and training*: Douglas V. Johnson II and Rolfe L. Hillman Jr., *Soissons, 1918* (College Station: Texas A&M University Press, 1999), 6.

125   *As the* New York Times *reported*: Sterba, *Good Americans*, 68.

126   *"The paper I represent"*: quoted ibid., 61.

126   *"there is not one Irishman"*: quoted in Leskovar, *One Night in a Bad Inn*, 202; facts on Butte during Registration Day from Leskovar, *One Night in a Bad Inn*, 204.

127   *Most anxious of all the "foreign element"*: Details on Poles on Registration Day, from Wytrwal, *Behold! The Polish-Americans*, 223; Pula, *Polish Americans*, 59.

128   *"Even though glory"*: Wytrwal, *Behold! The Polish-Americans*, 228.

128   *In the surge of patriotism that swept through Polonia*: Material on Selective Service Act and declarants from William Bruce White, "The Military and the Melting Pot: The American Army and Minority Groups, 1865–1924" (PhD diss., University of Wisconsin, 1968), 316–317.

131   *"the flower of our neighborhood"*: quoted in Sterba, *Good Americans*, 73.

131   *Fingers were also pointed at aliens*: Discussion of conscientious objectors draws on Harries and Harries, *The Last Days of Innocence*, 96.

133   *Montana men, foreign- and native-born alike*: Statistics on Montana in the war from Leskovar, *One Night in a Bad Inn*, 223.

134   *On day one, an officer called the roll*: Story of officer sneezing at Camp Meade from White, *Military and Melting Pot*, 327.

134   *"the difficulties of teaching"*: quoted in Sterba, *Good Americans*, 15.

134   *"Never in my wildest flights"*: Irving Crump, *Conscript 2989: Experience of a Drafted Man* (New York: Dodd, Mead, 1918), 2.

135   *One Camp Upton officer shuddered*: Herschkowitz as "worst possible material" from Major Charles Whittlesey, quoted in Lee J. Levinger, *A Jewish Chaplain in France* (New York: Macmillan, 1921), 117.

135   *Sizing up the abilities and prospects*: Quoted in Ford, *Americans All!*, 67–68.

135   *"Many, many" of the men*: quoted in Jennifer D. Keene, *Doughboys, the Great War, and the Remaking of America* (Baltimore, MD: Johns Hopkins University Press, 2001), 13.

136   *"A right-smart number"*: from Alvin Cullum York, *Sergeant York: His Own Life and War Diary*, ed. Tom Skeyhill (Garden City, NY: Doubleday Doran, 1928), 180.

139   *"Slovak men!"*: Details on Binghamton Slovaks going off to war and farewell banquet from Imrich Mažár, *A History of the Binghamton Slovaks over a Period of Forty Years, 1879–1919* (Phoenix, AZ: Via Press, 2003).

140   *"Old scores from the pages of history"*: quoted in Ford, *Americans All!*, 68.

140   *Soldiers of German and Austrian ancestry passed around stories*: Harries and Harries, *The Last Days of Innocence*, 128.

140   *"They have had no chance to speak"*: quoted in Ford, *Americans All!*, 70.

141   *To its credit*: Information on Camp Gordon plan from White, *Military and Melting Pot*, 337, and Ford, *Americans All!*, chap. 3.

145   *Former army chief of staff*: quoted in Harries and Harries, *The Last Days of Innocence*, 194.

145   *"We made an attack one day"*: Story and quote from British instructor ibid., 135.

146   *"Russia," in the words of one historian*: quoted in Paul Dukes, *Superpowers: A Short History* (London: Routledge, 2000), 36.

146   *Nine months after Wilson had declared war on Germany*: Statistics on numbers of Americans in France from Harries and Harries, *The Last Days of Innocence*, 210.

147   *"We expected to see two million cowboys"*: quoted ibid.

149   *Tony Pierro was not the only one*: Ethnic stereotypes relies on Keene, *Doughboys*, 19–20.

149   *Ashad G. Hawie, a Syrian Christian:* Hawie wrote about his experiences as a Syrian immigrant and
       a Doughboy in *The Rainbow Ends* (New York: Theo. Gauss, 1942); this anecdote is from 83–84.

149   *Lieutenant Jacob Rader Marcus:* Quoted in Jennifer D. Keene, *World War I* (Westport, CT: Green-
       wood Press, 2006), 112.

150   *Particularly galling was the merger:* Details on the merger of the 1st and the 7th regiments and
       quotes are from Stephen L. Harris, *Duty, Honor, Privilege: New York's Silk Stocking Regiment and
       the Breaking of the Hindenburg Line* (Washington, D.C.: Brassey's, 2001), 73, 77–78, 87.

151   *"Perhaps the difference between soldier and civilian":* Quoted in Richard Slotkin, *Lost Battalions:
       The Great War and the Crisis of American Nationality* (New York, Holt, 2005), 121.

151   *"When they fired":* Quotes on training are from Harris, *Duty, Honor, Privilege,* 100.

152   *The men were marched:* Ibid., 101. Note, however, that Gerald F. Jacobson in his *History of the 107th
       Infantry* (New York: DeVinne Press, 1920) says the Glassy Rock practice was in March.

152   *Mastine, who had been nicknamed "Cuckoo":* Harris, *Duty, Honor, Privilege,* 101–102.

153   *Sam Dreben, late of Pershing's Punitive Expedition:* The various articles and books written about
       Dreben contain discrepant accounts of the timing of his enlistment and how or whether it
       related to his daughter's birth. I have gone with the version I found most plausible.

154   *"It is quite apparent":* quoted in Keene, *World War I* , 39.

155   *"The foreign element is taking hold":* Crump, *Conscript 2989,* 72.

155   *Boleslaw Gutowski, a Polish immigrant:* Quoted in Rose Szewc Papers, Polish American Collec-
       tion, Immigration History Research Center, University of Minnesota.

156   *Jack Herschkowitz, a Rumanian-born Jew:* Quoted in Sterba, *Good Americans,* 128.

CHAPTER 7: I GO WHERE YOU SEND ME

158   *The Americans' turn finally came:* Details on Yankee Division at Seicheprey from Sterba, *Good
       Americans,* 179, and Harries and Harries, *The Last Days of Innocence,* 239–240.

164   *Sending these men into battle:* quoted in Keene, *Doughboys,* 47.

169   *The marines' objective at Belleau Wood:* For the fighting at Belleau Wood, see Harries and
       Harries, *The Last Days of Innocence,* 271, and "The Battle for Belleau Wood," http://www
       .worldwar1.com/dbc/ct_bw.htm.

170   *Even the Germans admitted:* Quoted in Harries and Harries, *The Last Days of Innocence,* 272.

CHAPTER 8: JULY 4, 1918

I used Chapter 8 of *Strangers in the Land* by John Higham for background on wartime repression, the
activities of the Committee on Public Information, and the oppression of German-Americans. Kennedy's *Over Here* and *The Last Days of Innocence* by Harries and Harries were also useful.

     I am indebted to Norman Hofer for the story of Jacob Wipf and the Hofer brothers. I am also
indebted to Susan E. Cohn for the superlative research she has done and generously shared. Ms. Cohn
secured and supplied me with a copy of the trial transcript of *United States v. Recruits David J. Hofer,
Michael J. Hofer etc.* recorded by the War Department, dated September 30, 1918. She also gave me
copies of the papers she obtained from the National Personnel Records Center in St. Louis, Missouri,
pertaining to the case. Additional information comes from the headquarters of the 91st Infantry
Division, Judge Advocate's Office, dated June 15, 1918, Rocky Mountain Region of the National
Archives and Records Administration; Daniel Hallock, "Persecution in the Land of the Free," http://
www.plough.com/articles/persecutionintheland.html; and Darrell R. Sawyer, "Anti-German Senti-
ment in South Dakota During World War I," *South Dakota Historical Collections* 38 (1976): 440–514.
I also found two books very useful: John Stahl, *Hutterite CO's in World War One,* trans. Karl Peter and
Franziska Peter (Hawley, MN: Spring Prairie Printing, 1996); and Gerlof D. Homan, *American Mennonites and the Great War* (Waterloo, ON: Herald Press, 1994).

176   *On April 4, 1918, a thirty-one-year-old baker:* Kennedy, *Over Here,* 68.

178   *A German pastor was imprisoned:* Work, *Darkest Before Dawn,* 152.

180   *When Jacob Wipf and the Hofer brothers:* Details on conscientious objectors from Harries and Harries, *The Last Days of Innocence,* 132.

181   *"Part of the time they had me on my back":* Harries and Harries, *The Last Days of Innocence,* 133.

181   *Those who refused to obey orders:* Stahl, *Hutterite CO's in World War One,* 93.

182   *From the testimony of Jacob Wipf: United States v. Recruits David J. Hofer, Michael J. Hofer etc.,* September 30, 1918.

184   *For over a year now:* George T. Blakey, *Historians on the Homefront: American Propagandists for the Great War* (Lexington: University Press of Kentucky, 1970), 57.

185   *The day's showcase event:* Quotes and descriptions of Mount Vernon celebration from George Creel, *How We Advertised America* (New York: Arno Press, 1972), 204.

186   *"When they arrived at Alcatraz":* Stahl, *Hutterite CO's in World War One,* 33–34.

189   *"Guards were stationed at each end":* Zenas A. Olson, *Following Fighting "F": Being an Intimate History of Company "F", 361st Infantry of the Ninety-first Division* (La Chapelle-Montligeon, France: Imprimerie de Montligeon, 1919), 33, 37, 38.

189   *Now, in the frenzy to get American soldiers overseas:* Statistics on soldiers shipping out from Edward M. Coffman, *The War to End All Wars: The American Military Experience in World War I* (New York: Oxford University Press, 1968), 227.

191   *Leonardo Costantino, an affable Italian immigrant:* Costantino's log was transcribed and typed by his daughter-in-law, Nancy Marzo Costantino, under the title "Daddy's Log." Leonardo's granddaughter Diane Dituri Lopez gave me a copy of the log and her permission to quote from it.

192   Quote from John Dos Passos, *Three Soldiers* (1921; repr., New York: Barnes & Noble Books, 2004), 108.

CHAPTER 9: THESE FOUGHT IN ANY CASE

196   *For the first time in an American war:* Information on rabbis in the army from Sterba, *Good Americans,* 192.

198   *July 18, the jump-off set:* Details on Marshal Foch's plan and the battle of Soissons from Harries and Harries, *The Last Days of Innocence,* 315.

199   *Kocak's platoon arrived at the drop-off site:* Details on 5th Marines at Soissons from George B. Clark, *Devil Dogs: Fighting Marines of World War I* (Novato, CA: Presidio Press, 1999); Johnson and Hillman, *Soissons, 1918.* Additional information comes from records of the 5th Marines at NARA, Record Groups 117 and 120.

200   *Kocak realized that the only way to advance:* Details on Kocak at Soissons from Roosevelt, *Rank and File,* 181. Additional details on this battle are from Harries and Harries, *The Last Days of Innocence,* 320, and Johnson and Hillman, *Soissons, 1918,* 87.

201   *Kocak made swift precise use of his bayonet:* The details on Kocak's use of the bayonet are from Roosevelt, *Rank and File,* but Great War authority Tom Gudmestad questions this. From extensive reading of soldiers' letters and firsthand accounts, Gudmestad concludes that the bayonet was little used in the field, "except to toast bread and dry socks." In the absence of any other account of Kocak's actions that day, I rely on Roosevelt.

204   *The commander of the 2nd Division:* Johnson and Hillman, *Soissons, 1918,* 84.

204   *The 2nd Division as a whole:* Harries and Harries, *The Last Days of Innocence,* 320.

205   *Like the Senegalese fighters:* Johnson and Hillman, *Soissons, 1918,* 66.

205   *Six other American divisions were deployed:* Daniel W. Strickland, *Connecticut Fights: The Story of the 102nd Regiment* (New Haven: Quinnipiack Press, 1930), 136

206   *"Goddamn replacements":* Elton E. Mackin, *Suddenly We Didn't Want to Die* (Novato, CA: Presidio Press, 1993), 87.

207   *On both sides of the road*: Dos Passos, *Three Soldiers*, 147.

209   *As they swept through the fields*: Captain Strickland's account of the battle and quotes are from *Connecticut Fights*, 183, 137,185.

210   *The attack began at dawn on July 22*: My account of Max Cieminski's death is necessarily speculative but I have tried to ground my speculation in as many detailed sources as I could. I drew on the account of the battle by Strickland in *Connecticut Fights* and compared this with three typescript accounts written by officers of the 102nd Infantry and archived at NARA, Record Group 120. I also drew on details from Max's burial case file at NARA. In addition, I had a number of conversations with Max's great-nephew, John Riggs, who has taken a particular interest in his ancestor's service and death. After studying the relevant documents, Riggs has concluded that the bullet that entered the base of Max's scull was fired by a German soldier intent on delivering a "coup de grâce" to the wounded American. In a letter of March 28, 2008, Riggs writes: "In a way it probably was murder. When a person was wounded and not picked up right away after a charge, or if wounded badly and a position overrun, the Germans would shoot you in the back of the head. They were famous for killing the seriously wounded on the battlefield this way. If Uncle Max was laying there without an arm, stomach opened up and bleeding out, they probably thought they did him a favor. . . . This would have confused my family when they got the body because he was actually wounded over 3 times then. Probably more than one piece of shrapnel hit him: his arm missing and coup de grâce to the head and other wounds among the many."

212   *"The earth booms"*: Erich Maria Remarque, *All Quiet on the Western Front* (1928; repr., New York: Ballantine Books, 1996), 105–106.

213   *Two of those fiery steel splinters*: My discussion of Max's wounds and death is based on a book by Louis A. La Garde, *Gunshot Injuries: How They Are Inflicted, Their Complications and Treatment* (London: John Bale Sons and Danielsson, 1916), chaps. 2, 3, and 9; A. M. Fauntleroy, *Report on the Medico-Military Aspects of the European War* (Washington, DC: Government Printing Office, 1915); and interviews and correspondence with Dr. Donald O'Neill.

218   *Major George R. Rau*: Major Rau's death is a composite of typescript in NARA, Record Group 120 and Strickland, *Connecticut Fights*, 197.

219   *"They marched in wearied silence"*: Francis Patrick Duffy, *Father Duffy's Story: A Tale of Humor and Heroism, of Life and Death with the Fighting Sixty-ninth* (New York: Doran, 1919), 206.

CHAPTER 10: THE JEWS AND THE WOPS AND THE DIRTY IRISH COPS

222   *"Yesterday was New York 'Old Home Day'"*: Duffy, *Father Duffy's Story*, 114–115.

223   *"It is strange to see this organization performing"*: Kenneth Gow, *Letters of a Soldier*, quoted in Harris, *Duty, Honor, Privilege*, 156.

223   *"At first they gazed"*: John F. O'Ryan, *The Story of the 27th Division*, vol. 1 (New York: Wynkoop Hallenbeck Crawford, 1921), 204.

224   *"We have passed through a lot of cities"*: quoted in Harris, *Duty, Honor, Privilege*, 129.

224   *The few inhabited villages*: Descriptions of the primitive condition of French villages is from Keene, *World War I*, 113.

227   *Meanwhile, Peter Thompson was picking up jokes*: Thompson's summer in France and quote about "wearing the kitchen stove": from Leskovar, *One Night in a Bad Inn*, 240.

229   *But that's exactly where Affatato and Valente*: Description of Belgium and battle from Harris, *Duty, Honor, Privilege*, 139, 155–156.

230   *"These buttons were a hated symbol"*: Ibid., 116

230   *"I can't describe the awfulness of the war"*: Ibid., 185.

231   *Officially, Private Antonio Pierro*: Movements of the 320th Field Artillery are from NARA, Record Group 120.

238   *The big offensive*: Information on the battle of St. Mihiel from Harries and Harries, *The Last Days of Innocence*, 340.

239   *There was much weeping and embracing*: French attitude toward Americans at St. Mihiel from Harries and Harries, *The Last Days of Innocence*, 347.

CHAPTER 11: THE ARC OF FIRE

242   "Vogliamo sperare": Tommaso Ottaviano's great-nephew John Ricci very kindly gave me photocopies of the letters and postcards that Tommaso wrote home to his mother and siblings during the war. The translations are my own.

244   *"You love your comrade so in war"*: Johan Huizinga, *The Waning of the Middle Ages* (Garden City, NY: Doubleday, 1954), 64.

245   *"The contribution to the American military record"*: O'Ryan, *The Story of the 27th Division*, vol. 1, 156.

245   *"[Selig is] a nice chap"*: Charles F. Minder, *This Man's War: The Day-by-Day Record of an American Private on the Western Front* (New York: Pevensey Press, 1931), 140.

247   *As the plan took shape*: Statistics and background on Argonne Offensive from Edward G. Lengel, *To Conquer Hell: The Meuse-Argonne, 1918* (New York: Holt, 2008), 4.

249   *"There was nothing to do but sit and listen"*: Ibid., 88.

249   *Meyer and the couple of hundred men*: Details on Company H, 58th Infantry, from NARA, Record Group 120.

CHAPTER 12: BREAKING THE LINE

256   *"the hardest task imposed on any unit"*: O'Ryan, *The Story of the 27th Division*, vol. 1, 318.

258   *Company C marched through the evening*: Jacobson, *History of the 107th*, 51.

259   *"After four minutes that curtain of exploding shells"*: Quoted in Harris, *Duty, Honor, Privilege*, 224.

261   *"Our company has been honored above all other companies"*: Kenneth Gow, *Letters of a Soldier* (New York: Herbert B. Covert, 1920), 389–390.

263   *"It was a slaughter; we ran into a trap"*: Quoted in Mitchell A. Yockelson, *Borrowed Soldiers: American Soldiers under British Command, 1918* (Norman: University of Oklahoma Press, 2008), 172.

266   *Mike Valente's moment came later*: Details on Valente rampage from interview with his grandson, Ralph Madalena, and from Harris, *Duty, Honor, Privilege*, 252. Mr. Madalena told me that his grandfather was wounded on the right wrist, while Harris maintains it was the left arm.

267   *"We feel that all this murdering"*: quoted in Neil Hanson, *Unknown Soldiers: The Story of the Missing of the First World War* (New York: Knopf, 2006), 176.

267   *"I am ashamed not only of my deeds"*: Quoted in Hedges, *War Is a Force*, 176.

268   *When the numbers were finally tallied*: Statistics on Hindenburg battle from Harris, *Duty, Honor, Privilege*, 237, 294.

268   *One of the men of the 107th*: Story of Corporal Kim from Yockelson, *Borrowed Soldiers*, 176.

270   *Sixty-two years after the battle*: F. H. Doane, "To My Buddy Jim—Killed, Sept. 29, 1918; Buried, Somme: Plot A, Row 32, Grave 3," *New York Times*, September 29, 1980. Doane's slain friend Jim was Private James Spire, who entered the service from New York State.

270   *In the days after the battle*: The graveyard near Bony is now the Somme American Cemetery and Memorial maintained by the American Battle Monuments Commission. Many of the 1,844 Americans buried here died in the breaking of the Hindenburg Line.

270   *Rabbi Levinger recalled*: Levinger, *A Jewish Chaplain in France*, 54–55, 142.

CHAPTER 13: BLANC MONT

271   *"The real man of the whole crowd"*: Quoted in Sterba, *Good Americans*, 188.

273   *The word that rippled down the line*: Details on marines at Blanc Mont from George Clark, *Devil Dogs*, 288–290; 296.

277   *"In some instances officers were leading in what appeared"*: NARA, Record Group 120; Clark came across the same memo and reprints it in *Devil Dogs*, 314.

277   *"Marines never retreat"*: Mackin, *Suddenly We Didn't Want to Die*, 201. Other details on battle from Mackin, 184, 191, and Clark, *Devil Dogs*, 316.

279   *The only account of Kocak's role*: Roosevelt, *Rank and File*, 185. Additional details from Kocak's military record and from his burial case file at NARA.

280   *"No one but those present"*: Hunt's account of Blanc Mont is at NARA, RG 117.

281   *"No one had the gall"*: Clark, *Devil Dogs*, 325.

281   *Inevitably the two groups of men began trading insults*: Mackin, *Suddenly We Didn't Want to Die*, 211.

282   *It was during those first weeks*: Note that Rabbi Zielonka in his 1928 article asserts that Dreben enlisted after his baby daughter died, feeling "that all joy had been taken from his life."

282   *A new, intensely virulent strain of influenza*: Details on influenza from http://www.cdc.gov/ncidod/EID/vol12no01/05-0979.htm.

283   *If Sam Dreben had not been an actual person*: There was in fact talk in the early 1940s of a Hollywood movie based on the life of the Fighting Jew, variously titled *Fighting Sam* and *Sergeant Sam Dreben, USA*. At one point John Ford was lined up to direct, but in the end the picture was never made.

286   *"Great work. Keep it up"*: quoted in Leibson, *The Fighting Jew*, 130.

292   *"This is not easy to talk about"*: Bob Herbert, "Sacrifice of the Few," *New York Times*, October 12, 2006, A27.

295   *"I walked up to the machine-gun nest"*: Minder, *This Man's War*, 326.

CHAPTER 14: WHY SHOULD I SHOOT THEM?

297   *That new front, a bleak, dangerous place*: The battle at Bois des Loges and Tommaso Ottaviano's involvement are from interviews with Pamela Rhodes and John Ricci and Tommaso's letters written to his mother. Additional details from NARA Record Group 117 and 120. I also relied on *The 78th Division in the World War*, compiled and edited by Thomas F. Meehan (New York: Dodd, Mead, 1921); *History of the 307th Field Artillery, Sept. 6, 1917 to May 16, 1919* (no author, date, or publisher listed); and *A History of the 310th Infantry, Seventy Eighth Division, USA*, by Raymond W. Thompson (New York: Schilling Press, 1919).

298   Information on enemy artillery from "The 78th Division at the Kriemhilde Stellung, October 1918" by Rexmond C. Cochrane, U.S. Army Chemical Corps Historical Studies, Gas Warfare in World War I, U.S. Army Chemical Corps Historical Office, Army Chemical Center, Maryland, 1957 (archived at U.S. Army Military History Institute), 20.

301   *At some point his eyes fell on Arnold Pratt*: Peter's heroic actions in Belgium from Leskovar, *One Night in a Bad Inn*, 293, 300–301.

302   *Any doubt that something big was brewing*: Information on massive release of poison gas, Italians singing songs and telling stories before battle, and battle details from Rock Marcone, "At Bois des Loges, Italian Immigrants in the U.S. Army's 78th Infantry Division Proved Their Mettle," *Military History*, June 2005, 18–22.

306   *The retreat from the Bois des Loges*: Cochrane, "The 78th Division at the Kriemhilde Stellung, October 1918," 56.

309   *During the first week of November:* Note that both Leibson, in *Sam Dreben,* and Meister, in "Fighting Jew—Forgotten Hero," say the regiment was marching in Alsace; but regimental histories and papers at NARA Record Group 120 indicate that the 141st was not in Alsace but rather in Lorraine. It's quite possible the whole story is apocryphal.

CHAPTER 15: POSTWAR

312   *The enormity of the killing:* Statistics on war deaths and casualties are from "Statistical Summary of America's Major Wars," http://www.civilwarhome.com/warstats.htm, and from Keene, *Doughboys,* ix. Leskovar notes that twice as many American servicemen died in eighteen months of fighting in the First World War as in the decade of Vietnam; see *One Night in a Bad Inn,* 223.

317   *When he strode into his aunt's house:* Quotes and information on Peter's return from Leskovar, *One Night in a Bad Inn,* 364.

322   *The division's welcome-home parade:* Details on parade of 27th Division from Harris, *Duty, Honor, Privilege,* 334–337.

323   *In total, over 280,000 immigrant soldiers became citizens:* Statistics are from Sterba, *Good Americans,* 200. But Lynn G. O'Neil and Omer S. Senturk in their master's thesis, "Noncitizens in the U.S. Military" (MA thesis, Naval Postgraduate School, March 2004), give a considerably lower estimate, writing that "More than 123,000 military members were naturalized by virtue of service in World War I." Their source is William S. Bernard, ed., *American Immigration Policy: A Reappraisal* (New York: Harper & Brothers, 1950), 148.

325   *the initials of his two American military honors:* The Purple Heart was established by George Washington in 1782, but it lapsed after the American Revolution and was not resuscitated until 1932. World War I veterans who had won army wound ribbons, wound chevrons, or meritorious service citation certificates could apply to have the Purple Heart awarded retroactively. This is why Epifanio Affatato's gravestone carries the initials PH.

328   *In fact, medical studies have established:* Information about gas and health effects from Denis M. Perrotta, "Long-Term Health Effects Associated with Sub-clinical Exposure to GB and Mustard," July 18, 1996; http://www.gulflink.osd.mil/agent.html. Additional background on gas from William L. Langer, *Gas and Flame in World War I* (New York: Knopf, 1965); Hammond, *Poison Gas,* 1999; and Vilensky, *Dew of Death,* 2005.

328   *Most chose to stay:* Wytrwal, *Behold! The Polish-Americans,* 235.

331   *American Jews were quick to point out:* Statistics on Jews in the army from *New York Times,* February 24, 1919.

333   *Ten thousand Jewish American war veterans:* Sterba, *Good Americans,* 208–209.

333   *Congress, responding to the ugly national mood:* Jennifer D. Keene, *The United States and the First World War* (Harlow, England: Pearson Education, 2000), 49.

334   *It was in this poisonous atmosphere:* Details on Dreben after the war from Leibson, *The Fighting Jew,* 143–144.

335   *"I am a loyal member of the Legion":* Ibid., 150.

335   *A crowd of a hundred thousand:* Details on interment of the Unknown from Hanson, *Unknown Soldiers,* 346–348.

339   *Before he went to war:* Details on Max Cieminski's insurance and payout from interviews and correspondence with John Riggs.

342   Kocak grave site and reburial from NARA burial case file.

INDEX

Affatato, Carmine, 24–29, 37, 44, 45, 95, 103

Affatato, Charles, 323, 324, 325

Affatato, Domenick, 323, 324, 325

Affatato, Domenico, 24, 25, 27, 40–41, 95, 163

Affatato, Edward, 323, 324, 325

Affatato, Epifanio, xi, xvii–xviii, 44–48, 78, 125
   in Flanders, 229
   in France, 163, 224, 225–26, 228, 229, 241,
      255–62, 264–65, 268, 269, 270, 327
   heroism of, 265, 268, 270, 322, 323
   induction and training of, 162–63, 164
   journey of, 24–29, 36–38, 40–41, 362n
   medals won by, 322
   postwar life of, 322–25, 327
   start of war and, 95–97
   superstitions shed by, 226
   U.S. citizenship papers of, 103–4
   wounding of, 264–65, 268

Affatato, Filomena Mancuso, 323, 324

African Americans, 110, 150–51, 247, 248, 333

Aguinaldo, Emilio, 76

Aire River, 290, 291, 298, 300, 308

Aisne-Marne operation, 205–20

Aisne River, 219, 273, 287

Alabama National Guard, 149–50

Alcatraz, Calif., 184, 186–87, 321–22

Alexander II, Czar of Russia, 94, 179

All Quiet on the Western Front
      (Remarque), 213

Alpine Landing, N.Y., 188, 189

America, Americans
   becoming, 63–64, 67–68, 77–78, 93, 101–2,
      120, 313–14
   character of, 66
   forced integration as, 154–55
   hyphenated, 101, 113, 154
   Teddy Roosevelt Sr.'s views on, 67–68, 101,
      102

American Cemetery 88, 306

American Expeditionary Forces (AEF), xvi,
      xxii, 132–33, 203, 239, 240, 317, 328
   Pershing made commander of, 130, 365n
   size of, 147

American Scene, The (James), 65–66

American Truth Society, 102

Amish, 178, 180

Anabaptists, 178

Anaconda Copper Mining Company, 86, 104–5

Ancient Order of Hibernians, 105, 126

anti-Austrian sentiments, 202

anti-German sentiments, 173–84, 202

anti-immigrant sentiments, xvii–xviii, xx, 34,
      66–68
   draft and, 130–31

anti-Semitism, 94, 150, 236, 333. *See also* pogroms

Argonne, xv, xix, 247–53, 255, 257, 273,
      289–93, 296–308, 328

Argyle Coal Company, 59, 60, 116

Arlington National Cemetery, Unknown Sol-
      dier at, 335–36

Army, U.S., 122–56
    carelessness with equipment in, 225
    chaplains in (see chaplains)
    draft for (see draft, U.S.)
    forced integration in, 154–55
    Foreign-speaking Soldier Subsection of
        (FSS), 141–42
    lack of preparedness of, 123–24
    morale problems in, 140–41, 203, 259
    National, 142, 159, 162, 164
    in Philippine-American War, 75–76
    Regular, 123, 142, 148, 161
    training and equipping ineptitude of, 145,
        164–65
ASARCO, 108
Assembly of God church, 344
Atlanta, Ga., 117–18, 165–66
Atlantic Fleet, 73
Australian Corps, 247, 256, 269
Austria-Hungary, 17, 70, 91, 104, 146, 202,
        272, 328
    at Caporetto, 146, 147
    Jews in, 94
    Slovak views of, 107, 138, 139
    start of war and, 79–80, 83, 84–85
    surrender of, 306
    in Triple Alliance, 97
    U.S. declaration of war against, 128–29,
        138, 147
Austrian Americans, 140
Authe, 308

Baker, Newton, 180
Bar-sur-Aube, 282, 283
Barzini, Luigi, 96
Base Hospital 34, 305–6
Basilicata (formerly Lucania), xvi, 7–12, 52,
        315–16
bayonets, 151, 244, 259–60
    Kocak's use of, 201, 368n
Belarus, 30, 92
Belfast, 19–20, 42, 114, 317
Belgic (ship), 190, 193
Belgium, 43, 91, 98, 288
    Germans in, 80, 82, 85, 87, 94, 146, 175, 306
    in Grand Offensive, 247, 300–302
Belleau Wood, Battle of (1918), 168–71, 197,
        199, 204, 275, 276
Bellizzi, Charles, 195
Benedict XV, Pope, 103
Berlin, 81, 306

Bessemer, Mich., 87–91, 143, 144, 216, 218
Bessemer Brewing Company, 89, 90
Bethincourt, 250, 251
Bible, 174, 178, 346
Binghamton, N.Y., 106–7, 137–39, 272, 273,
        342, 343
blacks
    in French armed forces, 203–4, 205
    See also African Americans
Black Tom Island explosion, 114–15
Blanc Mont, 248, 271–95, 306, 309, 334
Bliss, Fort, 171–72
Blue Riband, 31
Bois de Breteuil, 210, 216
Bois de Fays, 289
Bois des Loges, xv, 298–300, 302–6, 308, 341
Bolsheviks, xviii, xx, 142, 146, 157, 314, 333
bombing, 203, 224, 232, 308, 327
Bony, 258, 270, 370n
boot camps
    fighting at, 149–51
    See also names of specific camps
Bosnia, 79, 80
Bouresches, 169, 207, 208
Brandenburg, Broughton, 26, 29, 33–34
Brattestø, John, 16, 55
Brattestø, Magnus Andreas. See Christofferson,
        Andrew
Brattestø, Tollef, 55, 84
Brest, 161, 282
Brevda, Mr., 4–6, 30
British Expeditionary Force, 114, 157–58, 317
British Military Medal, 23, 322
Brooklyn, 36, 131, 324, 331
    Italians in, 41, 44–47, 95, 97, 103–4, 162, 163
Brown, Winks, 318
Browning guns, 46, 139
Bryan, Eugene C., 142
Buell, Ralph Polk, 255–56, 258, 263–64, 268
Buffalino, Mrs., 53
Buffalino family, 52, 53
Buffalo, N.Y., 62, 128
Buffs (regiment), 229
Burges, Richard F., 153–54, 283–86, 334
Butte, Mont., 20, 42, 81, 86–87, 104–6, 133
    antidraft sentiments in, 126–27
    postwar, 317–18

Cahan, Abraham, 50, 93–94, 126
Cain, James M., 257
Calabria, 9, 24–25, 97, 103, 104, 225, 226

California, 76, 118, 227, 336
Campania (ship), 32, 34–35
Camp Bowie, 153–54, 283
Camp Custer, 143–44
Camp Devens, 165
Camp Dix, 164, 329
Camp Forrest, 161
Camp Funston, 180
Camp Gordon, xviii, 133, 136–37, 139–42,
    147, 152, 159, 231, 272
    FSS at, 141
    "Camp Gordon plan," 141–42
Camp Lewis, 133, 149, 173, 175, 181, 186,
    188
Camp Meade, 134, 135
Camp Merritt, 188–89
Camp Pike, 144, 206, 340
Camp Sevier, 171
Camp Upton, xviii, 134–35, 156, 159, 162–63,
    165, 196, 322
Camp Wadsworth, 150–52, 160, 161–62
Canto Degli Emigranti, Il ("The Song of the
    Emigrants"), 12–13
Caporetto, 146, 147
Carnegie Steel Company, 11
Carranza, Venustiano, 108
Carras, Nicholas, 237–38
Carrizal, Battle of (1916), 110
Catholics, 101, 333
    chaplains for, 196–97, 208, 239, 258
    Italian, 9, 103, 140
    Kaszub, 88, 145, 208, 216–17
    Polish, 17–18, 59, 60, 61, 81, 127–28, 140
    Slovakian, 70, 106–7, 140
cavalry, U.S., 108, 110, 166–67, 346–47
Champagne, 236, 248, 273–87, 309
Champigneulle, xv, 298, 299
chaplains, 196–97, 208, 216–17, 219, 239, 258
    Jewish, 196–97, 234, 268, 270
    marine, 287–88
Chase, William Merritt, 42
Château-Thierry, 168, 198, 209, 232
Chauchat machine gun, 236
Chicago, Ill., 128, 185–86
Chinook, Mont., 56, 84, 86, 171, 344
chlorine gas, 98, 152, 269
Chmielewski, Chester, 61
Chmielewski, Frank, 17–19, 58–62, 116,
    122–23
    postwar life of, 329, 330
    start of war and, 81–83, 95
Chmielewski, Joseph (nephew), 330

Chmielewski, Joseph (uncle), xi, xvii, 17–19,
    60–63, 116–17, 122–23, 191
    enlistment of, 129–30
    in Great War, 78, 307–9
    postwar life of, 329–30
    start of war and, 81–83
    training of, 161–62
    in transport to Europe, 190
Chmielewski, Mary Yablonsky, 61–62, 116, 329
Chopin, Frédéric, 17
Christian Science Monitor, xxi
Christofferson, Andrew, xii, xvii, 13–16, 54–56,
    99, 362n
    enlistment and training of, 170–71
    in Great War, 78, 293–94, 307, 309–11
    journey of, 35, 41
    postwar life of, 344–46
    religious conversion of, 55–56, 85, 344
    start of war and, 84–86
Christofferson, Juline Ostrem, 16, 35, 344
Christofferson, Nellie, 345
Christofferson, Norman, 344–45
Christofferson, Ruth, 345
Christ Stopped at Eboli (Levi), 7
Cieminski, Boleslaw, 338, 339
Cieminski, Mary. *See* Kondziela, Mary Ciem-
    inski
Cieminski, Maximilian, xii, 87–91, 143
    death of, 213–18, 337–40, 369n
    drafting and training of, 143–45
    in France, xviii–xix, 159, 160, 205–11,
        213–18, 234
    missing trigger finger of, 89, 207
    stake in war of, 87–88, 95
Cieminski, Mrs. Paul, 90
Cieminski, Paul, 88, 89, 90
Ciorlano, 57, 243
citizenship, U.S., 74, 322–23, 372n
    amended law and, 167, 188, 237–38
    draft and, 128–31, 133, 165
Civilian Public Service (CPS), 321
Civil War, U.S., 100, 124
Clapp, Eugene H., 54, 104, 125
Clark, George B., 278–79
Clark, Sergeant, 264
coal mines, 45, 58, 59, 60, 116, 329
Coburn, Harold D., 183
Columbus, N. Mex., 108–9
Committee of the Foreign Born, 185–86
Committee on Public Information (CPI), 177,
    184–86
con artists, 25–26

Congress, U.S., xx, xxii, 83, 333–34, 338
    draft and, 120, 123, 124
    entry into war and, 119–21
    naturalization laws and, 167
Congressional Medal of Honor, 326, 327, 342,
        343
Congress Poland (Russian Partition), 18–21, 92
Connecticut Fights (Strickland), 218, 369n
Conrad, Joseph, 17
conscientious objectors (COs), 100–101,
        131–32, 175, 178, 180–81, 319–22
    court-martials of, 181–84, 187, 188
    noncombative service and, 180
copper mining, 42, 43, 86–87, 104–5, 116, 126
Cosgrove, Bertram M., 236–37
Costantino, Leonardo, 191, 194, 226–27,
        238–39, 251, 292–93, 301
courts-martial, 72, 181–84, 187, 188
Creel, George, 177, 184–86
Croix de Guerre, 318, 323, 343
Crump, Irving, 134–35, 155
Cubberley, Ellwood P., 66
Cuisy, 250, 251
Cukela, Louis, 148, 201–5, 207
Cunard Line, 31, 32, 42, 98–100
Cunel, 302, 306
Czechoslovakia, 342, 343
Czechoslovak Legion, 138

Danforth, William, 174
D'Angelo, Pascal, 25
Daniels, Josephus, 343
Davenport, Charles, 66
Day of Sorrow (1919), 333
Debs, Eugene V., 177
Declaration of Neutrality (1914), 83–84
Decorah Posten, 84–85, 86, 99
democracy, 119, 124, 312, 317
Democrats, 112–13, 118, 326
development battalions, 141, 162
Dewey, Thomas, 343
Dickebusch Lake, 229
Doane, Francis H., 265, 270
Dominican Republic, 111–12
Domnitz, Aaron, 49
Dos Passos, John, 191–93, 207–8, 336
draft, U.S., 120, 123–33, 162
    exemptions from, 128–29, 131, 163
    Hutterites and, 131, 173–75, 180
    opposition to, 126–27, 130–31
    physical exams and, 132, 143

Dreben, Helen Spence, 153, 282, 334–35
Dreben, Samuel, xii, xx, 69, 73–78, 364n, 367n
    at Blanc Mont, 282–86, 288, 334
    daughter's death and, 282, 334, 371n
    Hollywood qualities of, 284, 371n
    in Lorraine, 309, 372n
    marriages of, 153, 336
    in Philippine-American War, 75–76
    postwar life of, 334–37
    in Punitive Expedition, 109–10, 334
    as soldier of fortune, 74, 77, 108, 153
    synagogue of, 234
    in Texas National Guard, 153–54
Drum, Hugh A., 252–53
DuBois, W. E. B., 45
Duffy, Francis, 219, 223, 230
Duvid (Sam Goldberg's maternal uncle), 92

education, 19, 94, 117, 152
Edward, Prince of Wales, 322
8th Irish Fusiliers, 114
18th Field Artillery, 171–72
81st "Wildcat" Division, 171, 293–94, 307,
        309–11
82nd "All-American" Division, 142, 152,
        159–61, 299, 316
    320th Field Artillery of, xv–xvi, 142,
        159–60, 231–33, 238, 290, 291–92
    325th Infantry of, 159–60
85th Division, 144
87th Division, 144, 145
Eisenhower, Dwight David, 327
election of 1916, U.S., 112–13, 118
Ellis Island, 34–41, 46, 48, 115, 189
Ellis Island Marine Hospital, 38
El Paso, Tex., 74, 77, 108, 153, 233–34, 282,
        334–35
Emergency Quota Act (1921), 333–34
Empress of Russia (ship), 189–91, 193
enemy aliens, 147, 152, 172, 202
    draft exemption of, 128–29, 131
    Germans as, 177–78
    P. Kocak as, 147, 272–73
England, 68, 74–75, 159–60
English language, 137, 160, 164
    Cieminski's scant knowledge of, 143, 144
    Cukela's use of, 202–3
    school for, 152
Epieds, 207, 209–11
Epstein, Alexander (Zender), 3, 5, 30, 50, 91, 92, 93
Epstein, Harold, 331, 332

Epstein, Ida Rubinstein, 331, 332–33
Epstein, Jacob, 50
Epstein, Julius, 331, 332
Epstein, Leonard, 331, 332–33
Epstein, Meyer, xii, xvii–xix, 189
    background of, 2–6, 333
    dietary laws and, 289
    in Great War, 78, 234–39, 289, 332
    induction and training of, 165
    journey of, 30, 32, 35, 42
    in New York, 42, 48–51, 67, 91–93, 125,
        331, 332–33
    postwar life of, 330–33
    sinking of *Lusitania* and, 98–99
    start of war and, 91–95
Epstein, Sarah, 2–3, 50
Epstein, Yehuda, 2–3, 5, 6, 30, 35, 49–50, 331
Espionage Act (1917), 176–77
ethnic slurs, xix–xx, 137, 149, 196, 223
eugenics, 66–67, 363
Everett, William, 347–48
Ewert, J. Georg, 186–87

Falkenhayn, Erich von, 107
Feland, Logan, 278, 280
Fifficktown, Pa., 58–62, 116–17, 122–23, 129,
    191, 329
5th Marine Regiment, 147–48, 168–70,
    198–205, 207
    Blanc Mont and, 272, 274–81
15th New York Infantry (Harlem Hellfighters),
    150–51
58th Infantry Regiment, 234–37, 249, 289,
    330–31
Finnish miners, 86, 126–27
Fiore, Rosa, 24, 40, 163
First Amendment, 177
First Army, Russian, 80, 82
First Army, U.S., 302
1st Division, 147, 148, 198, 204
1st Infantry Regiment, 111, 130, 150
Fitts, Lieutenant, 227
Flanders, 96, 130, 152, 158, 229–31, 247, 301, 318
    musical entertainment in, 223–24
Flying Circus, 203
Foch, Ferdinand, 168, 198, 200, 228, 246–48
food, 45, 62, 74–75, 85, 141
    on ships, 31, 32, 33, 191
    in war, 226, 229, 249, 289, 294, 303
Ford, Henry, 46
Foreign-speaking Soldier Subsection, 141–42

Forenza, 7–12, 52, 53, 231, 315–16, 317
42nd "Rainbow" Division, 219, 222, 230, 235
Forverts (Daily Forward), 50, 91, 93–94, 98–99,
    119, 126
4th "Ivy" Division, 234–37, 249, 251, 253,
    330–31
4th Marine Brigade, 168–69
Fourth of July, 147, 184–88
France, 43, 72, 82, 98, 114, 129, 130, 142, 145–
    49, 157–61, 163, 164, 167–72, 194–220,
    222–28, 231–44, 247–311, 315, 317, 341
    divisions of, 142
    Germans in (*see* Germany, Germans: in
        France)
    Germany's declaration of war on, 80
    Grand Offensive in (*see* Grand Offensive)
    London Pact and, 97
    number of Americans in, 158–59
    peasants in, 224–25
    readiness for war of, 68
    training incompetence in, 145
    in Triple Entente, 94
    U.S. arrival in, 147, 161, 282
    U.S.–British forces combined in, 228–31
    U.S. troops lent to, 247, 274
    See also names of specific places and battles
Franco-Prussian War, 309
Franz Ferdinand, Archduke of Austria-Hungary,
    79–80, 82, 85
French army, Blanc Mont and, 274, 275, 277
French Foreign Legion, 198
French Moroccan Division, 198, 203
Frohman, Charles, 99

Galicia, 91–92, 96, 333
Garibaldi, Giuseppe, 6
gas alarms, 148
"gas house," 152
gas masks, 148, 151, 152
General Order 151, 237
George V, King of England, 87, 159–60
Georgia, training camp in. *See* Camp Gordon
Georgia, USS, 72, 280
German Americans, 324, 334
    at Camp Gordon, 140, 142
    conscientious objectors and, 100–101,
        131–32, 178, 180, 181
    election of 1916 and, 112, 113, 118
    as enemy aliens, 177–78
    fear of, 294–95
    reactions to war of, 100–103

German Americans *(cont.)*
  return home of, 100, 365*n*
  stereotypes about, 245
German language, 143, 144, 175–76, 178, 179, 321
Germany, Germans, 18, 42, 85–88, 96–103, 112–15, 118–21, 145–49, 173–87, 315, 322, 326
  back of the head shootings by, 369*n*
  in Battle of the Somme, 113–14
  in Belgium, 80, 82, 85, 87, 94, 146, 175, 306
  bombing by, 203, 224, 232, 308
  at brink of revolution, 306–7
  at Caporetto, 146
  chlorine gas used by, 98
  emigration from, 21
  in Flanders, 229
  Forverts backing of, 94, 99
  in France, xv, 107–8, 145, 146, 148, 149, 157–58, 168–70, 198–211, 214–16, 219, 225, 228, 232–33, 235, 237–41, 244–54, 256–70, 273–80, 283–300, 303–11, 324, 327, 369*n*
  as Huns, 85–86, 173, 202
  Jews in, 94
  Lusitania sunk by, 98–100
  Mexican alliance of, 108, 118–19, 167, 346
  passenger ships of, 26–29, 31, 33
  in Poland, 91–92
  postwar, 330–31, 333
  as prisoners of war, 225, 239–40, 246, 267, 291
  readiness for war of, 68
  retreat of, 306
  Russia's fighting with, 82, 88
  Russia's peace with, 146
  statistics on, 100, 365*n*
  surrender of, 246
  in Triple Alliance, 97
  U-boats of (*see* U-boats, German)
  U.S. declaration of war on, xxii, 124, 126, 128, 138, 146, 180
  vulnerability in line of, 198
  war declared by, 80
Giosi, Antonio, 243, 296, 305, 306
Goldberg, Asriel, 21, 40, 63, 165, 346
Goldberg, Leib, 63
Goldberg, Samuel, xii, 20–21
  enlistment and training of, 165–67
  furlough of, 233–34
  in Great War, xviii–xix, 346–47
  journey of, 32, 34–35, 38, 40
  postwar life of, 346–47
Goldberg, Sarah, 34–35, 38, 40, 63, 346
Gornitz, Dora, 35, 48, 49, 50

Gow, Kenneth, 152
Grand Offensive, 247–310
  attack on Hindenburg Line in, 247, 255–70, 274
  Blanc Mont in, 248, 271–95, 306, 309, 334
  Meuse-Argonne offensive in, 247–53, 255, 296–308
Grandpré, 300, 302, 308
Grant, Madison, 67, 68–69
graves, 335–40
  in Flanders, 229, 230
  in France, 196, 216–17, 267, 270, 287–88, 306, 337–38, 342, 343, 370*n*
  of Jews, 196
  in United States, 318, 325, 327, 335–36, 339–40, 372*n*
Graves Registration Service, 217, 288
Great Britain, 30–31, 129, 241, 254, 315, 334
  in Battle of the Somme, 113–14, 115, 124
  divisions of, 142
  entry into war of, 80
  in Flanders, 96, 130, 247
  in Grand Offensive, 247, 257, 262
  guns used by, 139, 190
  Irish views of, 87, 102, 105, 126, 149, 229–30
  London Pact and, 97
  Military Medals awarded by, 268
  sinking of *Lusitania* and, 99–100
  in Triple Entente, 94
  U.S. divisions lent to, 228–31, 247, 256, 274
Greece, Greeks, 129, 141
Greene, H. A., 183–84
Gregory, Thomas W., 176, 177
grenades, 145, 151–52, 227, 244, 260, 263
  E. Affatato's throwing of, 265, 270
  Valente's use of, 266, 267
Gross, Chaim, 50
Guggenheim family, 108
Guillemont Farm, 256, 262, 269
guns, 46, 152, 345
  rifles, 139, 151, 201, 291, 308
  ship, 190–91
  See also machine guns
Gutentag, Morris, xx–xxi
Gutowski, Boleslaw, 155–56
Gutowski, Stanislaw A., 141

Haggerty, James A., 206–7, 208
Haig, Douglas, 113–14, 228, 246–47
Haller, Jozef, 129
Hamilton, George W., 275–78, 280

Harbord, James, 204
Harries, Meirion, 102
Harries, Susie, 102
Haugesund, 13–16
Havre, Mont., 344–46
Hawie, Ashad G., 149–50
Hayward, William, 150–51
Hedges, Chris, xix, 260
Hendrix, James V., 338
Herschkowitz, Jack, 135, 156
high explosives, 211–13, 218, 232, 233, 303
Hill, James J., 56
Hillman, Rolfe L., Jr., 124
Hills, Orlando G., Jr., 209, 211, 214, 215
Hilt, Reynolds B., 181
Hindenburg, Paul von, 82
Hindenburg Line, 157, 240–41, 273
    attack on, 247, 255–70, 274, 288, 323, 324,
        325, 327
    mustard gas and, 269
Hirsch, Baron Maurice de, 50
Hitler, Adolf, 67
Hoey, Peter, 258
Hofer, David, 173–75, 179–87, 188, 319–22
Hofer, Jacob, 321
Hofer, Joseph, 173–75, 179–84, 186–87, 188,
    319–22
Hofer, Maria (Joseph's wife), 320, 321
Hofer, Maria (Michael's wife), 320
Hofer, Michael, 173–75, 179–84, 186–87, 188,
    319–22
Homestead Act, 84
homesteading, 16, 41, 55, 56, 84
Hoover, Herbert, 326
horses, 166, 231–32, 291–92
House of Representatives, U.S., 121
Huerta, Victoriano, 77
Hughes, Charles Evans, 112, 113, 118
"Hugh Selwyn Mauberley" (Pound), 195
Hunt, LeRoy P., 280
Hutter, Jakob, 178
Hutterites, 173–75, 178–84
    as conscientious objectors, 101, 131, 175,
        178, 180–84, 188, 319–22
    courts-martial of, 181–84, 188
    imprisonment of, 184, 186–87, 319–22
    origins of, 178–79

Immigrant Act (1924), 333–34
immigrants
    cessation of flow of, 80

Fourth of July dedicated to, 184–86
    grapevine of rumor, misinformation, and
        wishful thinking of, 129–30, 140
    quotas on, 333–34
    as refugees, 21
    strange ideas about, 65–68
immigrant soldiers
    amended naturalization law and, 167, 188
    author's selection of, 360n
    poor physical condition of, 154
Ingalls, Albert G., 230, 269
insurance policies, 144, 296, 339, 341
International Workers of the World (IWW;
    Wobblies), 45, 81, 86, 105, 126, 177
Inverclyde, Lord, 31
Iraq War, 292
Ireland, Irish, 19–20, 21, 42–43, 63, 317
    British as viewed by, 87, 102, 105, 126, 149,
        229–30
    draft opposed by, 126–27
    election of 1916 and, 112, 113, 118
    in 69th Infantry, 219, 222, 230
    stereotypes about, 149, 245
iron mining, 90
Italian Americans, 206, 290, 313, 326
    as blacks, 65
    in Brooklyn, 41, 44–47, 95, 97, 103–4, 162,
        163
    at Camp Gordon, 140, 141
    draft and, 25, 129, 132–33
    ethnic slurs and, 137, 149
    start of war and, 95–97
    stereotypes about, 149, 245
    in 310th Infantry, 302–3
Italy, Italians, 6–13, 16, 21, 40, 51–54, 56–59,
    334, 361n
    at Caporetto, 146, 147
    departures from, 24–29
    immigrant remittances to, 9–10, 103
    neutrality of, 96–97
    on ships, 28–29, 33, 34
    unification of, 6, 7
    wartime returnees to, 103
    in World War II, 326–27

Jackson, David T., 277
Jacobson, Gerald R., 257, 263, 266
James, Henry, 65–66
James, Luther R., 284, 286
Jastzemsky, Father, 59
Jewish Welfare Board, 234

Jews, 1–6, 20, 21, 81, 142, 314, 361*n*
  at Camp Gordon, 141
  in cavalry, 166–67
  as chaplains, 196–97, 234, 268, 270
  dietary laws of observant, 289
  draft and, 129, 131
  education of, 94
  fall of the czar and, 119
  in Newark, 63
  in New York, 21, 49–51, 66, 91–93, 131,
    222, 290, 331
  Rosh Hashanah and, 233–37
  start of war and, 91–92
  stereotypes about, 149, 245
  as veterans, 322–23, 332, 333
  See also anti-Semitism
Johnson, Douglas V., II, 124
Jonsson, Patrik, xxi
Justice Department, U.S., 176, 177

Kalisz, 91–92
Kansas, 319–21
Kasovich, Israel, 33
Kaszubia, Kaszubs, 88–89, 90, 143
Kaszubian language, 88, 143
Keegan, John, 158
Kelly, Francis J., Jr., 278–79
Kennedy, David, 177
Kerensky, Alexander, 146
Kilmer, Joyce, 219
Kim, Alexander, 269
Kinloch, Lieutenant, 161
Knoll (Picardy), 256, 262–68, 270
Koblenz, 330–31
Kobrin, Leon, 49
Kocak, John, 137, 342
Kocak, Julia, 106, 137, 147, 342, 343
Kocak, Matej, xiii, xx, 69–74, 77–78, 137–40,
    145, 364*n*
  as American, 77–78
  bayonet use of, 201, 368*n*
  in Binghamton, 106–7
  death of, 279–80, 287–88, 342–43
  in Dominican Republic, 111–12
  in France, 147–48, 168–70, 197–205, 207,
    271–76, 279–80, 283, 287–88, 342
  furlough of, 137–39
  going AWOL of, 72, 280
  as warrior, 69–70
Kocak, Mr. (cousin Paul's father), 138, 140, 272
Kocak, Mr. (Matej's father), 70, 71, 72

Kocak, Paul (Matej's brother), 106, 137, 342
Kocak, Paul (Matej's cousin), 106, 138, 139–40,
    147, 272–73
Kocak, Roza, 72, 343
Kondziela, August, 89, 143
Kondziela, Marguerite. *See* Riggs, Marguerite
    Kondziela
Kondziela, Mary Cieminski, 89, 90, 143,
    144–45, 339, 340
König Albert (ship), 26–29, 36–37
Kościuszko, Tadeusz, 17
Ku Klux Klan, xx, 333, 335

labor unions, 86, 93, 329. *See also* International
    Workers of the World
Larimore, N. Dak., 54–56, 84
Laskin, Emily, 348
Laskin, Samuel, 361*n*
La Società, 104
Leavenworth, Fort, 319–21
Lejeune, John A., 274, 277–78
Lemberg (Lviv; Lvov), 91, 96
Lengel, Edward G., 248
Lenin (Vladimir Ilyich Ulyanov), 146
Leskovar, Christy, 301–2
Levi, Carlo, 7, 11
Levin, Samuel, 20, 171–72
Levinger, Lee J., 196–97, 268, 270
Lewis, Isaac Newton, 46
Lewis, John L., 329
Lewis machine gun, 46, 139
Liège, 85, 247
Lithuania, Lithuanians, 3, 18, 82, 123
Little Italies, 51–52
Liverpool, 30, 75, 159, 193
Lodz, 20–21
London, 74–75, 97, 159–60
London Pact (1915), 97
Lone Tree Trench, 265, 270
Long Beach, N.Y., 325–26, 327
Long Island National Cemetery, 325, 327
Lorraine, 236, 309, 372*n*
Lost Battalion, 290–91
Louvain, 85
Ludendorff, Erich, 157, 168, 198
Lugg, T. B., 287–88
Lukens, Carl J., 162
Lusitania (ship), 30–32, 41–42
  sinking of, 98–100, 102
Luther, Martin, 178
Lutheranism, 15–16, 55

Lymansville, R.I., 57–59
lynchings, 176

machine guns, 175, 203, 206, 232, 238, 257
    German use of, 145, 169, 200–201, 204,
        208–11, 215, 219, 225, 235, 237, 250,
        253, 262, 263, 264, 266–67, 276, 278,
        284, 286–87, 291, 299, 300, 303–4
    Lewis, 46, 139
    in Mexican Revolution, 77
    U.S. training for use of, 161–62, 236
Mackin, Elton E., 277, 278, 281, 282
Madalena, Ralph, 327
Magdalena (girlfriend of Antonio Pierro), 316,
        317, 348
Magyarization, policy of, 70
Mancuso, Filomena. *See* Affatato, Filomena
        Mancuso
Marcone, Pasquale, 303
Marcus, Jacob Rader, 150
Marine Corps, U.S., 71–74, 106, 107, 108, 137
    breaking ranks and retreat of, 276–78, 280
    in Dominican Republic, 111–12
    in France, 107–205, 271–83
    qualifications for, 168
    self-confidence of, 168, 202
    Solo Club of, 201–2, 205, 207
Marne salient, 198–99, 204, 205, 207, 219, 235,
        240
Mastine, Joseph, 151, 152, 266–67
Mauretania (ship), 31
Mažár, Imrich, 139
McCormack, John, 185, 186
McKinley, William, 75
McPherson, Leslie Allen, 304
McRae, James H., 305
medical inspection, 27–28
Mennonites, 101, 131, 178
Metz, 240, 308
Meurthe et Moselle district, 232
Meuse-Argonne Cemetery, 342, 343
Meuse-Argonne sector, 247–53, 255, 296–308
Meuse River, 306, 309
Mexican Revolution, 73, 74, 77, 81
Mexico
    German alliance with, 108, 118–19, 167, 346
    U.S. relations with, 73, 74, 77, 108–12, 130,
        167, 334
Mezzogiorno, Il, 7–13, 21, 24–25, 28, 29, 41, 52
Mickiewicz, Adam, 17
Military Intelligence Division, U.S., 141–42

Miller, Archie, 52
Minder, Charles, 245–46, 294–95
mining
    accidents in, 105, 116, 127, 318
    See also coal mines; copper mining; iron
        mining
Minskoff, Sam, 51, 93
Moellering, Edward, 347
Montana, 20, 56, 84–87, 178
    enlistment in, 133
    Hutterites in, 179, 321
    See also Butte; Chinook
Monte Cassino, 110, 327
Montfaucon, 248–53, 273, 342
Montgomery, W. S., 287
morale problems, 140–41, 158, 203, 259, 276
Moranville, 310
Morro Castle, SS, 73, 74
Moselle River, 232, 240, 330
Mount Vernon, 185
Mourners' Parade, 333
Munson, Edwin S., 163
music, 12–13, 34, 176, 222–24, 309
    patriotic, 185, 186, 345
mustard gas, 269, 302, 328
Mustico, Angelo, 224

Nancy, 232, 235, 236
Nantes, 305–6
Naples, 25–26, 53, 362*n*
Naród Polski (Polish Nation), 81–83
National Guard, 109, 110, 123, 126, 142, 219
National Registration Day (June 5, 1917), 13–27
naturalization laws, 167, 188, 237–38
Navy, U.S., 73, 176, 322
Nebraska, 16, 41, 55
Needham, Gordon A., 215
Nelms, James A., 277
Neumann, Nellie C., 310–11
New England National Guard. *See* 26th "Yan-
        kee" Division
New Jersey, 37, 63, 188, 325, 338
New Mexico, 108–9, 119, 167, 233, 346
New York City, 24, 35–42, 44–52, 113, 119, 150
    arrival in, 36–42, 53
    Black Tom Island explosion and, 114–15
    draft in, 125, 126, 131, 165
    Lower East Side in, 35, 48–51, 66, 91–93,
        131, 222, 290, 331
    marine recruiting office in, 72, 107
    parades in, 322, 324, 325, 333

New York City (cont.)
    peace demonstration in, 102
    See also Brooklyn; Staten Island
New York City Department of Sanitation, 323, 324, 325
New York Harbor, 114–15, 188, 331
New York National Guard, 111, 125, 130
    1st Regiment of, 111, 130, 150
    7th "Silk Stocking" Regiment of, 150–51
New York Times, 80, 81, 85–86, 125, 127, 158, 186, 270, 336
Nicholas II, Czar of Russia, 82, 94, 119
Nichols, G. H. F., 259
1919 (Dos Passos), 336
90th Division, 304
91st "Wild West" Division, 149, 188, 189, 191, 226–28, 239, 253, 292–93, 317
    in Belgium, 300–302
    as Powder River Gang, 227–28
    2nd Battalion of, 251, 252
92nd Marine Company, 139
93rd Division, 247, 248
noncombative service, 180
nonresistance, doctrine of, 178
Norris, George W., 120
North German Lloyd, 26–29, 31
North Providence, R.I., 125, 163–64, 243, 340, 341
Norway, Norwegians, 13–16, 21, 54–56
    start of war and, 84–86
    See also Christofferson, Andrew

O'Connor, Sergeant, 264
Ogdensburg, N.Y., 111, 267
Oglethorpe, Fort, 166–67
Oise-Aisne Cemetery, 338
O'Leary, Jeremiah, 102, 113, 118
Olson, Zenas A., 189–90
101st Infantry, 215–16
102nd Infantry, 160, 205–11, 213–18, 339, 369n
105th Infantry, 256, 257
106th Infantry, 256–59
107th Infantry, 150–52, 161, 162–63, 222, 224–26, 322
    in attack on Hindenburg Line, 255–62, 266–68, 327
    in Flanders, 229–31
139th Infantry, 309
141st Infantry, 153–54, 281–87, 309
One Night in a Bad Inn (Leskovar), 301–2
O'Ryan, John, 223–24, 245, 256–59, 261

Ostrem, Juline. See Christofferson, Juline Ostrem
Ottaviano, Antonia, 56–59, 125, 163–64, 340, 341
    Tommaso's correspondence with, 242–44, 246, 296–97, 300, 303, 341
Ottaviano, Ascanio, Jr., 163–64, 341
Ottaviano, Ascanio, Sr., 57
Ottaviano, Carlo, 163
Ottaviano, Domenico, 163, 164
Ottaviano, Giacomo, 163
Ottaviano, Otto, 341
Ottaviano, Tommaso, xiii, 56–58
    death of, 305–6, 340–42
    draft registration of, 125
    in Great War, xviii–xix, 238–39, 242–44, 246, 296–300, 302–6, 308, 341
    induction and training of, 163–64, 165

Padgett, Edward R., 135
Pale of Settlement, 1–6, 21, 30, 32, 49, 74, 91–94, 119, 236, 334, 361n
    Hutterites in, 179
    roundups of Jews in, 92, 93–94
    start of war in, 91–92
    tzedakah in, 3, 5–6, 333
    See also names of specific places
Palo, Matt E., 224
Paris, 80, 81, 334
    German advance on, 168, 169, 198
    U.S. troops in, 147
Parsons, Lloyd C., 307
Passing of the Great Race, The (Grant), 67, 68–69
peasants, French, 224–25
Peck, DeWitt, 277
Pellegrini, Angelo M., 33
Pellicoro, Peter I., 288
Pershing, John J., 203, 332, 335, 343
    citizenship and, 237–38
    in France, 130, 147, 168, 219–20, 228, 237–40, 246, 247, 252, 253, 274, 334
    open warfare and, 151, 219–20
    in Punitive Expedition, 109–10, 111, 167, 334, 365n
    U.S. divisions lent by, 228, 247, 274
Pétain, Philippe, 252
Philippine-American War, 75–76
phosgene gas, xv, 269, 299
photographs, 57–58, 165, 169, 317, 345
Piasecki, Peter F., 128
Picardy, 255–57, 327

Pierro, Antonio, xiii, xv–xviii, 145
  background of, xvi, xxii, 6–13, 362*n*
  at Camp Gordon, xviii, 133, 136–37, 140,
    152, 159, 231
  drafting of, 132–33
  draft registration of, 125
  in England, 159–60
  ethnic slurs and, 137, 149
  in Great War, xv–xviii, xxii, 78, 142, 159–60,
    231–33, 238, 290, 291–92, 299, 347
  journey of, 41
  postwar life of, 315–17, 347–48
  in Swampscott, xvi, 53–54, 67, 104, 125,
    132
Pierro, Daniele, 10–12, 104, 317
Pierro, Guarino, 10–12, 104
Pierro, Maria, 316–17
Pierro, Michele, 10–12, 104, 125
Pierro, Nicola, 10–12, 104, 315–16, 317
Pierro, Nunzia dell'Aquila, 8, 10, 11, 52–53,
    316, 317
Pierro, Rocco, 8, 10, 52–54, 65, 67, 104
Pierro, Rosa, 10–12
Pierro, Vito, 10–12, 104, 317
Piltz, Father Francis, 61, 62
pogroms, xx, 2, 20–21, 94, 333, 334
poison gas, xv, 98, 152, 233, 238, 269, 299, 302,
    328, 365*n*
Poland, Poles, 17–19, 58–63, 143, 166, 334,
    361*n*
  at Camp Gordon, 140, 141
  draft registration and, 127–30
  election of 1916 and, 118
  in Fifficktown, 58–62, 116–17, 122–23
  morale problems of, 140–41
  postwar, 328–30, 333
  start of war and, 81–83, 91, 92, 93
  stereotypes about, 245
  See also Chmielewski, Joseph; Cieminski,
    Maximilian
Polish Army (Haller's Army), 129
Polish-Lithuanian Commonwealth, 17
Polonia, Wis., 88–89, 128, 338
Pound, Ezra, 195
Prager, Robert Paul, 176
Pratt, Arnold, 301
prisoners of war, 240
  German, 225, 239–40, 246, 267, 291
Progresso Italo-Americano, 95–97, 103, 104
Protestant Reformation, 178, 179
Protestants, 154, 196–97
Prussia, 17, 82, 328

Punitive Expedition, 109–10, 111, 130, 167, 334
Purple Heart, 372*n*

railroads, 25, 39–40, 46, 48, 56, 173–75, 319
  to boot camp, 135–36
  Des Moines, 95
  in France, 147–48
  troop transit, 188–89, 309
  Villa's attack on, 108
Raskin, Alexander, 135–36, 327–28
Rau, George R., 218, 219
Red Scare era, 142, 314, 333–34
Reed, James, 124
Reggio di Calabria, 24, 362*n*
religious conversion, 55–56, 85, 196, 344
Remarque, Erich Maria, 213
Republicans, 112, 113
Retz Forest, 199–203
Rheims, 95–96, 158, 169, 198, 207, 272, 273,
    287
Rhine River, 330, 331
Rhodes, Pamela Ottaviano, 341, 342
Ricci, Basil, 341
Ricci, John (nephew), 341–42
Ricci, John (uncle), 341
Richardson, Tracy, 77, 285
Richthofen, Baron von, 203
Riggs, John, 339–40, 369*n*
Riggs, Marguerite Kondziela, 90–91, 144, 218,
    339, 340
Riker, Matt, 149–50
Rizzi, Giuseppe Nicola (Joe; Woppy), xix, xx
Rockefeller, John D., 81
Roosevelt, Theodore, Jr., 142, 168, 279, 343
Roosevelt, Theodore, Sr., 67–68, 101, 102, 113,
    134
Rosh Hashanah, 233–37
Rubinstein, Ida. See Epstein, Ida Rubinstein
Runyon, Damon, 337
Russia, Russians, 17, 83, 91–94, 315, 328, 334
  at Camp Gordon, 141
  German fighting with, 82, 88
  Germany's peace with, 146
  London Pact and, 97
  Serbia supported by, 79–80
  Slavicization of Germans in, 179
Russian Army, 119, 146, 334
  Hutterites and, 179
  Jews in, 1, 74, 91, 171, 361*n*
  See also First Army, Russian; Second Army,
    Russian

Russian Partition. *See* Congress Poland

Russian Revolution (1917), xviii, 119, 146, 157

saber drill, 166–67

St. Anthony's Church, 59, 60, 61, 116

St. Etienne, 275, 278, 280, 283, 286

St. Juvin, 298–300, 308

St. Mihiel, 242, 247, 290

St. Mihiel salient, 148, 158, 238–41

St. Quentin Canal, 248, 256, 262–63, 266, 268, 269

Samsonov, Alexander, 82, 92

Sanger, Margaret, 67

Sant'Apollinare, 110, 327

Sarajevo, 79, 80, 85

Savage, Arthur, 139

Savage, Murray, 291

Savage Arms Company, 139

Scala Coeli, 24, 29, 163, 323, 362*n*

Schucker, Louis E., 310, 311

Schwegler, Sergeant, 264

seasickness, 33, 35

Second Army, Russian, 82

2nd Battalion, 251, 252

2nd Division, 148–49, 168–69, 198, 204, 274.
    See also 5th Marine Regiment; 6th
    Marines

Sedan, 248, 302

Sedition Act (1918), 177

Seicheprey, 158, 160, 206

Selective Service Act (1917), 123, 125, 128–29

Senegalese soldiers, 203–4, 205

Serbia, 79–80

76th Depot Division, 165, 189

77th Division, U.S., xx, 134, 222–23, 271, 294–95
    Lost Battalion of, 290–91
    in Mourners' Parade, 333

78th Division, 164, 242–44, 296–300, 302–6, 308, 328

79th "Baltimore's Own" Division, 253, 257

Shertzer, Robert S., 181

ships, 26–37, 53, 362*n*
    first-class passage in, 29, 31, 34, 42
    troop transport, 147, 161, 163, 189–91,
        231–32, 322, 331
    See also steerage; *and names of specific ships*

shrapnel, 211–12, 213, 233, 264

Shrapnel, Henry, 212

Shuler, G. K., 169

Signal Corps, 166, 167

6th Division, 161–62, 190, 307–9

6th Marines, 148, 204, 274, 287–88

16th Machine Gun Battalion, 161–62, 190,
        307–9, 329

66th (C) Company, 5th Marine Regiment,
        147–48, 169–70, 200, 207, 271–76,
        278–79

69th Infantry (Fighting Irish), 219, 222, 230

Slovak Republic, 70

Slovaks, 106–7, 123, 128, 138, 147, 343
    at Camp Gordon, 140

Smith, Albert C., 163

socialism, socialists, 105, 125, 126, 177
    of *Forverts,* 93, 99

Soissons, 198–205, 207, 209, 240, 275, 276,
        279, 342
    heroism at, 200–201, 203–5, 280

Solo Club, 201–2, 205, 207

Somme, Battle of the (1916), 113–14, 115, 124

Sophie, Duchess of Hohenburg, 79–80, 82, 85

South Dakota, 176, 178
    Hutterites in, 173–74, 178, 179–80, 319,
        320, 321

South Fork, Pa., 58, 59, 60, 81, 82, 116, 122,
        123, 329, 330

Spence, Helen. *See* Dreben, Helen Spence

Spire, James, 270, 370*n*

Staten Island, 36, 50

Statue of Liberty, 36, 115, 189

steerage, 26, 32–35, 42, 75
    description of, 28, 32–34
    profits from, 29
    in troop transport ships, 190–93

Sterba, Christopher M., 365*n*–66*n*

stereotypes, 149, 245

Stratton, Harry, 162–63

Strickland, Daniel W., 209, 210, 211, 218, 369*n*

strikes, 124, 127

Stryckmans, Felix J., 185–86

Suddenly We Didn't Want to Die (Mackin), 277,
        278, 281

superstitions, 226

Sutphen, Joseph W., 188

Swampscott, Mass., xvi, 10, 41, 52–54, 104,
        317, 348
    draft in, 125, 132

synagogues, 234, 331

Syrian Christian, 149–50

Taft, William Howard, 46

Talese, Gay, 25

Talmud, 3, 6

Tannenburg, Battle of (1914), 82, 92
taverns, 61, 106
tear gas, 152
Teilhard de Chardin, Pierre, 260
10th Cavalry Regiment, 110
Texas, 109, 119. *See also* El Paso
Texas National Guard, 153–54
Texas Rangers, 110
Thiaucourt, 297–98
3rd Division, 168, 171–72
13th U.S. Cavalry, 108
30th Division, 228, 247–48
35th Division, xix
36th Division, 274
    141st Infantry of, 153–54, 281–87, 309
Thompson, Denis, 114, 317
Thompson, John (nephew), 114
Thompson, John (uncle), 318
Thompson, Mary, 19–20
Thompson, Nellie, 19
Thompson, Peter, xiii, xvii–xviii, 114, 187–89,
        191
    background of, 19–20, 362*n*
    in Belgium, 300–302
    in Butte, 42, 86–87, 104–6, 317–18
    draft registration of, 127
    enlisting of, 133
    in France, 226, 227–28, 238–39, 251–52,
        292, 317, 318
    journey of, 42–43
    postwar life of, 317–18
    wages of, 105, 365*n*
Thompson, Rose, 19–20, 317
Thompson, Sam, 105, 106
310th Infantry, 164, 242–44, 297–300, 302–6,
        308
320th Field Artillery, xv–xvi, 142, 159–60,
        231–33, 238, 290, 291–92
321st Infantry, 293–94, 310
325th Infantry, 159–60
328th Infantry, 142, 291
345th Infantry, 144, 206, 338, 339, 340
362nd Infantry, 188, 251–52, 301–2
364th Infantry, 191, 194, 251, 301
Three Soldiers (Dos Passos), 191–93, 207–8
Trader, E. J., 71
trench warfare, 148, 151–52, 157–58, 240, 244,
        246, 264
    high explosives and, 212–13
trigger fingers, removal of, 89, 144, 207, 361*n*
Triple Alliance, 97
Triple Entente, 94

Trugny, 209–11, 213–18, 337–38
Turkey, 146, 306
Turrill, Julius S., 199
12th Cavalry Regiment, 166–67, 233
26th "Yankee" Division, 158, 205–11, 213–19
    102nd Infantry of, 160, 205–11, 213–18,
        339, 369*n*
27th Division, 160–61, 195, 245, 322, 325
    chaplains with, 196–97
    in Grand Offensive, 247, 259
    lent to British, 228–31, 247
    musical entertainment in, 223–24
    as O'Ryan's Traveling Circus, 224
    See also 107th Infantry
tzedakah, 3, 5–6, 333

U-boats, German, 98, 100, 105, 118, 365
    transport ships attacked by, 147, 191, 232
Ukraine, 3, 20, 74, 91, 179, 333
United Mine Workers of America, 329
United States
    entry into war by, xviii, xxii, xxiii, 69, 101,
        118–21, 124, 126, 128, 138, 146–47, 180
    Mexico's relations with, 73, 74, 77, 108–12,
        130, 167, 334
    neutrality of, 83–84, 94, 107
    percentage of immigrants in, 124
    in Philippine-American War, 75–76
    postwar life in, 312–48
    prewar life in, 45–46
    reactions to start of war in, 80–88, 91–97
    as world power, 68
    See also America, Americans

Valente, Margarita, 325
Valente, Michael, xiii, xx, 159–63
    background of, 110–11, 365*n*
    enlistment of, 110–11, 130
    in Flanders, 224, 229
    in France, 159, 161, 224, 228, 241, 262,
        266–68, 325
    Medal of Honor won by, 326
    postwar life of, 325–27
    training of, 150, 151, 162, 163
    transport to Europe of, 160–61
    wounding of, 267, 268
Valente, Patricia A., 313
Valente, Paul, 111
Valiani, Maria, 48

Valiani, Mrs., 48
Van Deman, Ralph, 142
Vanderbilt, Alfred Gwynne, 99
Vanzetti, Bartolomeo, 40
Vaux-devant-Damloup, 309–10
Veracruz, 73, 74, 77, 108
Verdun, 232, 236, 238, 247, 248, 298, 309
    Battle of (1916), 107–8, 110, 148, 149, 248,
        249, 310
Verdun Boche, 148, 149
Vesle River, 219, 235
veteran's compensation, 329–30
Villa, Pancho, 74, 77, 108–11, 130, 167
Vittorio Emanuele III, King of Italy, 103
Vosges Mountains, 293–94, 307

wage increases, 87, 104–5
Wagner, Richard, 240
War Department, U.S., xix, 142, 159, 337
    chaplains appointed by, 196
    draft and, 126, 130–32, 162
    National Guard policy of, 150
War Is a Force That Gives Us Meaning (Hedges),
    xix
War Risk Insurance program, 144
Washington, George, 185, 372n
Weisz, Eugene C., 141
Wells, H. G., 46
wheat, 179, 203, 208–9
Whittle, John R., 270
Whittlesey, Charles, 335
Wilhelm II, Kaiser of Germany, 87, 102, 306
Wilhelm of Germany, Prince, 249
Williams, Lloyd, 169
Wilson, Edith, 185
Wilson, Leonard J., 201
Wilson, Margaret, 185
Wilson, Woodrow, 16, 46, 66, 115, 118–20, 312,
    314, 328, 345
    draft and, 123, 131
    election of 1916 and, 112–13, 118
    entry into war and, 119–20, 146
    Espionage and Sedition acts and, 176–77
    Fourth of July and, 185, 186
    Mexico policy of, 73, 108, 109

on noncombative service, 180
"Peace Without Victory" speech of, 128
training and equipping ineptitude of, 145
U.S. neutrality and, 83–84, 94
Wipf, Jacob, 173–75, 179–84, 186–87, 188,
    319–22
Wise, Frederick May, 169, 170
Woevre Valley, 309–10
Wood, Leonard, 145
World War I
    Allied victory in, 246–48
    bonding in, xix–xx, 221–22, 243–46
    civilian casualties in, 85, 98–99, 364n
    Eastern Front in, 80, 82, 83, 88, 91–93, 96,
        146, 157
    end of, 310–11, 313, 319
    gas used in, xv, 98, 152, 233, 238, 269, 299,
        365n
    heroism in, 69, 197, 200–201, 203–5
    military casualties in, 79, 91–92, 96, 108,
        114, 124, 146, 157, 158, 169–70, 204,
        213–18, 219–20, 230, 239, 268–69, 275,
        279–80, 291, 298, 299, 305–6, 311, 312,
        315, 364n, 372n
    number of immigrants fighting in, xvi
    shrapnel vs. high explosives in, 211–13
    start of, 42–43, 79–88, 91–97
    treaty in, 312, 313
World War II, 85, 311, 312, 314, 321, 324–27,
    330, 332, 345

Yablonsky, John, 61
Yablonsky, Josephine, 61
Yablonsky, Mary. See Chmielewski, Mary
    Yablonsky
Yablonsky, Walter, 60, 61
Yockleson, Mitchell, 365n
York, Alvin Cullum, 136–37, 142, 145, 151,
    291, 335
Ypres, 98, 130, 158, 229–31, 247, 301
    first battle of, 96

Zimmermann, Arthur, 118–19
Zimmermann telegram, 118–19

About the author
2 Meet David Laskin

About the book
6 From Inspiration to Narration

Insights,
Interviews
& More . . .

Read on
12 The Great War in History and
   Literature
15 A Brief Guide to Military Units
   and Ranks

Meet David Laskin

I HAVE A HUNCH that I'm not the only baby boomer who grew up with a sense of being somehow outside of history. Of course, my generation made its share of headlines and shaped more than its share of social and cultural trends, everything from the Hula-Hoop craze to ecstatic Beatles and Rolling Stones concerts to massive anti–Vietnam War protests. But I was raised in a nice suburban house outside of New York City feeling that the nation's, and the world's, recent history—Cold War, Korean War, Second World War and the attendant destruction of Europe's Jews, Great Depression, just to reel off the big events of the two decades preceding my birth—had no real bearing my life. We were different, we baby boomers; we would write our own history; we would invent new media that in turn would generate new messages. Let the West's long nightmare torment the sleep of the old. We were going to Hula-Hoop our way into a new millennium.

I now see how misguided I was. I didn't escape the shadow of history— I merely skirted it for a while. For the past six or seven years, I've been groping my way back into it through my writing—and now I'm deeply, inextricably entangled.

I'll be a bit more concrete. After growing up in that nice suburban house and graduating from high school in 1971 (Summer of Love past, but Vietnam still raging), I went to Harvard College and studied history and literature, specifically the history and literature of England and America from roughly 1600 to 1950. You'd think that that major would have plugged me into the current of history like a cartoon cat with its paw in an electrical outlet—but somehow I spent more time with Henry James and Henry Miller, George Eliot and John Ruskin, than with the events and ideas that crashed their civilization. The one historic thread that captivated me was the emergence of modernism out of the ashes of the Great War—and this thread was considerably strengthened by my reading of *The Great War and Modern Memory* by Paul Fussell. The idea that four years of all-out industrial war should have shaped the consciousness of a generation of writers and intellectuals became a kind of touchstone that I came back to again and again.

I had the great privilege after college of going to graduate school at Oxford, where I spent two years reading English poetry and making lifelong friends. I returned to the States in 1977 able to recite long swatches of John Donne, Shakespeare, Ben Jonson, and Wordsworth, but totally unfit to make my way in commercial culture ("You must be some kind of freaking snob," a prospective employer spat at me when he glanced at my resume). But somehow, after a brief stint in publishing, I found my feet as a writer, and I began, slowly, sometimes painfully, but more often pleasurably, working my way toward what I hope is my true ▶

> 66 The idea that four years of all-out industrial war should have shaped the consciousness of a generation of writers and intellectuals became a kind of touchstone that I came back to again and again. 99

Meet David Laskin *(continued)*

subject: the sweep of cataclysmic history through the lives of ordinary men and women.

My professional turning point came in 2004 with the publication of *The Children's Blizzard*, an account of a deadly winter storm that hit the upper Midwest on January 12, 1888. True, the cataclysm here was natural, not man-made—but historic circumstances (the facts that the American prairie was so thinly settled then, that most of the settlers were immigrants, and that almost all of them were poor and struggling) determined the severity of the event and ensured that its tragic legacy would endure in families to this day.

I took a step further in my next book, *The Long Way Home: An American Journey from Ellis Island to the Great War*. Writing about how the greatest (in numbers and, I would argue, in influence) generation of American immigrants was profoundly shaped by the Great War allowed me to explore ideas and issues that I had been brooding about for a long time. My college reading about modernism and the "lost generation," the harrowing war poems and memoirs that Fussell had analyzed in *The Great War and Modern Memory*, my increasing awareness of the irresistible gravitational pull of the great cycles of history, and my own heritage as the grandson of immigrants who passed through Ellis Island in the first decades of the twentieth century—all this came together in this book. I was deep into

> ❝ Writing about how the greatest generation of American immigrants was profoundly shaped by the Great War allowed me to explore ideas and issues that I had been brooding about for a long time. ❞

the research when I learned that my great-uncle Hyman Cohen, an immigrant from the Russian Pale, was drafted in 1917 and served in France with the First Division. Like Sergeant Kocak, Hyman fought at Soissons in the summer of 1918; he was gassed and evacuated to a convalescent station.

As it turns out, it was a lucky stroke that Hyman's story came to my attention too late for me to include in *The Long Way Home*—because it will find a place in the book I am working on now. In my new project I have turned from the stories of other families to my own. In *Three Roads* (the working title), I am writing about the three branches of my mother's family and the very different paths they followed at the turn of the last century: One branch came to America and prospered in successful businesses, one branch emigrated to Palestine and went to work as pioneers on collective farms, and one branch remained behind in Poland and perished in the Holocaust.

It has taken nearly half a century for me to recognize fully, painfully, how much history has impinged— and no doubt will continue to impinge— on my life and the lives of those I love. I am blessed that I have been given the opportunity to share this recognition in my writing. ᕦ

> ❝ In my new project I have turned from the stories of other families to my own. ❞

From Inspiration to Narration

THE IDEA FOR *THE LONG WAY HOME* came to me in one of those rare moments when a lightbulb pops on and something new flashes unbidden in the brain. It happened one evening when I was chatting over dinner in a Thai restaurant with my wife, Kate, while silently, distractedly mulling over what to write about next. I had been playing around with the idea of doing something about the First World War but I was having trouble finding an angle. Meanwhile, I was also batting around the notion of doing something about immigration. As the grandson of four immigrants, I have always marveled at the courage and determination it takes to uproot yourself from the country of your birth and start a new life in an alien place. My grandparents arrived at Ellis Island shortly after the turn of the last century, during the peak period of U.S. immigration. America, I knew, had been transformed by this unprecedented wave of newcomers. The world, of course, was transformed by a war fought in and by the countries where most of those immigrants came from.

That's when the lightbulb lit up. Why not fuse the two? Write about the impact of the first industrial war on the millions of immigrants of my grandparents' generation.

Flash! *The Long Way Home* was born.

I wish the rest of the process unfolded

the way it does in the movies: The inspired writer, aglow with determination, chains himself to the typewriter or computer, smokes a lot of cigarettes while tossing crumpled pages on the floor or furiously hitting the delete key, and a couple of frames later legions of adoring fans clamor to have their copies of the book signed. In truth, it took months just to wrestle the idea into a proposal that I could send off to my publisher. I knew that I wanted to open up the immigrant experience in the war by focusing on the stories of individual immigrant soldiers and their families, but the problem was deciding which immigrants to focus on. Initially I toyed with the idea of writing about famous foreign-born World War I veterans like Irving Berlin (who wrote a musical while at Camp Upton boot camp out on Long Island), Fiorello LaGuardia (who flew fighter planes on the Italian front), and Frank Capra (who enlisted from Los Angeles in October 1918, just a month before the war ended). But at some point it struck me that the book would be more powerful and universal if instead I recounted the experiences of ordinary guys who came to the United States for freedom and opportunity and ended up shipping back across the Atlantic in the khaki uniform of their adopted country.

But where to find them?

That question brought many a sleepless night after I started work on the book in the summer of 2006. I decided at the outset that Sam Dreben, known as the Fighting Jew, was in ▶

> **❝** I wanted to open up the immigrant experience in the war by focusing on the stories of individual immigrant soldiers and their families, but the problem was deciding which immigrants to focus on. **❞**

About the book

because he was not only a genuine war hero but an incredibly colorful soldier of fortune, gambler, machine gun aficionado, and world-class character. Combing through lists of World War I Medal of Honor winners, I came upon Michael Valente, the only Italian American to win the MOH in the war, and Matej Kocak, a Slovak-born Marine who won *two* MOHs. I was able to track down Valente's grandson and Kocak's cousin's daughter, and both were extremely helpful with background information, photos, and guidance. So now I had three "characters" for my epic. But the sleepless nights continued.

I don't recall the chain of connections, but a long morning of phone calls led me to a man who had a list of all surviving American Great War Doughboys (there were a dozen, as I recall, in 2006), and a bit of research revealed that two of them were immigrants. It never occurred to me when I started work on the book that I would be including men who were still alive—and so I was thrilled to make contact with Tony Pierro, who was 110 that summer, and Samuel Goldberg, 106. Interviewing these centenarians became my number one priority, and after a bit of strategizing with their families, I managed to meet them both in the same hot July week. It helped that Tony's home in Swampscott, Massachusetts, was only about an hour away from the assisted living facility where Sam lived outside of Providence, Rhode Island. I will never forget the look in Tony's eyes when he reminisced about his French girlfriend from 1918 or Sam's stories of eating lobster and ice cream on the Jewish high holidays in El Paso, Texas. Tony and Sam both refrained from addressing big issues—to what extent serving in the U.S. Army (or in Sam's case, the cavalry) made them Americans; whether they faced much discrimination; whether they believed that the war had really "made the world safe for democracy." But they did give me a sense of the texture of the past—what it felt and sounded and smelled like—that no book or article or even film can convey. I also learned that when you're interviewing centenarians, you better be prepared to speak really loudly.

That made five.

At some point I came up with the idea of posting ads in ethnic newsletters and Web sites, immigrant newspapers, and veterans organizations. I advertised in *Italian America* (the magazine of the Order Sons of Italy in America) and *The Jewish Veteran*, published by the Jewish War Veterans of the USA; I wrote articles for Polish newspapers and websites about the important role that Polish Americans played in the war. At the end of the articles I mentioned that I was writing a book and would be happy to hear from readers who had stories to share. I placed an item in the newsletter of the Norwegian-American Historical Association, which had helped me with research for *The Children's Blizzard*. Then I sat back and waited and hoped that stories would roll in.

I didn't have to wait long. John Chmielewski called from Bakersfield, California, to tell me about the service and strange disappearance of his Polish-born uncle Joseph. All three Affatato brothers contacted me one after another on the same day to describe the heroic actions of their father, Epifanio, on the morning of September 29, 1918. Len Epstein saw my item in *The Jewish Veteran* and eagerly shared stories, photos, and memories of his father, Meyer, born in the Russian Pale and drafted by the U.S. Army in the spring of 1918. The fact that so many stories came pouring in was a sign of the intense pride that people still feel for the service and sacrifices of their immigrant ancestors. In a matter of weeks I had more stories than I could possibly include in one book.

Now the challenge was choosing the best ones and knitting them together into a coherent narrative. My guiding criteria were depth and detail. Since I wanted to go as far back as possible into the European backgrounds and origins of these men and their families, I singled out stories that were rich in personal history. Meyer Epstein, orphaned as a young boy, lost on the way to his relatives' village and forced to go to work in a strange town, allowed me to open up the hardship that Jews suffered in the Pale of Settlement in the nineteenth century. With Andrew Christofferson, I was able to touch on the social and religious changes sweeping Norway from 1860 to 1910. Max Cieminski, a Kaszub, belonged to an ethnic group I had never heard of and one that turned out to be rich in folklore, pride, and faith.

As the book took shape, I gradually realized that I had an ▶

From Inspiration to Narration *(continued)*

even dozen heroes and that their ethnic breakdown—four Italians, three Jews, two Poles, a Norwegian, a Slovak, and an Irishman— mirrored the demographics of the day. These twelve occupied the foreground of my narrative. For background, I was able to work in anecdotes and details that came from the more fragmentary stories: the Jewish tailor who altered his uniform so it fit him like a glove; the boy from the same village as Andrew Christofferson who wrote about the importance of a conversion experience in the spiritual life of the villagers; Polish Americans who served in Haller's Army (a semi-independent unit based in France, led by General Jozef Haller, and made up largely of Polish immigrants). It was pure luck that Mike Valente and Epifanio Affatato fought in the same infantry regiment and became heroes on the same morning and the same battlefield and that my two career soldiers—Matej Kocak and Sam Dreben—both fought in the horrible bloodbath at Blanc Mont.

The stories are still coming. One of the most gratifying parts of being an author is hearing from readers who have enjoyed your book enough to send a message expressing their appreciation and sharing relevant details from their own lives. "I often wondered why it was that my maternal grandfather became such a vibrant American citizen," Seattle resident Donald Lorentz wrote me about his mother's father, Conrad August Westerberg, a Swedish immigrant who enlisted in the U.S. Army in 1917, "while my paternal grandfather, also a Swedish immigrant, remained tied to the Swedish community throughout his life. We suspected that World War I participation had a role in their different perspectives. After reading your book we are absolutely assured that such was indeed the case."

Allan Denberg wrote me recently to describe his father's wartime experiences as a Jewish immigrant from the Russian Pale: "My father could have been one of your subjects. Shortly after emigrating from Poland he was drafted. He was given the choice of serving and thereby automatic citizenship, or refusing, in which case he could *never* become a citizen. He chose to serve and saw action in France. . . . He told me that during the high holy days he was pulled from the line and boarded with a French Jewish family.

I guess the army really was sensitive to ethnic and religious considerations."

If you have a story of a foreign-born Doughboy—an ancestor, someone you've always admired, a person whose service deserves to be recognized—please send it to me and I'll post it at my Web site, www.thelongwayhomebook.com. You'll find contact information at the website as well. ◡

The Great War in History and Literature

THE FIRST THING an American researcher runs up against is that most works—and the best works—of Great War history and literature have been written by British writers. Wilfred Owen, Siegfried Sassoon, Ford Madox Ford, Robert Graves, Edmund Blunden, and Isaac Rosenberg were among many brilliant writers from the British Isles who penned memorable Great War verse, novels, and memoirs; Winston Churchill, John Keegan, Hew Strachan, Martin Gilbert, and John Terraine have published influential and highly readable histories—and all of them are British. Erich Maria Remarque's *All Quiet on the Western Front* is arguably the best novel to have come out of the Great War—maybe *any* war—and of course Remarque was German (though he emigrated to the States in 1939).

Among the Great War classics written by Americans, only John Dos Passos's *Three Soldiers*, e. e. cummings's *The Enormous Room*, and John Allan Wyeth's recently rediscovered book of verse, *This Man's Army*, come to mind. Maybe because the British fought so much longer and suffered so many more casualties than we did, maybe because the war coincided with or sparked a period of burgeoning creativity in Britain, maybe because the trauma of war cut so much deeper in Europe— whatever the reason, Britain's history and literature of the Great War are far richer than America's.

That said, I did come upon a number of inspiring, illuminating, and opinion-altering works of literature about the period written by Americans. Even better than *Three Soldiers* in conveying the spirit of the times is Dos Passos's breathtaking U.S.A. trilogy: *The 42nd Parallel, 1919*, and *The Big Money*. If you want to breathe the air (and fumes) of this great nation of ours in the first decades of the twentieth century, I can think of no more powerful and pungent source than Dos Passos's prose. *Through the Wheat* by Thomas Boyd, though several rungs down the literary ladder, offers a vivid soldier's eye view of trenches. *Suddenly We Didn't Want to Die* by Elton E. Mackin, *Along the Road* by Thomas H. Barber, and the short stories of marine lieutenant colonel John W. Thomason Jr. also contain searing accounts of battles and trenches written by men who fought and lived.

As for histories of the war written by Americans from an American perspective, I would recommend *The Doughboys* by Laurence Stallings and a book by the same title by Gary Mead, *The War to End All Wars* by Edward M. Coffman, the recent social histories of the war by Jennifer D. Keene, and Barbara Tuchman's classic *The Guns of August* for its evocation of how and why the great empires of Europe descended into this titanic conflict. Edward G. Lengel gives a thorough account of the unfolding of the Battle of the Argonne ▶

" If yo
to breath
(and fur
this grea
of ours i
decades
twentiet
I can thi
more po
and pun
source th
Passos's

u want
e the air
ues) of
nation
the first
of the
century,
k of no
verful
ent
an Dos
rose. **"**

Hell, and James Carl Nelson vividly
his grandfather in his recent volume,
s for accounts of what life was like
ted States in the years before, during,
ommend David Kennedy's classic
of Innocence by Meirion and Susie

l for the fine job they do recounting
of immigrant soldiers: *Lost Battalions*
the predominantly Jewish and
77th Division (as well as the African
Infantry); in *Good Americans*,
erpoints the wartime experiences
ine gun battalion with that of
er East Side; and in *Americans*
ribes how the army accommodated
oorn soldiers recruited to fight in

l to France or Belgium to visit the
vised to get a copy of *American*
rope published by the Center of
text, and photos take you yard by
re our soldiers fought and died. In
ispensible for the armchair traveler or
o wants to visualize what happened

once more for my final
npleting *The Long Way Home*, I heard
Long Way by the Irish writer Sebastian
perience of Irish (not Irish American)
ves the most searing and heartbreaking
t it felt like to be gassed, shot at, frozen,
d by life and fate and history. ❧

A Brief Guide to Military Units and Ranks

A NUMBER OF READERS have asked me to clarify the size and h[]
of the various military units during the war (platoon, company[]
regiment, and so on) and to explain the ranking system used fo[]
officers and enlisted men. What follows pertains only to the U.S.
Army, and specifically the infantry, during the First World War—[]
navy, marines, cavalry, and air force used slightly different systems[]
and many changes and refinements have been made subsequently.

The army of the American Expeditionary Forces was divided int[]
forty-three divisions, each of which consisted of approximately
25,000 to 27,000 enlisted men and 980 officers. Divisions were
numbered 1 through 8 (the Regular Army divisions), 26 through
42 (the National Guard divisions), and 76 through 93 (the National
Army divisions). (For an explanation of the distinctions between
these three branches, see page 142.) Many divisions also had
nicknames—the 4th was known as the Ivy Division; the 42nd as
the Rainbow Division. A division was commanded by a major
general. Each division was further subdivided into two infantry
brigades (also identified by number) with a brigadier general
heading up each brigade.

Each infantry brigade in turn consisted of two infantry regiments,
which were also numbered, and at the head of each regiment was
a colonel. So there were four infantry regiments per division. For
example, the 91st Wild West Division, in which Peter Thompson
and Leonardo Costantino served, comprised the 361st, 362nd,
363rd, and 364th infantry regiments; Peter was in the 362nd,
Leonardo in the 364th.

Every regiment consisted for three battalions (First, Second,
and Third), each commanded by a lieutenant colonel or a major,
and every battalion was further divided into four companies,
identified by a letter (A, B, C, or D) and led by a captain, or in
some cases, a first lieutenant. (Since officer casualties were so
high, the peacetime command structure changed quickly in
combat; officers frequently took charge of units that were
technically above their level.) Companies typically had about
150 men. So, to take another example from the book, Epifanio
Affatato belonged to Company C of the 107th infantry ▶

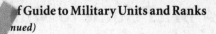

...egiment of the 54th infantry brigade of the 27th division; Mike Valente to Company D.

Below the company level came the platoon, consisting of about twenty-five to thirty men and led by a second lieutenant. The second lieutenant (known as a *shavetail* when newly commissioned) was the lowest commissioned officer grade. A platoon was further subdivided into four squads under the command of a corporal (see below).

Enlisted men—often referred to simply as *men* to distinguish them from officers—were also divided into ranks: at the bottom were privates, and one jump up were privates first class. If a private served bravely and showed leadership ability, he might be made a noncommissioned officer—which meant he had some authority over privates but was still officially subordinate to all commissioned officers. The two broad divisions of noncommissioned officers were corporal and sergeant. Matej Kocak was promoted to corporal in 1916 after nine years of service as a private with the marines; a year later, hours before his jump off in the battle of Belleau Wood, he was made a sergeant.